BUILDING TECHNOLOGY

IVOR H. SEELEY

B.Sc., M.A., Ph.D., F.R.I.C.S., C.Eng.,
F.I.Mun.E., F.I.Q.S., M.C.I.O.B., M.I.H.

*Professor and Head of Department of
Surveying and Dean of the School of
Environmental Studies, Trent Polytechnic,
Nottingham*

Second Edition

M

First edition 1974
Reprinted 1975, 1976, 1977, 1978, 1979
Second edition 1980
Reprinted 1981, 1982

Published by
THE MACMILLAN PRESS LTD
London and Basingstoke
Companies and representatives throughout the world

ISBN 0 333 30717 8
0 333 30718 6 pbk

Typeset in Great Britain by
STYLESET LIMITED
Salisbury, Wiltshire

Printed in Hong Kong

This book is dedicated to my wife for her continual interest in building matters and for producing such a good design for our new bungalow, where some of the constructional methods described in the book have been put into practice

CONTENTS

Contents

LIST OF ILLUSTRATIONS

LIST OF TABLES

PREFACE TO THE SECOND EDITION

THIS BOOK EXAMINES the general principles of building construction and applies them to practical examples of constructional work throughout all parts of simple domestic buildings. Consideration is also given to the choice of materials and implications of Building Regulations, British Standards, Codes of Practice and current government research.

The principal aim is to produce a comprehensive, simply explained and well-illustrated text which will form a basic construction textbook for first- and second-year students in a variety of disciplines, including architects, quantity surveyors, valuers, auctioneers, estate agents, building and mineral surveyors, builders, housing managers, environmental health officers, building control officers and clerks of works. It should also prove of value to those working for TEC certificates and diplomas in construction; degrees in architecture, building, quantity surveying and estate management; and the General Certificate of Education in building construction.

All drawings and building data are fully metricated and, although BS 1192 has been largely used as a basis for the preparation and presentation of drawings, it was decided to omit 'm' and 'mm' symbols from all descriptions as well as dimensions on the drawings, to avoid constant repetition of units of measurement. It is unlikely that any confusion will arise as a result of this. A metric conversion table is produced in an appendix to help readers who wish to make comparisons between imperial and metric measures. The abbreviations BRE, for Building Research Establishment, and DOE, for Department of the Environment, have been used extensively throughout the book. DOE Advisory Leaflets are sometimes referred to as Property Services Agency (PSA) leaflets.

This second edition has been considerably extended and updated to include the latest developments.

IVOR H. SEELEY

ACKNOWLEDGEMENTS

THE author acknowledges with gratitude the willing co-operation and assistance received from many organisations connected with the building industry — manufacturing, research and contractual.

The Controller of Her Majesty's Stationery Office has given permission to incorporate extracts from *The Building Regulations 1976* and the *First Amendment 1978*. Crown copyright material is also reproduced from BRS and BRE Digests by permission of the Director of the Building Research Establishment; in this connection it should be mentioned that copies of the digests quoted are obtainable from Her Majesty's Stationery Office, PO Box 569, London SE1, or any bookseller.

Extracts from British Standards and Codes of Practice are reproduced by permission of the British Standards Institution, 2 Park Street, London W1A 2BS. The Institution allows a discount of 40 per cent on sales of publications to students and educational bodies and BS Handbook 3 contains useful summaries of British Standards for building, which are updated periodically.

The quality and usefulness of the book has been very much enhanced by the large number of first-class drawings which do so much to bring the subject to life. Ronald Sears devoted many hundreds of hours to the painstaking preparation of the final drawings, for which he deserves special appreciation. The author also expresses his gratitude to Malcolm Stewart of the publishers, for his continual guidance and encouragement throughout the preparation and production of the book. Thanks are due to those readers, particularly D. J. Fisher and M. P. Fisby, who wrote to the author with helpful comments.

1 THE BUILDING PROCESS AND SITEWORKS

This book describes and illustrates the constructional processes, materials and components used in the erection of fairly simple domestic and associated buildings, and examines the principles and philosophy underlying their choice. Its primary aim is to meet the needs of students of a variety of disciplines whose studies embrace building technology.

This is a wide-ranging subject and no single textbook could cover every aspect. The student is advised to widen his reading by perusing technical journals, Building Research Establishment and other relevant government publications and trade catalogues. He will also obtain a better understanding of the processes, materials and site organisation by visiting buildings under erection. All building work is subject to control under the London Building Acts and Constructional By-laws in London and through the Building Regulations in the provinces. In addition, restrictions on the siting and appearance of buildings stem from the operation of the Town and Country Planning Acts.

All dimensions are in metric terms, and in this connection the metric conversion table in the appendix (page 202) may prove useful.

CHOICE OF SITE

Various matters should be considered when selecting a building site. The high cost of land and planning controls frequently combine to prevent the acquisition of the ideal site. Indeed many sites are now being developed which in years gone by would have been considered unsuitable. Some of the more important factors are now investigated.

Climate. This varies widely throughout the country being broadly warmer in the south, colder in the east and north, and wetter in the west. There are however local variations with some areas being more susceptible to mist and fog.

Aspect. The aspect of a site is important as it will determine the amount of sun received on the various elevations of the building. The ideal site is generally thought to be one below the summit on a gentle southerly slope, thus securing maximum sunshine and some protection from northerly winds. A house on a site with a north frontage should ideally have its main rooms at the rear. A further point to consider is the altitude of the sun varying from 61° 57' at the summer solstice to 15° 03' at the winter solstice. All habitable rooms should be planned to receive sunlight during their normal period of use.

Elevation. Elevated sites are generally preferable to low-lying ones, being drier and easier to drain, while hollows are likely to be cold and damp. Nevertheless, a hilltop site may need protection from winds by a tree belt.

Prospect. Ideally a site should command pleasant views and the adjoining land uses should be compatible. The site itself will be more attractive if it is gently undulating and contains some mature trees, but if the trees are too close to the building they will restrict light and air.

Available facilities. A housing site should ideally have ready access to schools and shops and other facilities such as parks, sports facilities, swimming pools, libraries and community centres, and good public transport services.

Services. Adequate and accessible gas and water mains, electricity cables and sewers are generally essential, although in rural areas it may be necessary to provide septic tank installations.

Subsoils. Subsoils merit special consideration because of their effect on the building work.

(1) Hard rock provides a good foundation but increases excavation costs and may cause difficulties in the disposal of sewage effluent on isolated sites.

(2) Gravel is probably the ideal subsoil being strong and easily drained.

(3) Sand drains well but if loose could be subject to movement; in practice it is usually combined with clay or gravel.

(4) Clay often has a good bearing capacity but does not drain well; with the shrinkable varieties it is necessary to take the foundations down to at least one metre below ground level because of the variations that occur with differing climatic conditions.

(5) Chalk provides a stable and. easily drained subsoil of good bearing capacity.

(6) Made ground is where soil or other fill has been deposited to make up levels and lengthy periods are needed for settlement; it may be necessary to use raft or piled foundations.

Water table. It is essential that the building should be erected well above the highest groundwater level. This level will vary according to the extent of the underground flow, evaporation and rainfall, and is normally highest in winter.

Subsidence. In areas liable to mining subsidence special and costly precautions of the type described in chapter 3 should be taken.

SITE INVESTIGATION

All potential building sites need to be investigated to determine their suitability for building and the nature and extent of the preliminary work that will be needed. Particular attention should be given to the nature of the soil and its probable load-bearing capacity by means of trial holes or borings, as there may be variations over the site. The level of the water table should also be established as a high water table may necessitate subsoil drainage and could cause flooding in winter.

The position and size of main services should be determined and it is advisable to take a grid or framework of levels over the site to indicate the amount of earthwork and the ease or otherwise of drainage. The nature and condition of site boundaries should be noted together with the extent of site clearance work, such as old buildings to be demolished and trees and shrubs to be grubbed up and removed.

By way of contrast there may be some attractive mature trees on the site which ought to be retained and they may even be the subject of tree preservation orders. Consideration should also be given to such matters as access, storage space and working conditions, which will all affect the cost of the job. Indeed, it is good practice to assess the approximate cost of site clearance work in building up an estimate of the likely cost of the complete job.

The local planning authority should be approached to ascertain whether there are any special or significant restrictions which could adversely affect the development of the site, and the position of the building line, so that the location of any new buildings can be established with certainty. The survey should include details of neighbouring development and the position with regard to facilities in the area. Enquiries will reveal the existence of any restrictive covenants such as rights of way, light and drainage, which may restrict the development.

Methods of Soil Investigation

A number of methods can be used to determine the soil conditions. For simple residential buildings on reasonable sites digging several holes about three spits deep, drilling holes up to 2 or 2½ m deep with a hand auger, or driving a pointed steel bar about 1½ m into the ground, are generally sufficient. With larger buildings or more difficult sites, the following methods are applicable

(1) Excavating trial holes about 1½ m deep outside the perimeter of the building.

(2) Drilling boreholes by percussion or rotary methods. The percussion method uses a steel bit with a chisel point screwed to a steel rod. The rotary method employs a hollow rod with a rotating bit and a core of strata is forced back up the hollow rod.

(3) Load testing often using a reinforced concrete slab about 1.20 to 1.50 m² and 300 mm thick, to which loads are added at about two-hour intervals and the amount of settlement is determined with a level. The safe bearing capacity of the soil is found by dividing half the load applied prior to appreciable settlement into the area of the slab.

Classification of Soils

Soils can conveniently be classified under five main headings

(1) rocks, which include igneous rocks, limestones and sandstones;

(2) cohesive soils, such as clays where the constituent particles are closely integrated and stick together;

(3) non-cohesive soils, such as gravels and sands, whose strength is largely dependent on the grading and closeness of the particles;

(4) peat, which is decayed vegetable matter of low strength with a high moisture and acidic content;

(5) made ground which may contain waste of one kind or another and can cause settlement problems.

GROUNDWATER DRAINAGE

On low-lying or damp sites, it is advisable to provide an effective system of groundwater drainage, as described in the BS Code of Practice on Building Drainage (see reference 1 at the end of this chapter).

Groundwater is the portion of the rainwater which is absorbed into the ground and drainage of it may be necessary for the following reasons

(1) to increase the stability of the ground;
(2) to avoid surface flooding;
(3) to alleviate or avoid dampness in basements;
(4) to reduce humidity in the immediate vicinity of the building.

The standing level of the groundwater (water table) will vary with the season, the amount of rainfall and the proximity and level of watercourses. Main groundwater drains should follow natural falls wherever possible and should be sited so as not to endanger the stability of buildings or earthworks.

Systems of Groundwater Drainage

The most commonly used systems embrace porous or perforated pipes or gravel-filled trenches (French drains) laid to one of the following arrangements

(1) *Natural.* The drains follow natural depressions or valleys on the site with branches discharging into the main pipe (figure 1.1.1).

(2) *Herringbone.* There are a number of main drains into which smaller subsidiary drains discharge from both sides. The subsidiaries run parallel to each other but at an angle to the main drains, and should not exceed 30 m in length (figure 1.1.2). The spacing of the subsidiary drains varies with the type of soil and where the main drains are 600 to 900 mm deep, the spacing of the subsidiaries might be within the following ranges

Sand	30 to 45 m
Loam	23 to 26 m
Sandy clay	11 to 12 m
Clay	8 to 9 m

(3) *Grid.* A main drain or drains are laid near the boundaries of a site into which branches discharge from one side only (figure 1.1.3).

(4) *Fan.* The drains converge to a single outlet at one point on the boundary of the site without the use of a main drain (figure 1.1.4).

(5) *Moat or cut-off system.* Drains are laid on one or more sides of a building to intercept the flow of groundwater and thereby protect the foundations (figure 1.1.5).

The choice of system will depend on site conditions. A common arrangement is 100 mm main drains with 75 mm branch drains, 600 to 900 mm deep with gradients determined by the fall of the land. The outlet of the groundwater drainage system will discharge into a soakaway or through a catchpit into the nearest ditch or watercourse (figures 1.1.6 and 1.1.7). Other forms of soakaway construction are detailed in BRE Digest 151.[17] Where these are not available groundwater drains may be connected, with the approval of the local authority, through a silt trap to the surface water drainage system. If connected to a foul drain a reverse-action intercepting trap would be required.

In clay or heavy soils, *mole drains* provide an alternative to pipes. These are formed by drawing a steel cartridge through the ground by a strong, thin blade attached to a mole plough driven by a tractor. Mole diameters range from 50 to 150 mm, at depths varying from 300 to 750 mm and spacings from 3 to 5 m. Their life is rather limited.

Groundwater Drains

A variety of materials are available as follows

(1) Clayware field pipes to BS 1196[2] of unglazed, porous pipes with square ends (butt jointed) — 65 to 300 mm nominal bores.

(2) Clay surface water drainpipes to BS 65[3], non-porous and laid with open joints. Fully and half-perforated pipes are also available — 75 to 900 mm nominal bores.

(3) Concrete porous pipes to BS 1194[4] with rebated, O.G. or butt joints — 75 to 900 mm nominal bores.

(4) Pitch impregnated fibre pipes to BS 2760[5], perforated and usually jointed with snap ring joints — 50 to 225 mm nominal bores.

(5) Unplasticised PVC underground drainpipes to BS 4660[6], non-porous with socketed joints — 100 and 160 mm nominal sizes.

Pipe trenches should be just wide enough at the bottom for laying the pipes and they are mainly laid with open joints to straight lines and suitable gradients (varying from 1 in 80 to 1 in 300). Pipes should be surrounded with suitable clinker, gravel or rubble to 150 mm above the pipes, covered with a layer of inverted turf, brushwood or straw to prevent fine particles of soil entering the pores of the filling or the pipes (figure 1.1.8). On occasions, a short length of slate or tile is provided over each open pipe joint as added protection. If the drain is also to receive surface water, rubble should extend up

LAND DRAINAGE SYSTEMS

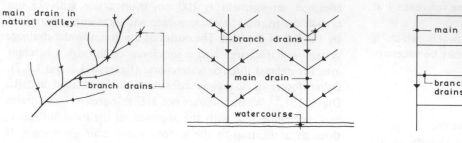

main drain in
natural valley

branch drains

1. 1. 1   N A T U R A L

branch drains

main drain

watercourse

1. 1. 2   H E R R I N G B O N E

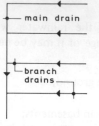

main drain

branch drains

1. 1. 3   G R I D

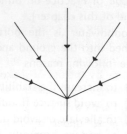

1. 1. 4   F A N

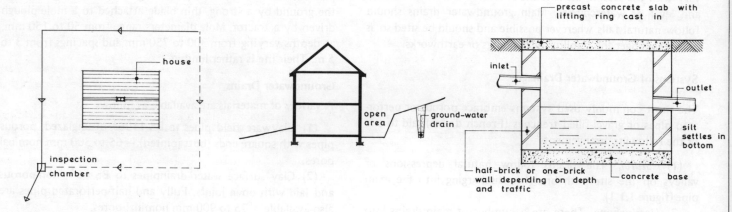

house

inspection
chamber

1. 1. 5   M O A T   O R   C U T - O F F   S Y S T E M

open
area

ground-water
drain

precast concrete slab with
lifting ring cast in

inlet

outlet

silt
settles in
bottom

half-brick or one-brick
wall depending on depth
and traffic

concrete base

1. 1. 7   C A T C H P I T

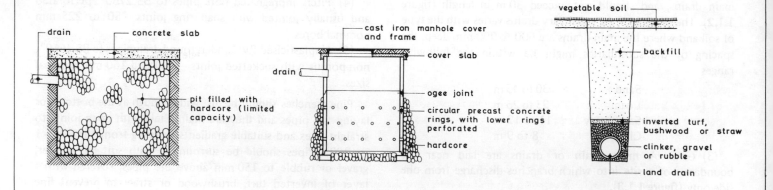

drain

concrete slab

pit filled with
hardcore (limited
capacity)

1. 1. 6   S O A K A W A Y S

cast iron manhole cover
and frame

cover slab

drain

ogee joint

circular precast concrete
rings, with lower rings
perforated

hardcore

vegetable soil

backfill

inverted turf,
bushwood or straw

clinker, gravel
or rubble

land drain

1. 1. 8   L A N D   D R A I N

Figure 1.1   G R O U N D - W A T E R   D R A I N A G E

the trench to ground level. Where drains pass near tree roots or through hedge roots, jointed socketed pipes should be used to prevent root penetration.

Problems sometimes arise with land drainage systems due to the following causes

(1) silting of pipes: this can be reduced by providing catchpits (figure 1.1.7);

(2) displacement of pipes by tree roots: this can be overcome by the use of socketed joints or collars;

(3) access of vermin to pipes: prevented by covering exposed ends with wire gratings;

(4) pipes under buildings damaged by settlement: they should be laid clear of buildings and if passage under a building is unavoidable they should be protected by a lintel or arch.

SETTING OUT

Establishing Levels on Site

Before building operations can be commenced on the site it is necessary to set out the building(s) and to establish a point of known level on the site which can be used to determine floor and drain invert levels. The basis for the levelling operations will be the nearest ordnance bench mark, whose value is obtained from the latest edition of the appropriate Ordnance Survey map. Levels are transferred from the Ordnance bench mark to the building site using a dumpy or precise level and levelling staff. The sights should be as long as possible and preferably be kept equidistant. A temporary bench mark is established at each change of instrument on a permanent fixed point such as a road kerb or top of a boundary wall, and this is suitably marked with a knife and detailed in the level book. The level point on the site should ideally consist of a bolt set in concrete or could alternatively be some permanent feature on the site. Flying levels are then taken back to the ordnance bench mark to check the accuracy of the levelling operations.

All normal precautions should be taken such as ensuring that the levelling staff is fully extended and is held truly vertical in both directions when taking levels. The instrument must be accurately set up and the levels properly recorded on either the collimation or rise and fall method. It is essential for the instrument to be in proper adjustment.

Foundation trench level pegs can be established individually with a dumpy or precise level from the temporary bench mark on the site. Another approach is to set up sight rails at the ends of each trench and to fix intermediate levels by means of boning rods (figure 1.2.1). Yet another method is to use a straight-edge about 2½ to 3 m long and not less than 25 x 150 mm in section, and a spirit level (figure 1.2.2). Peg A is driven in to

the correct level and another peg B is driven in almost the length of the straight-edge away. The straight-edge is placed across the two pegs with the spirit level on it and peg B is lowered until the correct level is achieved. After checks on the accuracy of the spirit level and straight-edge, peg C is driven in and the levelling process repeated.[7] A quicker method is to use a water level consisting of two tubes of clear glass or transparent plastics material, connected by a length of rubber or PVC tubing and almost completely filled with water. The water surfaces give two equal levels, as illustrated in DOE Advisory Leaflet 48.[7]

Setting Out Buildings

The first job is usually to establish the building line which demarcates the outer face of the front wall of the building. The building line is frequently determined by the highway authority and in urban areas is often around 8 m from the back of the public footpath. Often it can be fixed by measuring the appropriate distance from the highway boundary (back of the public footpath) along the side boundaries of the site, and stretching a line between the two pegs. On an infill site the line may be established by sighting between the existing buildings on either side of the site.

The next step will be to position the front of the building on the building line by checking the dimensions between the new building and the side boundaries. Flank walls will then be set out at right-angles to the building line often using a large builder's square or the 3:4:5 method (figure 1.2.3). The builder's square is a right-angled triangular timber frame with sides varying in length from 1.50 to 3.00 m. The square is placed against the building line and two pegs are driven in on the return side. By sighting across the two pegs a third peg can be driven in the same straight line, and a bricklayer's line stretched between them.

If a builder's square is not available, a right-angle can be set out based on a right-angled triangle whose sides are in the ratio of 3:4:5 (derived from the theorem of Pythagoras). A peg is first driven in at the corner of the building and a distance of 3 m is measured back along the building line. A peg is driven in at this point and the ring of the tape is placed over a nail driven into the top of the peg. The tape is held at the 12 m mark against the ring on the first peg and with the tape around the corner peg, the tape is stretched out to give the position of the third peg at the 7 m mark. The line extended through the third peg is at right-angles to the building line.

To secure permanent line markers, profiles are established at the corners of the building and at wall intersections (figure 1.2.4). They often consist of boards about 25 x 150 mm or

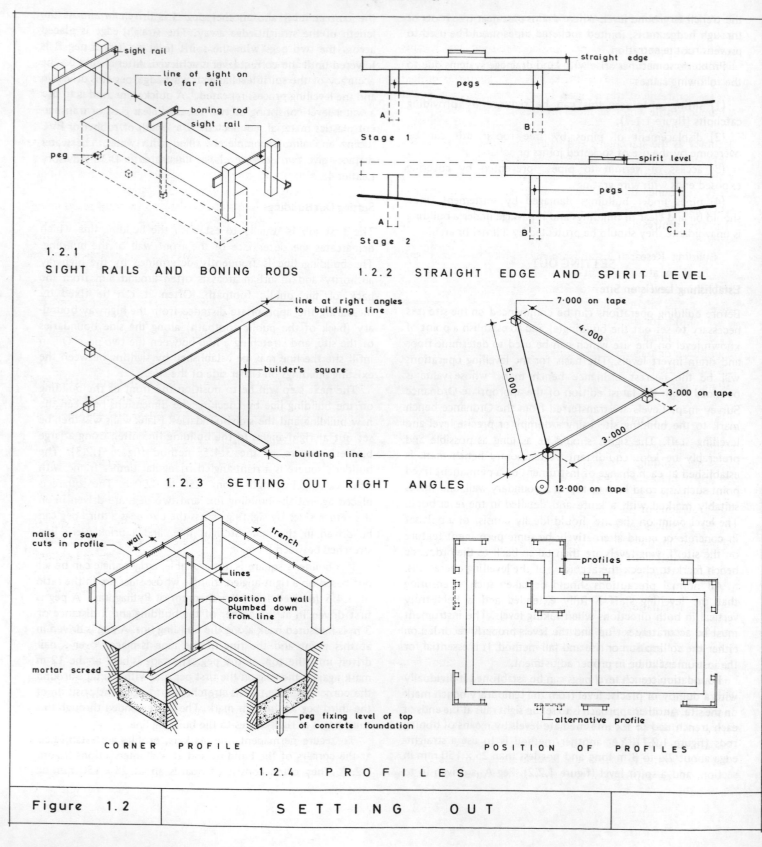

sight rail

line of sight on to far rail

boning rod

sight rail

peg

1.2.1

SIGHT RAILS AND BONING RODS

straight edge

pegs

A

B

Stage 1

spirit level

pegs

A

B

C

Stage 2

1.2.2   STRAIGHT EDGE AND SPIRIT LEVEL

line at right angles to building line

builder's square

building line

1.2.3   SETTING OUT RIGHT ANGLES

7·000 on tape

4·000

5·000

3·000 on tape

3·000

12·000 on tape

nails or saw cuts in profile

wall

trench

lines

position of wall plumbed down from line

mortar screed

peg fixing level of top of concrete foundation

CORNER   PROFILE

profiles

alternative profile

POSITION   OF   PROFILES

1.2.4   P R O F I L E S

| Figure   1.2 | S E T T I N G   O U T |

25 x 100 mm nailed to 50 x 50 mm posts or pegs driven firmly into the ground. Saw cuts or nails demarcate the width of walls and the spread of foundations and the bricklayer sets his lines to these and, when working below ground, plumbs down from them at the corners of the building. A radius rod worked from a centre pivot can be used to set out curved bays, whilst a framed profile would be made up for a cant bay.

The setting out of a steel-framed building requires extreme accuracy as the stanchions and beams are cut to length at the fabricator's works, and any error in setting out can involve expensive alterations on site. It is usual to erect continuous profiles around the building and to set out the column spacings along them. Wires stretched between the profiles will give the centres of stanchions around which templets are formed.

Building Research Establishment Digest 234[8] emphasises the need to attain predictable standards of accuracy in the setting out and erection processes, resulting from the increase in the partial or complete prefabrication of building structures, and methods for their achievement are described in BS 5606.[18] Errors in linear measurement may arise through reading, calibration or tension errors. However, if corrections are made for slope, the maximum error is likely to be about 25 mm in a 30 m length.

CONTROL OF BUILDING WORK

Almost all proposals for new buildings or alterations to existing buildings require approval by the relevant local authority before work can be commenced on site. In London, application is made to the appropriate district surveyor under the provisions of the London Constructional By-Laws[9], while in the provinces the operative constructional requirements are contained in the Building Regulations 1976 and subsequent amendments.[10] In addition approval is normally required under the Town and Country Planning Acts.

Building Regulations

Applications under Building Regulations involve the completion of application forms giving a description of the proposed building together with the various structural components, services and fittings. In addition plans are required of every floor and roof and sections of every storey to a scale of not less than 1:100, together with a block plan and key plan to show the siting of the building and the location of the site. The building control officer or other suitably designated official of the local authority advises the appropriate committee following which an approval or rejection notice will be sent to the applicant.

The building work is subject to inspection by an official of the local authority to ensure that the work is in accordance with the approved plans and the Building Regulations. The builder is required to supply the local authority with not less than twenty-four hours' notice in writing of the date and time at which building will commence; before the covering up of any excavation for a foundation, damp-proof course, concrete or other material laid over a site; and before any drain or private sewer is haunched or covered; and not more than seven days after the laying of any drain or private sewer. If the builder neglects to give any of these notices, he may be required to cut into, lay open or pull down so much of the building work as prevents the local authority from ascertaining whether any of these regulations have been contravened. Finally, the builder is required to give the local authority notice not more than seven days after completion of the building.

It should be noted that many building materials and components are the subject of British Standards, and there are also a number of Codes of Practice which cover workmanship and constructional techniques. Both sets of documents are issued by the British Standards Institution, which is a private body representing all interests with financial support from the government. The work of the Institution is mainly performed through technical committees. The use of these standards and codes ensures a good standard of materials and workmanship, and they embrace the results of latest research. In particular compliance with a British Standard is generally accepted as satisfying Building Regulations. Furthermore, their use reduces the work involved in specification writing and makes for greater uniformity in requirements, which helps the builder. Where a British Standard covers several grades or classes of material, it is essential that the required grade or class shall be stated. The Building Regulations are subject to periodic amendment and there is considerable pressure for them to be simplified.

Building Lines

The siting of new buildings is often affected by *building lines*. A building line is defined in section 73 of the Highways Act 1959 as a 'frontage line for building'. It may be prescribed by the highway authority on either one or both sides of a public highway, and no new building, other than a boundary wall or fence, shall be erected, and no permanent excavation below the level of the highway shall be made, nearer the centre line of the highway than the building line, except with the consent of the highway authority, who may impose conditions. The frontage line of existing buildings may in some cases form the building line. The main purpose of the building line is to ensure that all buildings are kept a reasonable distance from the highway to preserve amenity and obtain good sight lines.

Under section 74 of the same Act, a highway authority may prescribe lines called *improvement lines* on one or both sides of a street and they represent the lines to which the street will eventually be widened. This procedure is normally adopted when a public highway is narrow or inconvenient without any sufficiently regular boundary line, or it is necessary or desirable that it should be widened. Similar prohibitions apply to building work in advance of an improvement line as those in operation for building lines, except that boundary walls and fences are not excluded.

Planning Consent

Planning applications must be submitted where it is proposed to carry out development, which requires permission under the Town and Country Planning Acts. Such applications are usually required to be submitted in triplicate accompanied by plans.

'Development' is defined in the Town and Country Planning Act as 'the carrying out of building, engineering, mining or other operations in, on, over or under land, or the making of any material change in the use of any buildings or other land'.

Certain works do not constitute development and do not therefore require planning consent. Typical examples are works of maintenance, improvement or alteration of a building which do not materially affect its external appearance, the change of use of a building to another use within the same class as listed in the Town and Country Planning (Use Classes) Order, and roadworks by a local highway authority. On the other hand, the conversion of a single dwelling into two or more separate dwellings, the deposit of refuse or waste material and the display of advertisements generally require planning permission. Limited extensions to dwelling houses, boundary fences and walls within certain limits of height, painting the exteriors of buildings and erection of contractors' site huts are classified as 'permitted development' and no planning permission is needed, and these exclusions may be extended in the future.

A prospective developer often wishes to know whether the development he has in mind is likely to receive planning permission, before he purchases the land or incurs the cost of preparation of detailed plans. In these circumstances, the developer is able to submit an *outline application*, which entails merely the submission of an application form and a site plan. Consent may be given subject to subsequent approval by the local planning authority of any matters relating to siting, design and external appearance of the buildings. In this way the applicant can obtain officially the local planning authority's reaction to his proposals, without the necessity of preparing comprehensive plans.

All planning applications are recorded in a planning register, which is available for inspection by the public. The entries in the register will include details of the applicant and agent (if any), nature and location of proposed development, planning decision and result of appeal (where applicable). The local planning authority has three choices

(1) to grant permission unconditionally;
(2) to grant permission subject to conditions;
(3) to refuse permission.

Decisions must include reasons for refusal or for any conditions attached to a permission, and an aggrieved applicant has the right of appeal.

Other Statutory Requirements

It is often necessary to obtain an *industrial development certificate* before an industrial building can be erected, re-erected or extended, or an non-industrial building converted to industrial use. The certificates are issued by the Department of Trade and Industry and will only be issued if the Department considers that the proposed development is consistent with the proper and balanced distribution of industry. Where a certificate is refused by the Department of Trade and Industry there is no right of appeal and the local planning authority is unable to grant planning permission for the project. Under the provisions of the Control of Offices and Industrial Development Act 1965 it is frequently necessary to obtain an *office development permit* before applying for planning permission to erect or extend office buildings or to convert other buildings into offices.

New factory buildings are subject to various statutory controls, apart from Building Regulations, by virtue of the provisions of the Factories Act, the Clean Air Acts and the Thermal Insulation (Industrial Buildings) Act. New offices and shops have to comply with the Offices, Shops and Railway Premises Act. A number of statutes prescribe minimum provision in relation to means of escape in case of fire and fire-fighting appliances, while hotels and restaurants are examined under the Food Hygiene Regulations.

SEQUENCE OF BUILDING OPERATIONS

The building client may be a public authority, a private organisation or even an individual. Where an architect is engaged, the client supplies him with a brief of his basic requirements. The architect will inspect the site and consult with the local authorities having planning and building control functions, and with the various statutory authorities that will provide services to the site. He will then prepare preliminary sketch drawings for consideration by the client and from which a quantity surveyor can prepare an approximate estimate of cost. On the larger jobs, the quantity surveyor is likely to produce a cost plan as a basis for the effective control of the cost of the project throughout the design stage. The architect will proceed with the working drawings, and quotations will be obtained from sub-contractors and statutory undertakers, in addition to the various consents and approvals necessary. The specification and bill of quantities will be formulated and together with the drawings and conditions of contract will be sent to building contractors for the submission of tenders. The tenders are examined by the quantity surveyor who submits a report to the client through the architect and a contractor is selected. In some circumstances it is preferable to negotiate a price with a single contractor.

Following acceptance of his tender the contractor prepares a programme, orders materials and starts work on the site. His first job is to install certain vital temporary works, such as access roads, hoardings, site huts, storage compound and temporary services. Most of the temporary works are removed by the time the permanent work is complete. The contractor is responsible for the organisation of all the work on the site including that of nominated sub-contractors.

The building will be set out on the site in accordance with the dimensions and levels supplied by the architect, who will then check the setting out work. Foundation trenches are excavated and levelled for approval by the building control officer of the local authority, following which pegs are driven into the bottoms of the trenches with their tops delineating the finished surface of the foundation concrete, which is then poured. Bricklaying follows starting at the quoins and stringing lines between them, vegetable soil is stripped and concrete oversite laid. Brickwork is levelled to receive the damp-proof course, and after approval by the building control officer, the erection of walls continues with the fixing of external door frames and windows as the work proceeds.

The carpenter builds in first-floor joists and follows with the roof timbers. The roofer covers the roof to complete the carcassing. Partitions are erected and floor finishes follow after the insertion of services. Then comes the internal joinery in doors and staircases, followed by plastering to walls and ceilings, fireplaces and joinery fittings. The final stages of the job embrace the completion of services, decorations and cleaning of floors and windows, and removal of surplus materials from the site.

The architect is responsible for ensuring that the work is built in accordance with the contract documents, and on larger jobs will be assisted by a clerk of works who stays on the site. The quantity surveyor will make periodic valuations of the completed work and materials brought onto the site and the architect will then issue certificates authorising the client to make payments to the contractor. After completion of the work, the quantity surveyor will agree the final account with the contractor. At the end of the defects liability period, usually six months after completion, the architect will prepare a schedule of defects for which he considers the contractor is responsible. The contractor will not receive payment of the outstanding balance of retention money until these defects have been remedied.

PROBLEMS IN DESIGN AND CONSTRUCTION OF BUILDINGS

Principles of Design

Before we consider design and constructional problems it would probably be helpful to examine the elements or principles of design. A prime objective in building design is to secure an attractive building — one of high aesthetic value. It will also be appreciated that buildings serving different purposes tend to assume different forms. Hence public buildings often take a symmetrical form to emphasise formality, order and dignity, whilst individual houses are often quite informal. Character is derived from the composition as a whole.

The relationship of the various units (such as doors, windows, plinths and pilasters) to each other and to the whole building is termed proportion, and their relationship in size is described as scale. A well-designed building will always be well proportioned. Much of the beauty of older buildings stems from the use of local materials which generally weather to attractive colours. An effort should be made to harmonise with adjoining buildings by sympathetic choice of colours and to obtain an attractive street picture. The texture of the materials should also be considered; in general rougher surfaces have a more interesting texture with variations in colour which mellow over a period of time.

Designing a building is essentially a matter of making a long series of choices — choices about ends and choices about means.[11] Both client and designer are concerned with the choice of ends, embracing the purpose and character of the building and the client's special requirements. The means are solely the responsibility of the designer and will include the detailed planning, structural form, services and finishes; the completed design represents a set of instructions for the erection of the building.

The architect is concerned with providing a building which will satisfy the client's needs. In performing this task he will, however, be obliged to have regard to such factors as maximum use of land, cost, availability of labour and materials, technological constraints, planning and building regulations, relationship with other buildings, landscaping, services, circulation networks, and fire and noise prevention.

Even the design of a dwelling house can be quite complex if it is to effectively meet the changing needs of the occupants over a long period of time. In the last decade there has been a change of approach to the design of housing accommodation, stemming from the government report *Homes for Today and Tomorrow*.[12] Accommodation standards cease to be based on minimum room sizes, and depend on functional requirements and levels of performance, with minimum overall sizes for the dwelling related to the size of family. There should for instance be space for activities requiring privacy and quiet, for satisfactory circulation, for adequate storage and to accommodate new household equipment, in addition to a kitchen arranged for easy housework and with sufficient room in which to take at least some meals. The report includes recommendations for the provision of sanitary appliances, kitchen fitments, bedroom cupboards, electric-socket outlets and minimum heating standards. In 1980 the Government proposed modifications to the Parker Morris standards, described above, which were considered unduly restrictive.

Design Problems

Until the 1930s architects were seldom tempted to depart from traditional construction, and they tended to rely on inherited specifications which were based on the known effects of time, wear and weathering on local materials. Today the designer is faced with a wide range of both old and new materials and components, necessitating careful thought to design details to avoid unsatisfactory results.

The Building Research Establishment[13] has instanced cases where accelerated deterioration and/or unsightly appearance has resulted from the unsuitable placing of incompatible materials or inadequate attention to design details. Streaking by rainwater washings over walling, due to lack of suitable projecting features with adequate drips at the head of walls to buildings with flat roofs is a typical example.

Changes in appearance of materials used externally in buildings may result from any of the following three causes

(1) preventable faults in design and/or construction where

there is sufficient information available at the time to avoid them;

(2) unforeseeable behaviour of a material in a given situation where it could reasonably be argued that there was insufficient information generally available at the time, and where the change may not be wholly explicable in the light of existing knowledge;

(3) the process of 'natural' weathering, often as the result of microclimatic influences, frequently unpredictable in detail.

Some materials are commonly believed to be maintenance-free but this is not always so in practice. For instance, untreated teak and western red cedar often become unsightly particularly in urban situations. An annual application of a linseed oil/paraffin wax mixture containing a fungicide is the minimum necessary to preserve appearance. Aluminium used externally requires anodic treatment, followed by periodic washing, to maintain a satisfactory appearance.[19]

Building Maintenance

Building maintenance work uses extensive resources of labour and materials and in 1979 it was estimated that the annual building maintenance bill in Great Britain had reached £3500m. Hence it is vitally important that the probable maintenance and running costs of a building should be considered at the design stage, and due attention directed towards the maintenance implications of alternative designs. A reduction in initial constructional costs often leads to higher maintenance and running costs. The cheapest heating system in installation costs is often the most expensive to operate.[14]

Performance Standards

New buildings have to meet the performance standards prescribed by the Building Regulations or the London Constructional By-Laws. For instance the strength and stability of a building is covered by part D of the Building Regulations, and weather and water exclusion in part C, as it is vital that any building shall be wind and weathertight. With the greater provision of central heating, thermal insulation assumes greater importance and minimum thermal insulation requirements are contained in part F of the Building Regulations, whilst as the volume of noise increases and occupiers of buildings become more susceptible to noise, so the provision of adequate sound insulation (part G) becomes important. The inclusion of minimum standards of fire resistance are essential and these are detailed in part E of the Building Regulations. Occupants of buildings ideally want to enjoy ample natural lighting, and minimum standards for this are scheduled in part K of the Building Regulations. More detailed reference will be made to many of these regulations in later chapters.

Dimensional Co-ordination

BS 4011[15] recommends that the first selection of basic sizes for the co-ordinating dimensions of components should be in the order of preference, *n* by 300 mm; *n* by 100 mm; *n* by 50 mm and *n* by 25 mm, where *n* is equal to any natural number including unity. It is essential to relate components to a grid if they are to fit into the space planned for them within a building, without the need for trimming and cutting. Also, by using them in accordance with a grid, it will be possible to reduce the variety of components and increase standardisation and rationalisation in building. The grid lines can be based on the centre lines of walls and columns or on wall and column faces. Axial grids are most suitable for framed buildings and face grids are better suited for buildings with load-bearing walls. Key reference planes constitute load-bearing walls, columns and upper and lower surfaces of floors and roofs. The spaces between these key reference planes are termed *controlling zones* and the distances between these zones are called *controlling dimensions*. The recommended sizes of controlling dimensions are prescribed in BS 4330.[16] Intermediate key reference planes control the position of the heads and sills of windows and heads of door sets.

REFERENCES

1. BS Code of Practice CP 301: 1971 Building drainage
2. BS 1196: 1971 Clayware field drainpipes
3. BS 65 and 540: Clay drain and sewer pipes including surface water pipes and fittings; Part 1: 1971 Pipes and fittings, Part 2: 1972 Flexible mechanical joints
4. BS 1194: 1969 Concrete porous pipes for under-drainage
5. BS 2760: 1973 Pitch-impregnated fibre pipes and fittings, for above and below ground drainage
6. BS 4660: 1973 Unplasticised PVC underground drainpipes and fittings
7. DOE Advisory leaflet 48: Setting out on site (1975)
8. *BRE Digest 234*: Accuracy in setting-out (1980)
9. Greater London Council. London Building Acts 1930–1939, Constructional By-Laws (1972)
10. The Building Regulations 1976. SI 1676 HMSO (1976) and The Building (First Amendment) Regulations 1978 SI 723. HMSO (1978)
11. *BRE Digest 12*: Structural design in architecture (1969)
12. Ministry of Housing and Local Government. Homes for today and tomorrow (Parker Morris report). HMSO (1963)
13. *BRE Digests 45 and 46*: Design and appearance (1964)
14. I. H. Seeley. *Building Economics: appraisal and control of building design cost and efficiency.* Macmillan (1976)
15. BS 4011: 1966 Recommendations for the co-ordination of dimensions in building. Co-ordinating sizes for building components and assemblies
16. BS 4330: 1968 Recommendations for the co-ordination of dimensions in building. Controlling dimensions. PD 6444: Recommendations for the co-ordination of dimensions in building; Part 1: 1969 Basic spaces; Part 2: 1971 Co-ordinating sizes of fixtures, furniture and equipment
17. *BRE Digest 151*: Soakaways. HMSO (1973)
18. BS 5606: 1978 Code of practice for accuracy in building
19. I. H. Seeley. *Building Maintenance*. Macmillan (1976)

2 BUILDING DRAWING

Building drawing is important as it is often easier to explain building details by drawings or sketches than by written descriptions. Drawing thus forms an effective means of communication and drawings constitute an essential working basis for any building project. It is desirable to achieve maximum uniformity in the presentation of building drawings and this is assisted by implementing the recommendations contained in BS 1192: 1969 (Building drawing practice). This chapter is concerned with the various drawing materials in use and the basic principles to be observed in the preparation of drawings and sketches.

DRAWING INSTRUMENTS AND MATERIALS

Students are advised to purchase good quality drawing instruments which can quite easily last a lifetime. It is possible to purchase complete sets with varying ranges of instruments or to buy instruments singly. A popular and useful set of instruments is illustrated in figure 2.1.

Compasses. These are usually 125 or 150 mm in length and a lengthening bar (figure 2.1.6) permits the drawing of large radius curves. They should be designed to take a pencil point (figure 2.1.1), pen point (figure 2.1.4) or divider point (figure 2.1.5). Ideally they should be self-centring so that even pressure can be maintained on both points. Pen or ink points are often made from tungsten steel for use with plastic tracing film which is highly abrasive. In some cases the instruments are double knee-jointed and the points fitted with fine adjusting devices. The pencil should be sharpened to a chisel point and fitted tangentially to the circle.

Beam compasses. These are used for drawing circles and curves of extra large radius. They consist of attachments which are fastened to a wood lath or metal bar and have needle points and interchangeable pen and pencil points. They can be useful for plotting land surveys but are rarely needed for building drawings.

Spring bow compasses. These are used for drawing small radius

circles and curves. They are obtainable as separate dividers, and pen and pencil compasses (figures 2.1.7, 2.1.8 and 2.1.9).

Many sets of drawing instruments also include the larger *divider compasses* shown in figure 2.1.2. These are useful for setting out larger divisions of equal length.

Ruling pens. These consist of a handle and two blades of equal length connected by an adjusting screw for varying the thickness of line (figure 2.1.3). One blade is normally hinged or pivoted for ease of cleaning. The ink is fed in between the blades and the pen is held in a vertical position and drawn along the edge of a tee square or set square. The pen must be carefully cleaned after use.

In recent years many draughtsmen have used 'Graphos', 'Rapidograph', 'Page' and similar type pens which are a form of fountain pen that take Indian ink and have interchangeable heads or nibs for producing lines of different thicknesses. They are also useful for sketching and writing. Some manufacturers have produced compasses which will take interchangeable ink heads with ink containers.

Freehand pens. These are useful for producing small lettering such as explanatory or descriptive notes on drawings of building details. Larger lettering can be obtained with a Gillott No. 303 nib attached to a standard holder, while mapping pens with fixed or interchangeable nibs are useful for small lettering.

Drawing ink. This is usually black in colour and waterproof, although a wide range of coloured inks is also available. Indian ink is sold in bottles or tubes which must be kept sealed.

Stencils. These are used extensively by draughtsmen, especially for main headings. They are available in a wide range of sizes and in a number of different styles of letters and figures. Special pens are used with the stencils and have ink feeds. They are not suitable for examination purposes.

Pencils. They are produced in a variety of grades and HB is generally found to be most suitable for small-scale drawings and B pencils for large scale work and sketches. It is advisable to use good-quality pencils sharpened to a round point, preferably with a penknife.

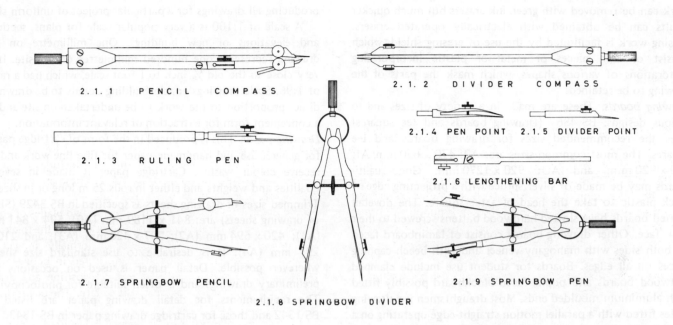

2.1.1 P E N C I L   C O M P A S S

2.1.2 D I V I D E R   C O M P A S S

2.1.3 R U L I N G   P E N

2.1.4 P E N   P O I N T   2.1.5 D I V I D E R   P O I N T

2.1.6 L E N G T H E N I N G   B A R

2.1.7 S P R I N G B O W   P E N C I L

2.1.8 S P R I N G B O W   D I V I D E R

2.1.9 S P R I N G B O W   P E N

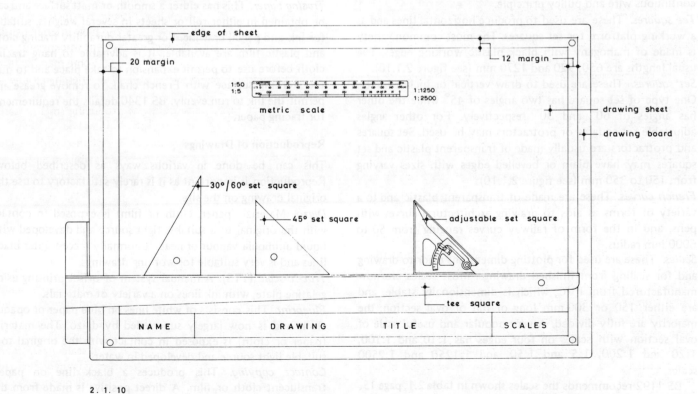

edge of sheet

20 margin

12 margin

metric scale

30°/60° set square

45° set square

adjustable set square

drawing sheet

drawing board

tee square

| NAME | DRAWING | TITLE | SCALES |

2.1.10

| Figure   2.1 | D R A W I N G   I N S T R U M E N T S | |

Erasers. The erasing of pencil work is generally done with soft erasers, although art gum may be used for large surfaces. Ink work can be removed with green ink erasers but much quicker results can be obtained with electrically operated erasers. Erasing work is facilitated by the use of erasing shields which consist of thin pieces of metal or plastic incorporating perforations of various shapes, which mask the parts of the drawing to be retained.

Drawing boards. These are made in a variety of sizes and to various designs. BS 4867 (Drawing boards and tee squares) gives the recommended sizes for drawing boards and tee squares. The most common sizes are: A2: 470 x 650 mm; A1: 650 x 920 mm; and A0: 920 x 1270 mm. Good-quality boards may be made of silver spruce with a projecting edge of black plastic to take the head of a tee square. The dovetail jointed boards have slotted hardwood battens screwed to their back face. Other drawing boards consist of laminboard faced on both sides with mahogany veneer and with beech capping pieces on all edges. Boards for student use include clamped softwood boards and poplar-faced blockboard possibly fitted with aluminium moulded ends. Most draughtsmen use drawing tables fitted with a parallel motion straight-edge operating on a continuous wire and pulley principle.

Tee squares. These are used to produce horizontal lines and as a working platform for set squares. The most common variety is made of mahogany with black plastic working edges. The usual lengths are 650, 920 and 1270 mm (see figure 2.1.10).

Set squares. These are used to draw vertical or inclined lines. One type of set square has two angles of 45° while the other has angles of 60° and 30° respectively. For other angles adjustable set squares or protractors may be used. Set squares and protractors are usually made of transparent plastic and set squares may have plain or bevelled edges with sizes varying from 150 to 350 mm (see figure 2.1.10).

French curves. These are made of transparent plastic and to a variety of forms as aids to drawing architectural curves with pens, and in the form of railway curves ranging from 50 to 6000 mm radius.

Scales. These are used for plotting dimensions prior to drawing and for scaling from finished drawings. Most scales are now manufactured from PVC, which is dimensionally stable, and are either 150 or 300 mm long of flat or oval section; the majority are fully divided. A very popular and useful scale of oval section with scales on four edges has 1:10 and 1:100, 1:20 and 1:200, 1:5 and 1:50, and 1:1250 and 1:2500 scales.

BS 1192 recommends the scales shown in table 2.1, page 15. Choice of scale is influenced by the need to communicate information accurately and adequately to secure maximum output, the nature of the subject and the desirability of producing all drawings for a particular project of uniform size.

A scale of 1:100 is a very popular scale for plans, sections and elevations of new buildings. One millimetre on the drawing represents one hundred millimetres on the site. It is very close to the old $\frac{1}{8}$ inch to 1 foot scale, which had a ratio of 1:96. Scale drawings enable building details to be drawn in direct proportion to the work to be undertaken on site and in a convenient form for extraction of relevant information.

Drawing paper. This is produced in the form of cartridge paper for general use and handmade paper for fine line work and to receive colour washes. Cartridge paper is made in several qualities and weights and either in rolls 25 m long or in sheets. Trimmed sizes of drawing sheets as specified in BS 3429 (Sizes of drawing sheets) are: 841 x 1189 mm (A0); 594 x 841 mm (A1); 420 x 594 mm (A2); 297 x 420 mm (A3); and 210 x 297 mm (A4). It is desirable to use standard size sheets wherever possible. Detail paper is used on occasions for preliminary drawings and is often suitable for photocopying. The requirements for detail drawing paper are listed in BS 1342 and those for cartridge drawing paper in BS 1343.

Tracing paper. This has either a smooth or matt surface and can be obtained in either roll or sheets in several weights, suitable for ink and pencil drawing. For greater durability tracing cloth and plastic film are available. It is advisable to hang tracing cloth before use to permit expansion to take place and to dust the working surface with French chalk to remove grease and permit the ink to run evenly. BS 1340 details the requirements for tracing paper.

Reproduction of Drawings

This can be done in various ways as described below. Reproduction is important as it is rarely satisfactory to use the original drawing on the site.

Diazo. Material (paper, cloth or film) is exposed in contact with the original to a suitable light source and developed with liquid ammonia vapour or heat. It normally incorporates black lines and is very suitable for working drawings.

True-to-scale (TTS). A manual system of offset printing using gelatine plate, with ink lines on a variety of materials.

Blueprint. This consists of white lines on blue paper or opaque cloth and is now largely superseded by diazo. The material (paper or cloth) is exposed in contact with the original to a suitable light source and developed in water.

Contact copying. This produces a black line on paper, translucent cloth or film. A direct positive is made from the translucent original by transmitted light and a negative positive

Table 2.1   *Recommended scales*

Type of drawing	Function of drawing	Scales for use with metric system	Notes
DESIGN STAGE			
Sketch Drawings	preliminary drawings, sketches or diagrams to show designer's general intentions		Scales will vary but it is recommended that preference be given to those used in production stage
PRODUCTION STAGE			
Location drawings Block plan	to identify site and locate outline of building in relation to town plan or other wider context	1:2500* 1:1250*	*Ordnance Survey indicate that due to large costs and labour involved these scales will probably have to be retained for a transitional period
Site plan	to locate position of buildings in relation to setting out point, means of access, general layout and drainage	1:500 1:200	
General location	to show position occupied by various spaces in building, general construction and location of principal elements, components and assembly details	1:200 1:100 1:50	
Component drawings Ranges	to show basic sizes, system of reference and performance data on a set of standard components of a given type	1:100 1:50 1:20	
Details	to show all the information necessary for the manufacture and application of components	1:10 1:5 1:1 (full size)	
Assembly	to show in detail the construction of buildings; junctions in and between elements, between elements and components, and between components	1:20 1:10 1:5	

Source: BS 1192: 1969 Building drawing practice

made by reflex copying of an opaque original to give a laterally reversed first copy and right reading subsequent copies.

Optical copying. A black line is produced on silver sensitised materials from a photographic negative, using a camera and film.

For smaller drawings of size A3 and below three other reproduction processes are available: namely, diffusion transfer, electrophotography and offset litho. These processes are described in BS 1192 (Building drawing practice), and BS 4212: 1967 Guide to the selection of processes for reproducing drawings.

LAYOUT AND PRESENTATION OF DRAWINGS

It is important that drawings shall be logically and neatly arranged to give a balanced layout. BS 1192 recommends that every drawing sheet shall have a filing margin, title and information panel. The title and information panel is located at the bottom right-hand corner of the sheet and incorporates the job title, subject of drawing, scale, date of drawing, job number and revision suffix, and name and possibly address and telephone number of architect or surveyor responsible for the job. Sometimes the names and initials of the persons drawing, tracing and checking the drawing are also added. The student will not need to give so many particulars and can reduce the size of the title and information panel on the lines adopted for the figures in this book. He particularly needs to incorporate his name, the subject of the drawing and the operative scales, as indicated in figure 2.1.10.

Setting out the various parts of a drawing may prove difficult to the student in his early days. It may therefore be helpful to draw rough outlines on tracing paper in the first instance. Ensure that the drawing paper has its smoothest surface upwards and pin the sheet onto the drawing board with a drawing pin at the top left-hand corner. Manipulate the sheet so that the top edge is parallel to the tee square and a second pin can then be inserted at the top right-hand corner of the sheet. With small sheets it is not advisable to use pins at the bottom corners of the sheet as they obstruct the tee square. Another alternative is to fix the sheets with sellotape.

The first step is to draw the border lines using a tee square in both directions. Thereafter all vertical lines are drawn with set squares. The title and information panel should follow, and the space remaining is available for drawing.

Apart from the desirability of securing a balanced arrangement and a nice-looking drawing, it is also possible to save a considerable amount of time by carefully locating the various parts in a logical relationship one with another. Hence when preparing plans, sections and elevations of buildings or component parts, such as windows and doors, it is good practice to draw the plan(s) at the bottom of the sheet and the section(s) on the right-hand side. The elevation(s) can then be drawn by projecting many of the lines upwards from the plan(s) and horizontally across from the section(s). Many examples of this arrangement appear in figures throughout this book.

GENERAL ASPECTS RELATING TO DRAWINGS

Some of the more important general aspects relating to building drawings are now considered. Firstly it is vital that all instruments, scales, tee squares and set squares are kept clean with a suitable cloth or duster. Scale and plot a number of points before connecting them with lines in order to accelerate the drawing process. Always use a scale which is sufficiently large to permit the drawing of details that can easily be read, with adequate space for annotated notes. Draw all lines lightly in the first instance and work from centre lines wherever practicable.

Lines

BS 1192 recommends the use of lines of different thicknesses for specific purposes as follows.

Thick lines. These should be used for site outline of new buildings on block and site drawings; primary functional elements, such as load-bearing walls and structural slabs, on general location and assembly drawings; outlines requiring emphasis on component ranges and horizontal and vertical profiles on component details. Services are shown by thick chain lines.

Medium lines. These should be used for existing buildings on block drawings; general details on site drawings; secondary elements and components, such as non-loadbearing partitions, windows and doors, on general location and assembly drawings; outlines of components, on component ranges; and general details on component details.

Thin lines will be used for reference grids, dimension lines, leader lines and hatching on all types of drawing; centre lines are shown by thin chain lines.

Dimensions

Dimension lines for figured dimensions should always be drawn in positions where they cannot be confused with other information on the drawing. The terminal points to which dimension lines refer must be clearly shown and BS 1192 makes certain recommendations. Open arrows of the type in figure 2.2.1 should be used for basic or modular dimensions or for the sizes of spaces or components taken to grid lines, centre lines or unfinished carcase surfaces. Tolerances or gaps, and the work sizes of components which are specified for manufacture (so that allowing for the tolerances, the actual size lies between the required limits) shall be indicated by solid arrows of the type shown in figures 2.2.2 and 2.2.3.

Dimension figures. These should be written immediately above and along the line to which they relate, as shown in figure 2.2.1. Running dimensions should take the form illustrated in figure 2.2.4. In all cases dimension figures, when not written for viewing from the bottom, should be written only for viewing from the right-hand edge (see figure 2.2.5).

Linear dimensions may be expressed in both metres and millimetres on the same drawing. In order to avoid confusion metres should be given to three places of decimals, for example, 2.1 m should be entered as 2.100. Millimetres will be entered as the actual number involved, for example, a wall thickness of 215. In this way the need for symbols (m and mm) is largely avoided. Summing up, the approach to be adopted for metric dimensions is

(1) whole numbers indicate millimetres;
(2) decimalised expressions to three places of decimals indicate metres;
(3) all other dimensions should be followed by the unit symbol.

The sequence of dimensions of components must be in the order of (a) length, (b) width and (c) depth or height. New levels, such as floor levels, should be distinguished from existing ground levels by inserting them in boxes; for example, a finished floor level could be shown as FFL 150.750 .

Graphical Symbols

Materials in section on plans and vertical sections are best hatched to assist in interpreting the drawings. Hatching is preferable to colouring which is costly, laborious and conducive to error. Hatchings representing the more commonly used materials are contained in BS 1192 and these are reproduced in figure 2.2.6. The use of nationally recognised hatchings leads to uniformity in the presentation of drawings and enables them to be more readily and easily understood.

BS 1192 also contains a large number of graphical symbols representing components in connection with services, installations and fixtures and fittings, and a selected sample is shown in figure 2.2.7. The British Standard also contains a list

of abbreviations whose use can result in a reduction in the length of explanatory notes on drawings. A selected sample of these officially recognised abbreviations follows.

Aggregate	Agg	Hardcore	hc
Air brick	AB	Hardboard	hdb
Aluminium	al	Hardwood	hwd
Asbestos	abs	Inspection chamber	IC
Asphalt	asph	Insulation	insul
Boarding	bdg	Joist	jst
Brickwork	bwk	Mild steel	MS
Cast iron	CI	Pitch fibre	PF
Cement	ct	Plasterboard	pbd
Concrete	conc	Rainwater pipe	RWP
Copper	copp	Reinforced concrete	RC
Cupboard	cpd	Softwood	swd
Damp-proof course	DPC	Stainless steel	SS
Damp-proof membrane	DPM	Tongue and groove	T & G
Foundation	fdn	Vent pipe	VP
Glazed pipe	GP	Wrought iron	WI

Lettering

Lettering is needed on drawings to provide information which could not otherwise be obtained. The aim should be to produce uniform, neat and easily legible lettering. General notes may with advantage be collected in groups but more specific particulars should be inserted as near as possible to the items to which they relate. Care must however be taken not to obscure any part of the drawing with lettering or lines linking lettering and details.

Lettering can take a variety of forms from block letters to small case (italics), or even neatly handwritten notes. It should be appreciated that the lettering on a drawing can quite easily occupy one-third of the total time required to prepare the drawing. Hence if time was short in the examination, a candidate would be justified in completing drawings to later questions with neatly handwritten notes, provided that he had earlier lettered up a drawing adopting the more orthodox approach. Each individual adopts his own style of lettering — vertical or sloping, thin or broad, simple or more flamboyant. Generally, a student would be well advised to adopt a fairly simple style of lettering and to gain adequate practice to achieve a fair standard of competence. Small-case lettering is best suited for notes and block letters for headings.

In any event, in the early stages, the students should produce his lettering between faint parallel lines. The letters and figures should be well formed with uniform spacing between the letters, using a fairly soft pencil, such as a B grade, kept

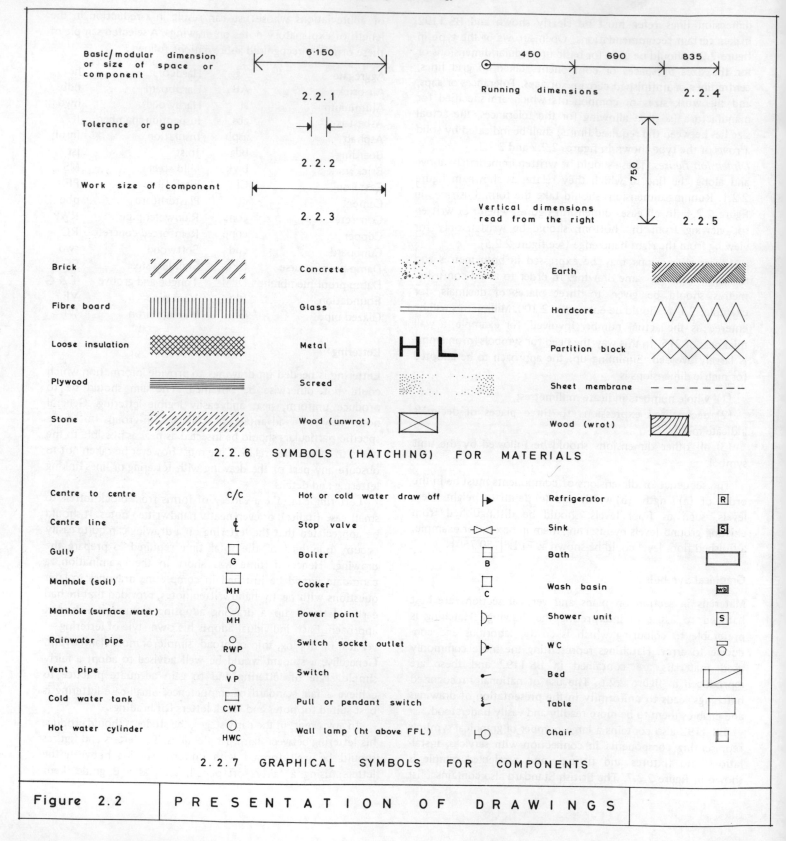

Basic/modular dimension or size of space or component   6·150   **2.2.1**

    450   690   835

Running dimensions   **2.2.4**

Tolerance or gap   **2.2.2**

Work size of component   **2.2.3**

750

Vertical dimensions read from the right   **2.2.5**

Brick	Concrete	Earth
Fibre board	Glass	Hardcore
Loose insulation	Metal **H L**	Partition block
Plywood	Screed	Sheet membrane
Stone	Wood (unwrot)	Wood (wrot)

2.2.6   SYMBOLS (HATCHING) FOR MATERIALS

Centre to centre	C/C	Hot or cold water draw off		Refrigerator	R
Centre line	₵	Stop valve		Sink	S
Gully	G	Boiler	B	Bath	
Manhole (soil)	MH	Cooker	C	Wash basin	wb
Manhole (surface water)	MH	Power point		Shower unit	S
Rainwater pipe	RWP	Switch socket outlet		WC	
Vent pipe	VP	Switch		Bed	
Cold water tank	CWT	Pull or pendant switch		Table	
Hot water cylinder	HWC	Wall lamp (ht above FFL)		Chair	

2.2.7   GRAPHICAL SYMBOLS FOR COMPONENTS

Figure 2.2	P R E S E N T A T I O N O F D R A W I N G S

continuously sharpened. BS 1192 recommends heights of 5 to 8 mm for headings and 1.5 to 4 mm for notes. Three styles of lettering are illustrated in figure 2.3.1. The first is a simple broad, vertical style for headings, the second a thinner, sloping and more elaborate style also suitable for headings, and the third is small case for notes.

SKETCHES

In recent years some examining bodies have introduced sketches into technology questions as a substitute for scale drawings. This approach is based on the philosophy that more drawings can be produced in a given period by sketching than by scale drawing. Hence the examination candidate has a greater opportunity to demonstrate his knowledge of the subject and of his ability to apply this knowledge to practical construction problems. Furthermore, the drawing process is simply a means of communication and the main aim of a building technology examination is not to test the ability of the candidate in producing scale drawings.

Some students have expressed concern at the move towards sketches, using the argument that they never have been and never will be artists. There is something of a fallacy here. Sketching requires practice in the same way as scale drawing does and the majority of students can acquire a reasonable skill over a period of time. There are several basic rules to be observed

(1) all sketches should be fairly large: it is difficult to show sufficient details on small sketches;

(2) sketches should be in correct proportions and one useful approach is to plot a few leading dimensions to a suitable scale before starting to sketch;

(3) use a soft pencil (a B grade pencil is often used); faint lines can be thickened up subsequently;

(4) develop the ability to move the hand down or across the paper working from the shoulder or elbow, and not merely the fingers;

(5) start by sketching simple objects like garden sheds, summerhouses and garages: valuable experience can be obtained by taking a notebook on one's travels and sketching buildings or building details;

(6) to give the sketches a measure of solidity, establish a source of light and shade all faces which are hidden from it.

Some typical sketches are shown in figures 7.5, 9.5, 12.1 and 13.3.

WORKING DRAWINGS

The majority of working drawings consist of plans, sections and elevations drawn by orthographic projection, whereby they are all in flat planes. The working drawings form one of the most effective ways of conveying the designer's requirements to the contractor, to assist him in constructing the work on the site.

Plan. This represents a view from above of an object on a horizontal plane. With buildings, plans are normally drawn of each floor at about one metre above floor level, looking down at the floor and cutting through walls, doors and windows.

The drawing represents a sectional plan of the walls, doors and windows, as if they were cut open. A roof plan looks down on the roof which is opened up to show the roof members.

Sections. These are taken vertically through a building cutting through the foundations, walls, partitions, floors, roof, windows and doors, to show the form of construction. Some features such as internal doors, wall tiling, skirtings and picture rails will appear in elevation where they are seen in the background.

Elevations. These represent external faces of a building including windows and doors.

BS 1192 describes how, in theory, with orthographic projection one or more views of an object are obtained by dropping perpendiculars from all significant points on the object to one or more planes of projection. These perpendiculars are known as projectors and the principal planes of projection are assumed to be at right angles to one another.

Figure 2.3.2 illustrates plans, sections and elevations of an electricity switch house and is a good example of orthographic projection. These drawings are drawn to a scale of 1:200 because of the limitations of page size, although 1:100 is a more common scale for working drawings of this type. Drawings are projected one from another as far as possible using a tee square and set square. An alternative scale is 1:50 which permits the inclusion of greater detail and provides more space for descriptive notes and dimensions.

Other forms of projection are used on occasions to make drawings more easily understood.

Isometric projection. This is a common form of projection in which length, breadth and height of the object are shown on the one drawing. An example of a house drawn to isometric projection is illustrated in figure 2.4.3 from the plan in figure 2.4.1 and elevation in figure 2.4.2. All vertical lines are drawn as verticals and horizontal lines at 30° to the horizontal. A layman can far more readily appreciate and comprehend the

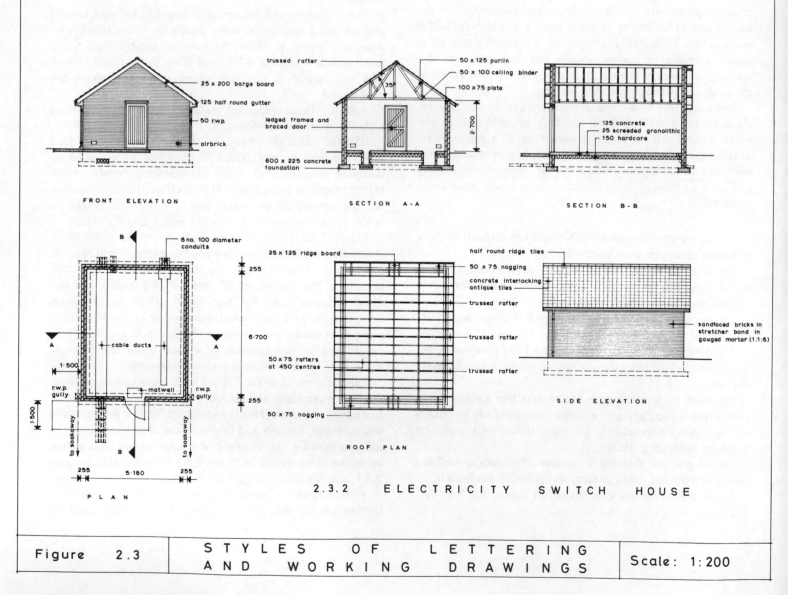

ABCDEFGHIJKLMNOPQRSTUVWXYZ

ABCDEFGHIJKLMNOPQRSTUVWXYZ

abcdefghijklmnopqrstuvwxyz   1234567890

2.3.1   STYLES   OF   LETTERING

FRONT ELEVATION

- 25 x 200 barge board
- 125 half round gutter
- 50 r.w.p.
- airbrick

SECTION A-A

- trussed rafter
- 35°
- ladged framed and braced door
- 600 x 225 concrete foundation
- 50 x 125 purlin
- 50 x 100 ceiling binder
- 100 x 75 plate
- 2·700

SECTION B-B

- 125 concrete
- 25 screeded granolithic
- 150 hardcore

PLAN

- B
- 8 no. 100 diameter conduits
- cable ducts
- 1·500
- r.w.p. gully
- matwell
- r.w.p. gully
- to soakaway
- to soakaway
- 255
- 5·180
- 255
- A
- A
- B

ROOF PLAN

- 25 x 125 ridge board
- 50 x 75 nogging
- trussed rafter
- trussed rafter
- 50 x 75 rafters at 450 centres
- trussed rafter
- 50 x 75 nogging
- 255
- 6·700
- 255

SIDE ELEVATION

- half round ridge tiles
- concrete interlocking antique tiles
- sandfaced bricks in stretcher bond in gauged mortar (1:1:6)

2.3.2   ELECTRICITY   SWITCH   HOUSE

| Figure   2.3 | STYLES   OF   LETTERING AND   WORKING   DRAWINGS | Scale: 1:200 |

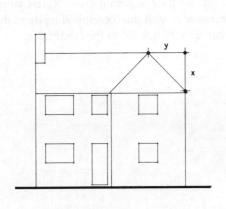

2.4.2   E L E V A T I O N

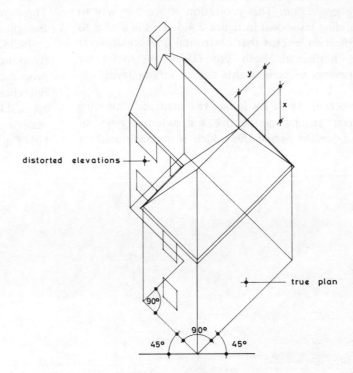

distorted elevations

true plan

90°

90°

45°   45°

2.4.4   A X O N O M E T R I C   V I E W

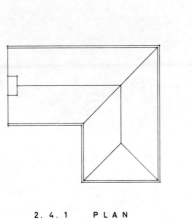

2.4.1   P L A N

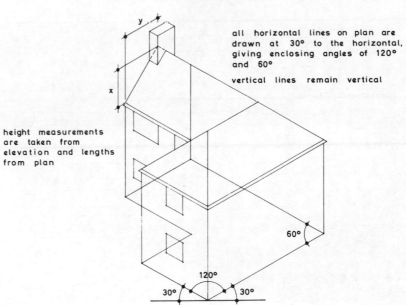

all horizontal lines on plan are
drawn at 30° to the horizontal,
giving enclosing angles of 120°
and 60°

vertical lines remain vertical

height measurements
are taken from
elevation and lengths
from plan

60°

120°

30°   30°

2.4.3   I S O M E T R I C   V I E W

| Figure 2.4 | I S O M E T R I C   A N D A X O N O M E T R I C   P R O J E C T I O N S | |

nature and appearance of the building from the isometric sketch than he can from the plan and elevation.

Axonometric projection. This projection is another way to draw the building, as shown in figure 2.4.4. This is similar to isometric projection, except that horizontal lines are drawn at 45° to the horizontal. Both projections are useful for preparing drawings of components and of service layouts in buildings.

Pictorial projection. There are in fact two methods of drawing in perspective: angular perspective for external views of buildings and parallel perspective which is normally used for interiors. Both methods provide projections from a spectator situated some distance from the building or room being drawn. These processes are not considered in detail as they are thought to be outside the scope of this book.

Building Research Establishment Current Paper 18/73 (Working drawings in use) draws attention to the need for the proper classification of drawings with appropriate titles and references to other drawings for further information where applicable, to assist in their dissemination. Referencing of materials and components with the corresponding items in the bill of quantities can be very helpful to the contractor.

3 FOUNDATIONS

It is imperative that the foundations of a building be properly designed to spread the dead and superimposed loads over a sufficient area of soil. In this context 'soil' is that part of the earth which lies below the topsoil and above the rock, having being formed by the erosion of the earth's crust by water, atmospheric means, and intense pressure over many thousands of years. This involves an understanding of soil types and their characteristics and awareness of the different types of foundation that are available. Chapter 1 dealt with site investigations and these included an examination of soil conditions and the level of the water table. These aspects are becoming even more important as some building sites now occupy land which has been avoided in the past. Building Research Establishment Digest 64[1] suggests an initial approach to the local authority with its intimate knowledge of soil and general conditions in the area. Older editions of Ordnance Survey maps may provide useful information on features that cause difficulty, such as infilled ponds, ditches and streams, disused pipes and sites of old buildings and tips. A polygonal pattern of cracks about 25 mm wide on the ground surface during a dry summer, indicates a shrinkable soil. Whereas larger cracks, roughly parallel to one another generally indicate deeper-seated movements and may be caused by mining, brine pumping, or landslips.[1]

IDENTIFICATION AND CHARACTERISTICS OF SOILS

Table 3.1 is extracted from Building Research Establishment Digest 64[1] and classifies soil types; it shows how they are identifiable in the field, and details possible foundation difficulties. Quantitative tests in the field and laboratory will be necessary where comprehensive information on soil conditions is required. In this connection reference to CP 2001[2] and BS 1377[3] should prove useful.

CP 101[4] provides a basis for field identification of soil in terms of the predominant size of soil particle and the strength features which have an important influence on foundation behaviour; the following descriptions are extracted from the Code.

Non-cohesive Soils

Gravel. A natural deposit consisting of rock fragments in a matrix of finer and usually sandy material. Many of the particles are larger than 2 mm in size.

Sand. A natural sediment consisting of the granular and mainly siliceous products of rock weathering. It is gritty with no real plasticity. The particles normally range between 0.06 and 2.00 mm in size.

Well-graded sand. A sand containing a proportion of all sizes of sand particles with a predominance of the coarser grades.

Compact gravel and sand. Deposits require a pick for removal and offer high resistance to penetration by excavating tools.

Loose gravel and sand. Deposits readily removable by hand-shovelling only.

Uniform or poorly-graded sand. The majority of particles lie within a fairly restricted size range.

Cohesive Soils

Clay. A natural deposit consisting mainly of the finest siliceous and aluminous products of rock weathering. It has a smooth, greasy touch, sticks to the fingers and dries slowly. It shrinks appreciably on drying and has considerable strength when dry.

Stiff clay. A clay which requires a pick or pneumatic spade for its removal and cannot be moulded with the fingers at its natural moisture content.

Firm clay. A clay which can be excavated with a spade and can be moulded by substantial pressure with the fingers at its natural moisture content.

Very soft clay. Extruded between fingers when squeezed in fist at its natural moisture content.

Soft clay. A clay which can be readily excavated and can be easily moulded with the fingers at its natural moisture content.

Table 3.1   Soil identification

Soil type	Field identification	Field assessment of structures and strength	Possible foundation difficulties
Gravels	Up to 76.2 mm (retained on No. 7 BS sieve) Some dry strength indicates presence of clay	Loose — easily removed by shovel 50 mm stakes can be driven well in	Loss of fine particles in water-bearing ground
Sands	Pass No. 7 and retained on No. 200 BS sieve Clean sands break down completely when dry. Individual particles visible to the naked eye and gritty to fingers	Compact — requires pick for excavation. Stakes will penetrate only a little way	Frost heave, especially on fine sands
Silts	Pass No. 200 BS sieve. Particles not normally distinguishable with naked eye Slightly gritty; moist lumps can be moulded with the fingers but not rolled into threads Shaking a small moist lump in the hand brings water to the surface Silts dry rapidly; fairly easily powdered	Soft — easily moulded with the fingers Firm — can be moulded with strong finger pressure	As for fine sands
Clays	Smooth, plastic to the touch, sticky when moist. Hold together (when dry). Wet lumps immersed in water soften without disintegrating Soft clays either uniform or show horizontal laminations Harder clays frequently fissured, the fissures opening slightly when the overburden is removed or a vertical surface is revealed by a trial pit	Very soft — exudes between fingers when squeezed Soft — easily moulded with the fingers Firm — can be moulded with strong finger pressure Stiff — cannot be moulded with fingers Hard — brittle or tough	Shrinkage and swelling caused by vegetation Long-term settlement by consolidation Sulphate-bearing clays attack concrete and corrode pipes Poor drainage Movement down slopes; most soft clays lose strength when disturbed
Peat	Fibrous, black or brown Often smelly Very compressible and water retentive	Soft — very compressible and spongy Firm — compact	Very low-bearing capacity; large settlement caused by high compressibility Shrinkage and swelling — foundations should be on firm strata below
Chalk	White — readily identified	Plastic — shattered, damp and slightly compressible or crumbly Solid — needing a pick for removal	Frost heave Floor slabs on chalk fill particularly vulnerable during construction in cold weather Swallow holes
Fill	Miscellaneous material — for instance rubble, mineral, waste, decaying wood		To be avoided unless carefully compacted in thin layers and well consolidated May ignite or contain injurious chemicals

Source: BRE Digest 64[1]

Boulder clay. A deposit of unstratified clay or sandy clay containing subangular stones of various sizes.

Silt. A natural sediment of material of finer grades than sand. Most of the grains will pass a 75 micrometer test sieve. It shows some plasticity, is not very gritty and has appreciable cohesion when dry.

Soil Identification and Classification Tests

Soils may be subjected to a number of tests to establish their identity and classify them. Some of the more important tests will now be briefly described.

Particle size distribution. This test determines the proportion of gravel, sand, silt and clay in a particular soil. The soil is dried and sieved through a nest of sieves, and the weight retained on each soil is recorded. The results of the grading test are plotted on a graph. Sand has particles between 0.060 and 2.000 mm, silt between 0.060 and 0.002 mm and clay is less than 0.002 mm.

Liquid limit test. The object is to determine the moisture content at which the soil passes from plastic to liquid state. The moisture content is expressed as a percentage of the dry weight of the soil.

Plastic limit test. This determines the moisture content at which the soil ceases to be plastic (soil sample can be rolled into a thread 3 mm diameter without breaking).

Plasticity index. This refers to the difference between liquid and plastic limits.

Casagrande devised a soil classification chart and a group symbol of two letters for each soil, the first letter representing the size of the soil particles and the second letter its main characteristic or property, as indicated below.

First letter		Second letter	
Gravel	G	Fines	F
Sand	S	High compressibility	H
Silt	M	Intermediate compressibility	I
Clay	C	Low compressibility	L
Organic silts and clays	O	Poorly graded	P
		Well graded	W
		Clay	C

A poorly graded sand would have a group symbol of SP, if it were well graded it would be SW, and if it were a sandy clay it would be SC.

Other tests include a dry density test to determine the density of the dry soil in its natural position, the standard penetration and consolidation tests to determine the compressibility and the shear vane, unconfined compression and triaxial compression tests to determine the shear strength parameters; both the compressibility and shear strength parameters are needed to ascertain the allowable bearing pressure.

DESIGN OF FOUNDATIONS

The primary aim must always be to spread the loads from the building over a sufficient area of soil to avoid undue settlement, particularly unequal settlement. The Building Regulations 1976[5] lay down the following requirements for the foundations of a building. They shall

(1) safely sustain and transmit to the ground the combined dead load, imposed load and wind load in such a manner as not to cause any settlement or other movement which would impair the stability, or cause damage to, the whole or any part of the building or of any adjoining building or works;

(2) be taken down to such a depth, or be so constructed, as to safeguard the building by swelling, shrinking or freezing of the subsoil;

(3) be capable of adequately resisting any attack by sulphates or any other deleterious matter present in the subsoil.

CP 101[4] includes a table of permissible bearing pressures for different soils and these are incorporated in table 3.2. The dead, imposed and wind loads of a building are calculated in accordance with the principles laid down in CP 3, chapter V Parts 1 and 2.[6] For instance the imposed load on a floor to a house is likely to be about 1.50 kN/m^2. In this way the total load in kilonewtons per linear metre of wall can be calculated, and dividing this by the safe bearing capacity of the soil in kilonewtons per square metre will give the required width of foundation in metres. The Building Regulations[5] also contain a schedule giving the minimum width of strip foundation for varying total loads expressed in kilonewtons per linear metre of wall for use on various soils. For example, with gravel and sand the minimum width ranges from 250 to 650 mm for loads varying from 20 to 70 kN/m. For firm clay or sandy clay the corresponding widths are 300 and 850 mm. With soft silt, clay, sandy clay or silty clay, the foundation widths are 450 mm for a loading of 20 kN/m and 650 mm for 30 kN/m. For greater loadings it would be advisable to pile or use a raft foundation. The table to Building Regulation D7 is illustrated in diagrammatic form in figure 3.1.

Building Research Establishment Digest 67[8] describes how a typical two-storey semi-detached house of about 85 m^2 floor area, in cavity brickwork, with lightweight concrete or clay

Total load of loadbearing walling not more than:					
20 kN/m	30 kN/m	40 kN/m	50 kN/m	60 kN/m	70 kN/m

Rock which is not inferior to sandstone, limestone or firm chalk and which requires at least a pnematic or other mechanically operated pick for excavation is in each case to be equal to width of wall.

Gravel or sand which is compact and requires pick for excavation. A 50mm square wooden peg hard to drive beyond 150mm.

250mm	300mm	400mm	500mm	600mm	650mm

Clay or sandy clay which is stiff, and cannot be moulded with the fingers and requires a pick or pneumatic or other mechanically operated spade for its removal.

250mm	300mm	400mm	500mm	600mm	650mm

Clay or sandy clay which is firm but can be moulded by substantial pressure with the fingers and can be excavated with graft or spade.

300mm	350mm	450mm	600mm	750mm	850mm

Sand, silty sand and clayey sand which is loose and can be excavated with a spade. A 50mm square wooden peg can be easily driven.

400mm	600mm	If total load exceeds 30kN/m foundations do not fall within provisions of Regulation D 7			

Silt, clay, sandy clay and silty clay which is soft and is fairly easily moulded in the fingers and readily excavated.

450mm	650mm	If total load exceeds 30kN/m foundations do not fall within provisions of Regulation D 7			

Silt, clay, sandy clay and silty clay where a natural sample in winter conditions exudes between fingers when squeezed in fist.

600mm	850mm	If total load exceeds 30kN/m foundations do not fall within provisions of Regulation D 7			

Figure 3.1	MINIMUM WIDTH OF STRIP FOUNDATIONS (AS TABLE TO BUILDING REGULATION D 7)	

Table 3.2   *Typical bearing capacities of soils* (Source: CP 101[4])

Soil	Typical bearing capacity kN/m²
Non-cohesive	
Compact gravel or compact sand and gravel	> 600
Medium dense gravel or medium dense gravel and sand	200 to 600
Loose sand or loose sand and gravel	< 200
Compact sand	> 300
Medium dense sand	100 to 300
Loose sand	< 100
Cohesive	
Very stiff boulder clays and hard clays	300 to 600
Stiff clays	150 to 300
Firm clays	75 to 150
Soft clays and silts	75
Very soft clays and silts	< 75
Other soils	
Peat	Foundations should be carried down through peat and organic soil to a reliable bearing stratum below.
Made ground	It should be investigated with extreme care.

block partitions, timber floors and tiled roof, exerts a combined load approaching 1000 kN, excluding the weight of foundations. The loads at ground level are, in kilonewtons per linear metre of wall, approximately: party wall 50, gable end wall 40, front and back walls 25 and internal partitions less than 15. When the foundations have been designed, their weight must be added to the loadings already calculated so as to obtain the total bearing pressure on the soil beneath.

A load applied through a foundation always causes settlement as it compresses the soil beneath it. However, not even uniform ground uniformly loaded settles evenly, and the complex properties of soil make it difficult to assess the settlement of individual foundations or to predict the distortion of complete buildings.

Clays which shrink on drying and swell again when wetted often cause movement in shallow foundations. Shrinkage of clays occurs both horizontally and vertically, and so there is a tendency for walls to be drawn outwards, in addition to settling, and for cracks to open between the clay and the sides of the foundations. Water may enter the cracks during the following winter and soften the clay against or below the foundations.

Trees which are close to buildings on a clay soil can cause extensive damage, particularly fast-growing trees with widespread roots like poplars. The tree roots extract water from the soil causing the clay to shrink due to the reduction in moisture content. Buildings on shallow foundations should not be closer to single trees than their height on maturity and not closer to clumps or rows of trees than one-and-a-half times the mature height of the trees. Adequate time should be allowed after felling trees on new building sites to allow time for the clay to regain its water content.

With beds of sand, fine particles can on occasions be washed out of the bed reducing its stability. During severe winters, frost may penetrate soil to a depth of 600 mm or more. If the water table is close to the ground surface and the spaces between the soil particles are within a certain range, as with fine sands, silts and chalk, water can move into the frozen zone and form ice lenses of increasing thickness. The ground surface is lifted and this movement is termed *frost heave*.[7]

FOUNDATION TYPES AND THEIR SELECTION

There are a number of different foundation types available for use with domestic and small industrial and commercial buildings. Their selection is influenced by the type of building and the nature of the loadings, and the site conditions. The more common types of foundation are now described.

Strip Foundations

The majority of buildings up to four storeys in height have strip foundations, in which a continuous strip of concrete provides a continuous ground bearing under the loadbearing walls. A typical strip foundation is illustrated in figure 3.2.1. This type of foundation is placed centrally under the walls and

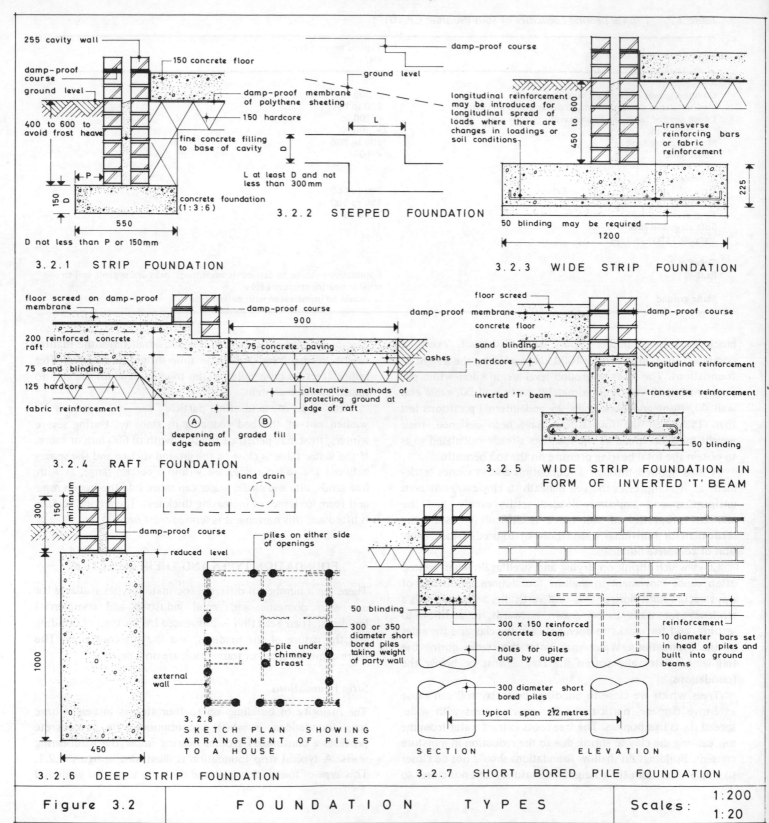

255 cavity wall

damp-proof course

ground level

150 concrete floor

damp-proof membrane of polythene sheeting

150 hardcore

400 to 600 to avoid frost heave

fine concrete filling to base of cavity

150

D

P

concrete foundation (1:3:6)

550

D not less than P or 150mm

3.2.1   STRIP FOUNDATION

damp-proof course

ground level

L

D

L at least D and not less than 300mm

3.2.2   STEPPED FOUNDATION

damp-proof course

longitudinal reinforcement may be introduced for longitudinal spread of loads where there are changes in loadings or soil conditions

450 to 600

transverse reinforcing bars or fabric reinforcement

225

50 blinding may be required

1200

3.2.3   WIDE STRIP FOUNDATION

floor screed on damp-proof membrane

200 reinforced concrete raft

75 sand blinding

125 hardcore

fabric reinforcement

damp-proof course

900

75 concrete paving

ashes

alternative methods of protecting ground at edge of raft

Ⓐ deepening of edge beam

Ⓑ graded fill

land drain

3.2.4   RAFT FOUNDATION

floor screed

damp-proof membrane

concrete floor

sand blinding

hardcore

damp-proof course

longitudinal reinforcement

inverted 'T' beam

transverse reinforcement

50 blinding

3.2.5   WIDE STRIP FOUNDATION IN FORM OF INVERTED 'T' BEAM

300

150 minimum

damp-proof course

1000

450

3.2.6   DEEP STRIP FOUNDATION

piles on either side of openings

reduced level

pile under chimney breast

external wall

3.2.8   SKETCH PLAN SHOWING ARRANGEMENT OF PILES TO A HOUSE

50 blinding

300 x 150 reinforced concrete beam

holes for piles dug by auger

300 or 350 diameter short bored piles taking weight of party wall

300 diameter short bored piles

typical span 2½ metres

SECTION

reinforcement

10 diameter bars set in head of piles and built into ground beams

ELEVATION

3.2.7   SHORT BORED PILE FOUNDATION

| Figure 3.2 | FOUNDATION TYPES | Scales: 1:200 1:20 |

is generally composed of plain concrete often to a mix of 1:3:6 by volume (1 part cement, 3 parts sand and 6 parts coarse aggregate, usually gravel), with the thickness being not less than the projection of the foundation and in no case less than 150 mm. The Building Regulations 1976[5] specify a concrete mix of 50 kg of cement to not more than 0.1 m^3 of fine aggregate (sand) and 0.2 m^3 of coarse aggregate. Internal walls may be supported by separate strip foundations or by thickening the concrete oversite.

On a sloping site the most economical procedure is to use a *stepped foundation* (figure 3.2.2), thus reducing the amount of excavation, backfill, surplus soil removal and trench timbering. The foundation is stepped to follow the line of the ground and the depth of each step is usually 150 or 225 mm (multiple of brick courses) and the lap of concrete at the step must not be less than the depth of the concrete foundation and in no case less than 300 mm. The damp-proof course may also be stepped in a like manner. Where the slope exceeds 1 in 10, it is desirable to use short bored pile foundations to overcome the sliding tendency.

Figure 3.2.1 also shows the linking of the damp-proof course in the external wall with the damp-proof membrane under the floor by means of a short vertical connecting damp-proof course, so that the ground floor is completely tanked. Vegetable soil is removed from over the whole area of the building and in figure 3.2.1 is replaced by suitable hardcore which forms a base for the solid concrete ground floor. Alternatively the damp-proof membrane may be laid under the floor screed.

Wide-Strip Foundations

Where the loadbearing capacity of the ground is low, as for example with marshy ground, soft clay silt and 'made' ground, wide strip foundations may be used to spread the load over a larger area of soil. It is usual to provide transverse reinforcement to withstand the tensions that will arise (see figure 3.2.3). The depth below ground level should be the same as for orthodox strip foundations. All reinforcement should be lapped at corners and junctions. If there is any danger of the foundation failing as a beam in the longitudinal direction, it may be necessary to use a reinforced inverted 'T' beam of the type illustrated in figure 3.2.5.

Deep-Strip Foundations

This type of foundation was first introduced to reduce the expense entailed in constructing orthodox strip foundations to depths of 900 mm or more in shrinkable clay soils, to counteract the variable soil conditions at different seasons. In reducing the width of the foundation trench, the quantity of excavation, backfill and surplus soil removal are also reduced. The deeper foundation also provides greater resistance to fracture from unequal settlement.

In more recent times concrete trench fill has been advocated as a more economical substitute for the traditional strip foundation with brickwork below ground in many situations. There can also be time savings due to the quicker completion with concrete trench fill. A typical deep strip foundation is illustrated in figure 3.2.6.

Raft Foundations

These cover the whole area of the building and usually extend beyond it. They consist primarily of a reinforced concrete slab up to 300 mm thick which is often thickened under load-bearing walls. The level of the base of the raft is usually within 300 mm of the surface of the ground and the reinforcement is often in the form of two layers of fabric reinforcement, one being near the top and other near the bottom of the slab (figure 3.2.4).

Raft foundations are best suited for use on soft natural ground or fill, or on ground that is liable to subsidence as in mining areas. The ground at the edge of the raft should be protected from deterioration by the weather and this can be achieved in one of three ways

(1) laying concrete paving around the building as shown in figure 3.2.4;

(2) deepening the edge beam as illustrated in figure 3.2.4.(A);

(3) laying a field drain in a trench filled with suitable fill as shown in figure 3.2.4.(B).

Flexible joints should be provided on services where they leave the raft. Design of the raft involves the calculation of the loads to be carried and careful assessment of the disposition and distribution of these loads.[4] The primary advantage over strip foundations is the ability of the raft foundation to act as a single unit, thus eliminating differential settlement.

Pad Foundations

These are isolated foundations to support columns. The area of foundation is determined by dividing the column load plus the weight of the foundation by the allowable bearing pressure of the ground. The thickness of the foundation must not be less than the projection from the column (unless reinforced) and must in no case be less than 150 mm. The size of the foundation can be reduced by providing steel reinforcement towards the bottom of the foundation running in both directions.

Grillage Foundations

These transmit very heavy loads by reinforced concrete or steel grillages.

Short-Bored Pile Foundations

These were devised to provide economical and satisfactory foundations for houses built on shrinkable clay. They consist of a series of short concrete piles, cast in holes bored in the ground and spanned, for loadbearing walls, by light beams usually of reinforced concrete. They have several advantages over strip foundations through speed of construction, reduced quantity of surplus spoil and ability to proceed in bad weather. Problems do however arise on stony sites or where there are many tree roots.

Holes are bored to a depth of 2½ to 3½ m by a hand or mechanically operated auger, keeping the holes vertical and on the centre line of the beams. The piles, generally about 300 to 350 mm diameter, should be cast immediately after the hole has been bored. A mix of 1:2:4 concrete is generally used. Short lengths of 20 mm reinforcing bar should be set in the top of each corner pile and bent over to be cast in with the beams. The reinforced concrete beams, often 300 x 150 mm in section, are usually cast in formwork but in some cases may be laid in trenches.

The distribution of the piles is influenced by the design of the building, the loads to be carried and the loadbearing capacity of the piles. With loadbearing walls, piles should be provided at corners and junctions of walls, and under chimney stacks with intervening piles located to give uniform loading and, so far as possible, to keep ground floor door and window openings midway between piles (figure 3.2.8). A short bored pile is detailed in figure 3.2.7.

Piled Foundations

These are frequently used with multi-storey buildings and in cases where it is necessary to transmit the building load through weak and unstable soil conditions to a lower stratum of sufficient bearing capacity. Building Research Establishment Digest 95[10] gives advice on choosing the correct type of pile for a particular job. Piles may be classified in several ways. With *end-bearing* piles, the shaft passes through soft deposits and the base or point rests on bedrock or penetrates dense sand or gravel, and the pile acts as a column. A *friction* pile is embedded in cohesive soil and obtains its support mainly by the adhesion or 'skin friction' of the soil on the surface of the shaft. Another method of pile classification relates to 'displacement' piles where soil is forced out of the way as the pile

is driven, and 'replacement' piles where the hole is bored or excavated in the soil and the pile is formed by casting concrete in the hole. Preformed solid piles of timber or reinforced concrete, and concrete or steel tubes or 'shells' with the lower end closed are examples of 'displacement' piles. The choice of pile depends on soil conditions, economic considerations and structural requirements. Sometimes piles are linked by beams to carry loadbearing walls.

Table 3.3 gives guidance on the choice of foundation for a variety of site conditions (extracted from *Building Research Establishment Digest 67*[8]).

EXCAVATION AND TRENCH TIMBERING

Preliminary Siteworks

BS Code of Practice, CP 101[4] recommends that bulk excavation and filling should if possible be carried out as soon as the site is cleared. The filling should be placed in thin layers, preferably not exceeding 150 mm deep, and consolidated at a moisture content which will ensure adequate compaction. As indicated in chapter 1, if the contours of the site are such that surface water will drain towards the building, land or other drains should be laid to divert the water from the vicinity of the building. The Building Regulations 1976[5] require the site of a building to be cleared of turf and other vegetable matter, and wherever dampness or position of the site of a building renders it necessary, the subsoil of the site shall be effectively drained or such other steps taken to protect the building against damage from moisture.

Excavation

Excavations should be cleared of water before concrete is deposited and those in clay, soft chalk or other soils likely to be affected by exposure to the atmosphere should, wherever practicable, be concreted as soon as they are dug. When this is not possible it is advisable to protect the bottom of the excavation with a 75 mm layer of lean concrete blinding, or to leave the last 50 to 75 mm of excavation until the commencement of concreting. Any excess excavation should be refilled with lean concrete; a mix of 1:12 is suitable. The method of excavation will depend on the nature and size of job and the soil conditions. The bulk of excavation work is now performed by machine. The principal machines are now briefly described.

Dragline. A bucket is dragged towards the machine, and it generally excavates below its own level.

Face shovel. This digs in deep faces above its own level.

Table 3.3   Choice of foundation

Soil type and site condition	Foundation	Details	Remarks
Rock, solid chalk, sands and gravels or sands and gravels with only small proportions of clay, dense silty sands	Shallow strip or pad foundations as appropriate to the loadbearing members of the building	Breadth of strip foundations to be related to soil density and loading (see figure 3.1). Pad foundations should be designed for bearing pressures tabled in CP 101.[4] For higher pressures the depth should be increased and CP 2004: 1972[4] (Foundations) consulted	Keep above water wherever possible. Slopes on sand liable to erosion. Foundations 0.5 m deep should be adequate on ground susceptible to frost heave although in cold areas or in un-heated buildings the depth may have to be increased. Beware of swallow holes in chalk
Uniform, firm and stiff clays: (1) Where vegetation is insignificant	Bored piles and ground beams, or strip foundations at least 1.0 m deep	Deep strip foundations of the narrow width shown in figure 3.2.6 can conveniently be formed of concrete up to the ground surface	
(2) Where trees and shrubs are growing or to be planted close to the site	Bored piles and ground beams	Bored pile dimensions as in table 3 of Building Research Establishment Digest 67[8]	Downhill creep may occur on slopes greater than 1 in 10. Unreinforced piles have been broken by slowly moving slopes
(3) Where trees are felled to clear the site and construction is due to start soon afterwards	Reinforced bored piles of sufficient length with the top 3 m sleeved from the surrounding ground and with suspended floors, or thin reinforced rafts supporting flexible buildings, or basement rafts		
Soft clays, soft silty clays	Strip foundations 1.0 m wide if bearing capacity is sufficient, or rafts	See figure 3.1 and CP 101: 1972[4]	Settlement of strips or rafts must be expected. Services entering building must be sufficiently flexible. In soft soils of variable thickness it is better to pile to firmer strata (see Peat and Fill below)
Peat, fill	Bored piles with temporary steel lining or precast or in situ piles driven to firm strata below	Design with large safety factor on end resistance of piles only as peat or fill consolidating may cause a downward load on pile (see Building Research Establishment Digest 63[7]). Field tests for bearing capacity of deep strata or pile loading tests will be required	If fill is sound, carefully placed and compacted in thin layers, strip foundations are adequate. Fills containing combustible or chemical wastes should be avoided.
Mining and other subsidence areas	Thin reinforced rafts for individual houses with loadbearing walls and for flexible buildings	Rafts must be designed to resist tensile forces as the ground surface stretches in front of a subsidence. A layer of granular material should be placed between the ground surface and the raft to permit relative horizontal movement	Building dimensions at right angles to the front of longwall mining should be as small as possible

Source: BRE Digest 67[8]

Drag shovel or backactor. This digs below its own level and toward itself and is primarily used for trench excavation.
Skimmer. This is for shallow excavation up to 1½ m deep and is particularly useful for levelling and roadwork.
Grab and clamshell. This is for moving loose materials.

Other machines include the *scraper* which operates like an earth plane and carries its scrapings with it; the *bulldozer* and *angledozer* for bulk excavation and grading; and the *rooter*, which is a tractor-drawn toothed scarifier for breaking up hard surfaces.

Problems with Wet Ground

Various methods have been employed to excavate below the water table. A common method is to excavate under the continuous protection of tightly closed sheeting and to form one or more sumps from which pumping proceeds continuously. For wide excavation, the strutting is expensive and inconvenient and difficulty arises in maintaining the floor of the excavation. There is also a danger of fine particles being extracted from the soil by the continuous pumping and possibly causing settlement of neighbouring ground.

Another method is the *wellpoint* system whereby cylindrical metal tubes of about 50 mm diameter are sunk into the ground until they have penetrated below the lowest level of the proposed excavation at about 1½ m intervals. A riser pipe connects each wellpoint with a header pipe on the surface, which in turn is coupled to a suction pump. The pump creates a vacuum in the system which draws the water from the ground and so lowers the water table. A wellpoint is capable of extracting large quantities of water with a minimum of air and fine particles.

Where the ground is impervious it may be possible to use *electro-osmosis*, whereby water is extracted from the ground and the ground is stabilised. Steel tubing is used for cathodes with smaller diameter tubing as the anode. A potential of 40 to 180 volts is applied and the groundwater flows to the cathode (negative pole) from which it is extracted.

Where the ground is unstable due to groundwater it may be necessary to enclose the excavation with *steel-sheet piling*. The piling is made in a variety of sections of differing strengths to resist a range of pressures. The pile sections are usually driven by a double-acting steam-operated pile hammer.

Some work in waterlogged ground has been made possible by *freezing* which has both solidified the loose ground and prevented water flowing into the working area. Freezing is normally undertaken by drilling a series of vertical boreholes of about 150 to 175 mm diameter at approximately one metre intervals around the perimeter of the work. The boreholes are lined with 100 to 150 mm diameter freezing tubes closed at the bottom end. An inner tube is then inserted with the bottom left open. Cooled brine solution is fed into the inner tube and returned to the next cooling tube or the refrigeration plant from the outer tube. The constant passage of the brine solution with a temperature lower than the freezing point of water gradually freezes the groundwater.

Other processes include the use of compressed air, soil stabilisation and grouting of the soil. All these methods are very costly and waterlogged sites should only be used as a last resort.[11]

Timbering of Excavations

The amount of timbering required to the sides of excavations is largely dependent upon two main factors — the depth of the excavation and the nature of the soil to be upheld. Vibration and loads from traffic or other causes, position of water table, climatic conditions and the time for which the excavation is to remain open also affect the decision. Figure 3.3.2 lays down broad guidelines for timbering requirements under different site conditions.

In relatively shallow trenches in firm soil it may be possible to dispense with timbering or, as it is sometimes termed, planking and strutting. The most that would be required would be pairs of 175 x 38 mm poling boards, spaced at about 1.80 m centres, and strutted with a single 100 x 100 mm strut. Alternatively adjustable steel struts may be used as illustrated in figure 3.3.1.

Most of the timber used is softwood, often red or yellow deal, possibly using pitch pine for heavy struts. The various members and their uses are now described.
Poling boards. These are boards 1.00 to 1.50 m in length (depending on the depth of excavation) and they vary in cross section from 175 x 38 to 225 x 50 mm. The boards are placed vertically and abut the soil at the sides of the excavation.
Walings. These are longitudinal members running the length of the trench or other excavation and they support poling boards. They vary in size from 175 x 50 to 225 x 75 mm.
Struts. These are usually square timbers, either 100 x 100 or 150 x 150 mm in size. They are generally used to support the walings which, in turn, hold the poling boards in position. Struts are usually spaced at about 1.80 m centres to allow adequate working space between them.
Sheeting. This consists of horizontal boards abutting one another to provide a continuous barrier when excavating in loose soils, and so permitting the timbering to closely follow the excavation. A common size for the sheeting in 175 x 50 mm.

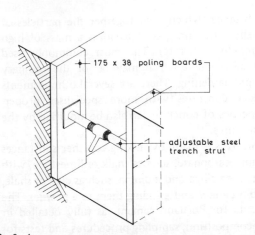

3.3.1
TIMBERING TO SHALLOW
TRENCH IN FIRM SOIL

- 175 x 38 poling boards
- adjustable steel trench strut

TYPE OF SOIL	DEPTH OF EXCAVATION		
	UP TO 1·50m SHALLOW	1·50 TO 4·50m MEDIUM	OVER 4·50m DEEP
SOFT PEAT	C	C	C
FIRM PEAT	A	C	C
SOFT CLAYS AND SILTS	C	C	C
FIRM AND STIFF CLAYS	A+	A+	C
LOOSE GRAVELS AND SANDS	C	C	C
SLIGHTLY CEMENTED GRAVELS AND SANDS	A	B	C
COMPACT GRAVELS AND SANDS WITH OR WITHOUT CLAY BINDER	A	B	C
ALL GRAVELS AND SANDS BELOW WATER TABLE	C	C	C
FISSURED OR HEAVILY JOINTED ROCKS (SHALES, ETC)	A+	A+	B
SOUND ROCK	A	A	A

A. NO SUPPORT NEEDED
B. OPEN SHEETING
C. CLOSE SHEETING OR SHEET PILING
+   OPEN OR CLOSE SHEETING MAY BE REQUIRED
    IF SITE CONDITIONS ARE UNFAVOURABLE

SOURCE:

CP 2003 : Earthworks (11)

3.3.2 TIMBERING REQUIREMENTS UNDER VARYING
CONDITIONS

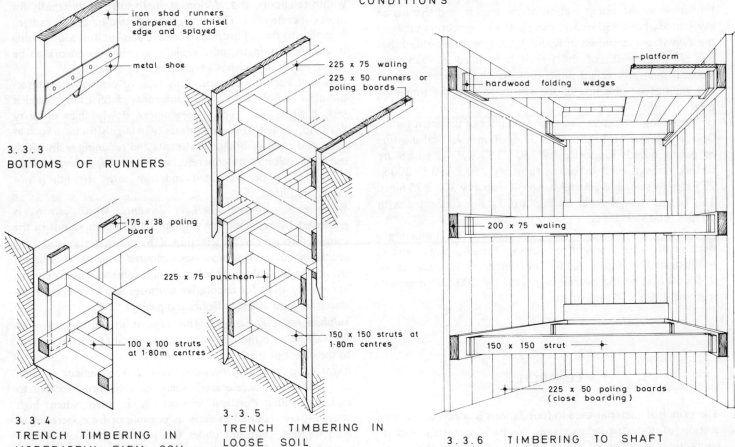

- iron shod runners sharpened to chisel edge and splayed
- metal shoe

3.3.3
BOTTOMS OF RUNNERS

- 225 x 75 waling
- 225 x 50 runners or poling boards
- 175 x 38 poling board
- 225 x 75 puncheon
- 100 x 100 struts at 1·80m centres
- 150 x 150 struts at 1·80m centres

3.3.4
TRENCH TIMBERING IN
MODERATELY FIRM SOIL

3.3.5
TRENCH TIMBERING IN
LOOSE SOIL

- platform
- hardwood folding wedges
- 200 x 75 waling
- 150 x 150 strut
- 225 x 50 poling boards (close boarding)

3.3.6 TIMBERING TO SHAFT

Figure 3.3 TIMBERING TO EXCAVATIONS

Runners. These are poling boards in continuous formation with tapered bottom edges, possibly shod with metal shoes as shown in figure 3.3.3. A common size for these members is 225 x 50 mm. They are particularly suitable for use in loose or waterlogged soils which will not stand unsupported.

In *moderately firm ground* the timbering is likely to take the form illustrated in figure 3.3.4. It consists of a series of poling boards which are quite widely spaced at about 600 mm centres, supported by walings and struts in the normal way. In shallow trenches the poling boards would probably only be needed at about 1.80 m centres, with each pair of poling boards individually strutted with a single 100 x 100 mm strut, and no walings, on the lines indicated in figure 3.3.1.

Timbering in *loose soil* may take one of two main forms. One method uses continuous horizontal sheeting supported by pairs of poling boards and struts at about 1.80 m centres. Another method employs a continuous length of poling boards or runners supported by walings and struts. If the trench exceeds 1.50 m in depth, it is necessary to step the timbering so that the timbering to the lower stage fits inside the timbering to the upper section of the trench, and there is an overlap of about 150 mm at the point of connection between the two stages as shown in figure 3.3.5. It will be noticed that the effective width of the trench is reduced at each step. Vertical members called puncheons may be inserted between the walings on the line of the struts, to give the timbering greater rigidity.

Timbering for *shafts* can take a number of different forms. One arrangement is to use 225 x 38 or 50 mm vertical sheeting or poling boards supported by 150 x 150 walings which are further strengthened by beams (possibly 250 x 150 to 300 x 225 mm in size) and vertical hangers (probably 300 x 75 mm). The beams, hangers and walings are usually connected with 20 mm mild steel bolts.

Another and rather simpler arrangement for timbering a shaft or pit, such as might be necessary for a heating chamber, is illustrated in figure 3.3.6. The timbering is made up of 225 x 50 mm poling boards supported by 200 x 75 mm walings and 150 x 150 mm struts.

CONCRETE

The principal material used in foundations is concrete and this section of the chapter examines the basic materials used in concreting, selection of mixes, site operations and the use of reinforced and prestressed concrete.

Cements

Cements are substances which bind together the particles of aggregates (usually sand and gravel) to form a mass of high compressive strength (concrete). The most commonly used cement is Portland cement which may be of the ordinary variety or be rapid-hardening. There are several other cements that will produce concretes with more specialised properties.[12] The properties of concrete can also be modified by the addition of admixtures.[9,12]

Portland cement. This is made by mixing together substances containing calcium carbonate, such as chalk or limestone, with substances containing silica and alumina, such as clay or shale, heating them to a clinker and grinding them to a powder. The basic requirements for Portland cement, as fully detailed in BS 12,[13] cover composition, sampling procedures and tests for fineness, chemical composition, compressive strength, setting time and soundness. The cement combines with water to form hydrated calcium silicate and hydrated calcium aluminate. The initial set takes place in about 45 minutes and the final set within ten hours, and develops strength sufficiently rapidly for most concrete work. The setting time of rapid-hardening cement is similar to that of ordinary Portland cement, but after setting it develops strength more rapidly, enabling formwork to be struck earlier. It also has advantages in cold weather.[14]

Low-heat Portland cement. This type of cement is manufactured to comply with the requirements of BS 1370,[15] and it sets, hardens and evolves heat more slowly than ordinary Portland cement. It is used primarily in large structures such as dams and massive bridge abutments and retaining walls which use large volumes of concrete, and where the generated heat cannot easily be dissipated and high early strength is not usually required.

White and coloured Portland cements. White cement is produced by reducing the content of iron compounds in the cement through careful selection of the raw materials and using special manufacturing processes. Coloured cements are obtained by adding suitable pigments to white cement. These cements are mainly used for decorative purposes and have been introduced to good effect in floors and pavings.[14]

Sulphate-resisting cement. This cement should comply with BS 4027.[29] It is more resistant than ordinary Portland cement to the effect of sulphates.

Extra rapid-hardening Portland cement. The cement is made by adding an accelerator, such as calcium chloride, to rapid-hardening Portland cement. It is used when high strengths are required as early as possible or for concreting in cold weather (but not under freezing conditions). Both the rate of hardening and of heat evolution are accelerated, as well as the setting process.

Waterproof and water-repellent Portland cements. Concrete made with these cements, mixed with a metallic soap, is less permeable to water than concrete made with ordinary Portland cement. Careful control is required during the mixing process as some types tend to entrain air, and low permeability is also dependent on a dense concrete.[14]

Hydrophobic Portland cement. This cement has been developed to prevent partial hydration of cement during storage in humid conditions, resulting in a reduction in strength and the formation of air-set lumps. Substances added during the grinding process form a water-repellent film around each grain of cement and so prevents deterioration during storage. During mixing the protective film is lost by abrasion and hydration takes place.[12]

Portland blastfurnace cement. This is made by grinding a mixture of ordinary Portland cement clinker and granulated blastfurnace slag. It is similar to ordinary Portland cement but evolves less heat and is rather more resistant to chemical attack by sulphates or sea water.[16]

Supersulphated cement. This consists of granulated blast-furnace slag, calcium sulphate and a small percentage of Portland cement or lime. Its prime advantage is the high resistance to chemical attack by sulphate-bearing waters and weak acids, It deteriorates rapidly if stored under damp conditions. Concrete based on this cement requires a longer mixing period and the surface of the finished concrete needs to be kept moist during curing.[16]

High-alumina cement. This cement differs in method of manufacture, composition and properties from Portland cements.[30] Its main advantages stem from its very high early strength and resistance to chemical attack. Heat evolution is rapid, permitting the concrete to be placed at lower temperatures than ordinary Portland cement concrete. When mixed with a suitable aggregate, such as crushed firebrick, it makes an excellent refractory concrete to withstand high temperatures. Its use for structural work in Greater London is forbidden.[35]

Pozzolanic cements. These cements are mixtures of a Portland cement and a pozzolanic material (one combining with lime to form a hard mass). They offer good resistance to chemical attack, but the rate of heat evolution and strength attainment is reduced.

Aggregates

Aggregates are gravels, crushed stones and sand which are mixed with cement and water to make concrete. The two most essential characteristics for aggregates are durability and cleanliness; cleanliness includes freedom from organic impurities.

Fine aggregate. This consists of natural sand, or crushed stone or crushed gravel sand that mainly passes through a 4.76 mm British Standard sieve with a good proportion of the larger particles.[19]

Coarse aggregate. This is primarily natural gravel, or crushed gravel or stone that is mainly retained on a 5.00 mm BS 410 test sieve. Both types of aggregate should comply with the grading requirements of BS 882.[18] Artificial coarse aggregates such as clinker and slag are used for lightweight concrete.[31] The maximum size of coarse aggregate is determined by the class of work. With reinforced concrete the aggregate must be able to pass readily between the reinforcement and it rarely exceeds 20 mm. For foundations and mass concrete work the size can be increased possibly up to 40 mm. The type of aggregate used directly influences the fire protection and thermal insulation qualities of the concrete.[12]

SITE TESTING OF MATERIALS

It is sometimes necessary to carry out site tests on materials to determine their suitability. The following tests relating to cement and sand serve to illustrate the approach.

Cement

(1) Examine to determine whether it is free from lumps and of a flour-like consistency (free from dampness and reasonably fresh).

(2) Place hand in cement and if of blood heat then it is in satisfactory condition.

(3) Settle with water as paste in a closed jar to see whether it will expand or contract.

Sand

(1) Handle the sand; it should not stain hands excessively, ball readily or be deficient in coarse or fine particles.

(2) Use a standard sieve test — if more than 20 per cent is retained on a 1.25 mm sieve, it is unsuitable for use.

(3) Apply a silt or organic test — a jar half filled with sand and made up to the three-quarters mark with water; shake vigorously and leave for three hours; the amount of silt on top of sand is then measured and this should ot exceed six per cent.[19]

CONCRETE MIXES

Concrete must be strong enough, when it has hardened, to resist the various stresses to which it will be subjected and it often has to withstand weathering action. When freshly mixed it must be of such a consistency that it can be readily handled without segregation and easily compacted in the formwork. The fine aggregate (sand) fills the interstices between the coarse aggregate, and both aggregates need to be carefully proportioned and graded. The strength of concrete is influenced by a number of factors

(1) proportion and type of cement;
(2) type, proportions, gradings and quality of aggregates;
(3) water content;
(4) method and adequacy of batching, mixing, transporting, placing, compacting and curing the concrete.

Concrete mixes can be specified by the volume or weight of the constituent materials or by the minimum strength of the concrete. A Building Research Establishment paper (Specification of Concrete)[20] outlines the disadvantages of nominal mix proportions by volume, such as 1:3:6 for plain concrete foundations, 1:2:4 for normal reinforced concrete and 1:1½:3 for strong reinforced concrete work. These mixes fail to specify adequately the cement content of the mix, as the actual cement content of a cubic metre of concrete made to a particular nominal mix varies with different aggregates and different water contents. Hence there is considerable merit in specifying mixes in terms of weight of aggregate per 50 kilogrammes of cement, as these mixes have more realistic cement contents and make allowance for different maximum sizes of aggregate and levels of workability. Nevertheless, the strength of concrete produced under site conditions varies widely. A more realistic approach is to specify a minimum strength for the concrete and for the proportions of cement, sand, coarse aggregate and water then to be selected to achieve it. This trend away from the specification of nominal mixes and towards the design of mixes on a strength basis is reflected in Code of Practice, CP 114[21] and CP 110[32]. Table 3.4 indicates some typical concrete mixes with their probable strength requirements. CP 110[32] recommends grade 7 or 10 plain concrete (7 to 10 N/mm²) and grade 15 to 25 for reinforced concrete. BS 5075[9] details five categories of admixture which may be used to modify one or more of the properties of cement concrete, such as accelerating and retarding qualities.

Table 3.4   Concrete mixes and strength requirements

Concrete made with Portland cement (BS 12) and natural aggregates (BS 882)

Nominal mix by volume	Volume of dry aggregate in m³ per 50 kg of cement		Minimum cube crushing strength at 28 days in MN/m² (N/mm²)	
	Fine	Coarse	Preliminary test	Works test
1:1:2	0.035	0.07	40	30
1:1½:3	0.05	0.10	34	25.5
1:2:4	0.07	0.14	28	21

Source: *Building Research Establishment Digest 13*[22]

The water/cement ratio is a most important factor in concrete quality. It should be kept as low as possible consistent with sufficient workability to secure fully compacted concrete with the equipment available on the site. The higher the proportion of water, the weaker will be the concrete. Water/cement ratios are usually in the range of 0.40 to 0.60 (weight of water divided by weight of cement). Allowance has to be made for absorption by dry and porous aggregates and the surface moisture of wet aggregates. Badly proportioned aggregates require an excessive amount of water to give adequate workability, and this results in low strength and poor durability. A common test for measuring workability on the site is the slump test, although for greater accuracy the compacting factor test or consistometer test should be used.[23]

The *slump test* consists of filling a 300 mm high open-ended metal cone with four consolidated layers of concrete and then lifting the cone and measuring the amount of slump or drop of the cone-shaped section of concrete. Slumps vary from 25 mm for vibrated mass concrete to 150 mm for heavily reinforced non-vibrated concrete.[34]

The apparatus for the *compacting factor test* consists of two conical hoppers mounted above a cylindrical container. The top hopper is filled with concrete and this is allowed to fall through a hinged trapdoor at the base into the second hopper below. The trapdoor of the second hopper is then released and the concrete allowed to fall into the container, when the top surface of the concrete is levelled off and the combined weight of the cylinder and partially compacted concrete is taken. Finally the cylinder is filled with concrete in 50 mm layers which are each compacted and the cylinder is weighed once again. The compacting factor is the ratio of the partially compacted to the fully compacted weight.

CONCRETING OPERATIONS

Concreting operations comprise batching, mixing, transporting, placing, compacting and curing the concrete.

Batching

Prior to use the cement needs to be stored in a damp-proof and draught-proof structure, while aggregates should be stored on clean, hard surfaces. For very small jobs the aggregates may be measured by volume using a timber gauge box, with its size related to the quantity of aggregate required for a bag of cement. Weigh batching gives much more accurate results. It is necessary to check constantly that the gauging apparatus is being used correctly and that allowance is made for the moisture content of the sand.

Mixing

Concrete can be mixed by hand or machine on the job or purchased ready-mixed, when it is delivered to the site from a central batching plant in lorries with mixing drums or with revolving containers. Hand mixing should only be used for limited quantities of concrete on the smallest jobs, with the proportion of cement increased by ten per cent. The materials are mixed dry on a clean, hard surface, water added through a rose and further mixing continued until a uniform colour is obtained. Concrete mixers are designed by their capacity of concrete and type (tilting and non-tilting).[33] The normal mixing time is at least one minute after all materials, including water, have been placed in the mixer.[19]

Transportation

The type of plant used to transport concrete from the mixer to the point of deposition depends on the size of the job and the height above ground at which the concrete is to be placed. Barrows, lorries, dumpers and mechanical skips are used for this purpose, and on very large jobs pumping may prove to be easier, quicker and cheaper.[24] Outputs claimed vary from 14 to 24 m^3/h per cylinder with a range of 500 mm horizontally and 50 m vertically through a 180 mm pipeline. Concrete should generally be placed in its final position within 30 minutes of leaving the mixer.

Placing

All formwork should be checked, cleaned and oiled before concrete is placed against it. Concrete should not be permitted to fall freely more than one metre, to prevent air pockets forming and segregation of the materials.

Compaction

The concrete must be adequately compacted to secure maximum density. It can be done by hand or by using vibrators. Vibrators are more efficient and can be used with drier mixes. There are three main types: external, internal and surface vibrators. Care must be taken to ensure that concrete is well compacted against forms and at corners and junctions. Construction joints, at the end of a day's work, should be carefully formed and be so positioned as not to be too conspicuous in the finished work. The 'laitance' film of cement and sand at the joint is removed to ensure a good bond between the new and old work. In large areas of concrete expansion joints should be incorporated to permit movement of the concrete with differences in temperature.

In cold weather the rate of setting and hardening of cement slows down and practically ceases at freezing point. Concrete should be at least 5°C when placed and should not fall below 2°C until it has hardened. Special precautions should be taken such as keeping aggregates and mixing plant under cover; covering exposed concrete surfaces with insulating material; using a richer mix of concrete and/or rapid-hardening cement; heating water and aggregate; placing concrete quickly and leaving the formwork in position for longer periods.[17]

Curing

The chemical reaction which accompanies the setting of cement and hardening of concrete is dependent on the presence of water; so exposed concrete should be covered with plastics or waterproofed sheets or building paper to protect it from the sun and drying winds for at least seven days.[17]

Formwork

Formwork or shuttering is used to retain concrete in a specific location until the concrete has developed sufficient strength to stay in position without support. The general requirements for formwork are as follows

(1) it should be sufficiently rigid to prevent undue movement during the placing of the concrete;
(2) it should be of sufficient strength to withstand working loads and weight of wet concrete;
(3) it must have sufficiently tight joints to prevent loss of fine material from concrete;
(4) erection and stripping shall be easily performed with units of a manageable size;
(5) it should produce concrete face of good appearance;
(6) it must permit removal of side forms prior to striking of soffit shuttering.

Formwork may be of timber, with possibly six or seven uses, or of sheet steel or metal-faced plywood. The use of telescopic units, adjustable props and special beam and column clamps, permit the easy erection and dismantling of steel formwork.

The period which should elapse before formwork is struck depends on the type of cement, nature of component and the temperature. For example, formwork to beam sides, walls and columns could be removed after one day with rapid hardening Portland cement and normal weather conditions (15°C), whereas six days would be required for ordinary Portland cement in cold weather (just above freezing). Similarly with slabs the periods would range from two to ten days, and with beam soffits from four to fourteen days, with props remaining for longer periods in both cases.

REINFORCED CONCRETE

Concrete is strong in compression but weak in tension, and where tension occurs it is usual to introduce steel bars to provide the tensile strength which the concrete lacks. For example, with a concrete beam or lintel, compression occurs at the top and tension at the bottom, so the reinforcement is placed about 25 mm up from the bottom of the beam and the ends are often hooked to provide a grip. The 25 mm cover prevents rusting of the reinforcement.

The steel must be free from loose mill scale, loose rust, grease, oil, paint, mud and other deleterious substances which impair the bond between the steel and concrete. The most common form of reinforcement is the mild steel bar to BS 4449 or BS 4482.[25] Medium and high tensile bars are also available, and deformed bars which are twisted and/or ribbed provide a better bond and greater frictional resistance than round bars and obviate the need for hooked ends. It is important that the reinforcement is fixed securely to avoid displacement during the placing of the concrete. The bars are tied together with soft wire at intersections, and spacers of small precast concrete blocks, concrete rings or plastic fittings ensure the correct cover of concrete. In raft foundations, reinforcement may take the form of steel fabric to BS 4483,[26] and this consists of a grid of small diameter bars, closely spaced and welded at the joints.

PRESTRESSED CONCRETE

The basic principle of prestressed concrete is that by using high-grade tensile steel as reinforcement, and stretching it and releasing it, stresses are set up in the member which act inversely to the applied loads. When a prestressed beam is subjected to bending and shear stresses by the application of a load, the concrete in what would be the tensile zone of an ordinary reinforced concrete beam is no longer carrying tensile stresses but only compression, which generally should reach zero when the full design load is applied. In this way the best qualities of both concrete and steel can be used effectively and economically. The deadweight is likely to produce only small compressive stresses and smaller size members can be used. Tensioning cables may be wires or bars and there are many proprietary systems. In *pre-tensioning* concrete is poured into moulds around stretched wires, whereas with *post-tensioning* the concrete is permitted to harden before the wires are stretched.[14]

BASEMENT CONSTRUCTION

Basement walls have to withstand earth pressures and also possibly water pressure. Earth pressures vary with the type of earth and are influenced by its angle of repose (the greatest angle at which soil will stand without slipping). Typical angles of repose are 16° for wet clay and 30° for loose dry earth or gravel. The horizontal pressure exerted by the earth on a basement wall is largely determined by the weight of the earth and its angle of repose.

Neither brickwork in dense bricks in cement mortar nor dense concrete are completely impermeable to water. It is therefore essential to provide a continuous waterproof membrane, preferably placed under the structural floor and behind the structural walls (figure 3.4.1). The membrane that gives the best guarantee of complete continuity is mastic asphalt, using three-coat work when there is water pressure. Where there is sufficient room to provide a minimum width of working space of 600 mm, the vertical mastic asphalt may be applied externally. On more restricted sites and in existing buildings, the vertical asphalt may be applied internally, with a loading wall as shown in figure 3.4.8 to resist pressure from groundwater. Three-coat asphalt is likely to finish about 29 mm thick for the horizontal work and 19 mm for the vertical work.

It is necessary to provide a good key for the asphalt by raking out brickwork joints or hacking concrete surfaces. Horizontal asphalt should be protected with a 50 mm fine concrete screed and vertical asphalt should be separated from loading walls by a 25 mm layer of grout. Two-coat asphalt fillets should be provided at all angles as shown in figure 3.4.1. Care must be taken to keep the site dry while the basement is under construction and to protect the asphalt from damage. Figure 3.4.5 shows the method of continuing the horizontal damp-proof membrane under a column or stanchion in the basement.

Care must be taken to ensure that watertight joints are formed wherever pipes pass through basement tanking. Figures 3.4.2, 3.4.3 and 3.4.4 show one method of treating pipes using an asphalt sleeve around the pipe. An alternative method is illustrated in figure 3.4.6 using a sheet lead collar sandwiched between the flanges and into the asphalt.

Access to basements is sometimes provided through external open areas and typical constructional details are illustrated in figure 3.4.7. DOE Advisory Leaflet 51[27] gives advice on the construction of watertight concrete basements. It

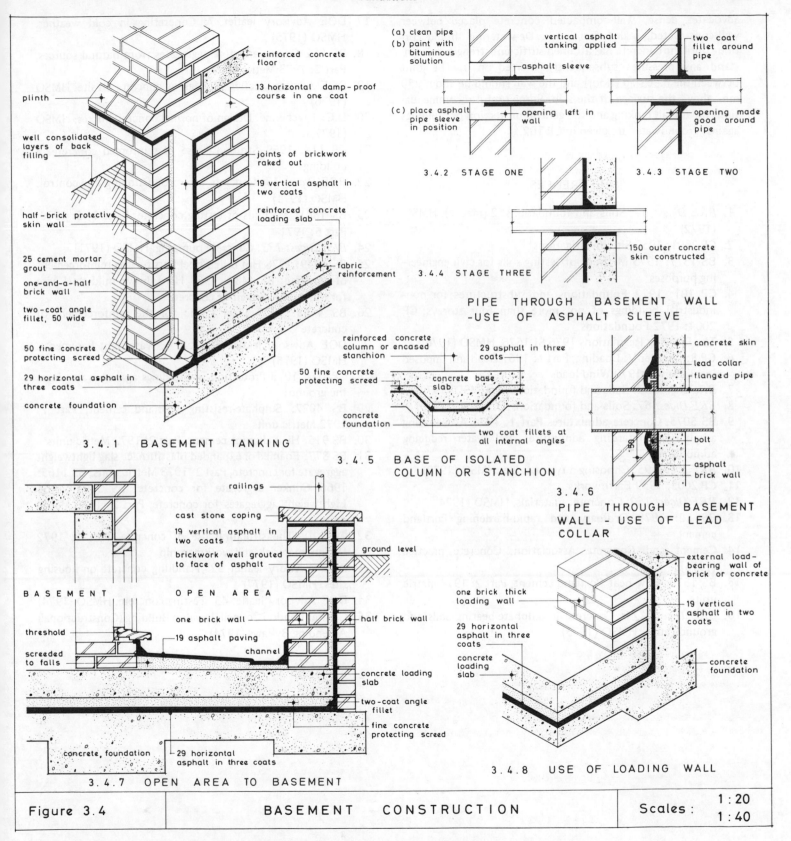

reinforced concrete floor

plinth

13 horizontal damp-proof course in one coat

well consolidated layers of back filling

joints of brickwork raked out

19 vertical asphalt in two coats

reinforced concrete loading slab

half-brick protective skin wall

25 cement mortar grout

one-and-a-half brick wall

fabric reinforcement

two-coat angle fillet, 50 wide

50 fine concrete protecting screed

29 horizontal asphalt in three coats

concrete foundation

3.4.1 BASEMENT TANKING

(a) clean pipe
(b) paint with bituminous solution
(c) place asphalt pipe sleeve in position

vertical asphalt tanking applied

asphalt sleeve

opening left in wall

two coat fillet around pipe

opening made good around pipe

3.4.2 STAGE ONE

3.4.3 STAGE TWO

150 outer concrete skin constructed

3.4.4 STAGE THREE

PIPE THROUGH BASEMENT WALL -USE OF ASPHALT SLEEVE

reinforced concrete column or encased stanchion

50 fine concrete protecting screed

concrete base slab

concrete foundation

29 asphalt in three coats

two coat fillets at all internal angles

3.4.5 BASE OF ISOLATED COLUMN OR STANCHION

concrete skin

lead collar

flanged pipe

bolt

asphalt brick wall

3.4.6
PIPE THROUGH BASEMENT WALL - USE OF LEAD COLLAR

railings

cast stone coping

19 vertical asphalt in two coats

brickwork well grouted to face of asphalt

ground level

BASEMENT   OPEN AREA

one brick wall

half brick wall

threshold

19 asphalt paving

channel

screeded to falls

concrete loading slab

two-coat angle fillet

fine concrete protecting screed

concrete foundation

29 horizontal asphalt in three coats

3.4.7 OPEN AREA TO BASEMENT

external load-bearing wall of brick or concrete

one brick thick loading wall

19 vertical asphalt in two coats

29 horizontal asphalt in three coats

concrete loading slab

concrete foundation

3.4.8 USE OF LOADING WALL

| Figure 3.4 | BASEMENT CONSTRUCTION | Scales: 1:20 1:40 |

advocates dense, well-compacted concrete placed between well-constructed, sealed formwork. De-watering should continue until the concrete has attained sufficient strength to withstand any pressures subsequently exerted on it. The joint between the base slab (floor) and the wall should be about 250 mm above the surface of the slab. Some useful recommendations on asphalt tanking and the general protection of buildings against groundwater are given in CP 102.[28]

REFERENCES

1. *BRE Digest 64*: Soils and foundations: 2 (metric). HMSO (1972)
2. CP 2001: 1957 Site investigations
3. BS 1377: 1975 Methods of testing soils for civil engineering purposes
4. CP 101: 1972 Foundations and substructures for non-industrial buildings of not more than four storeys; CP 2004: 1972 Foundations
5. The Building Regulations 1976 SI 1676. HMSO (1976)
6. CP 3: Chapter V: Loading; Part 1: 1967 Dead and imposed loads, Part 2: 1972 Wind loads
7. *BRE Digest 63*: Soils and foundations: 1. HMSO (1979)
8. *BRE Digest 67*: Soils and foundations: 3. HMSO (1970)
9. BS 5075. Concrete admixtures Part 1: 1974 Accelerating admixtures, retarding admixtures and water reducing admixtures
10. *BRE Digest 95*: Choosing a type of pile. HMSO (1979)
11. CP 2003: 1959 Earthworks
12. *BRE Digest 150*: Concrete: materials. HMSO (1974)
13. BS 12: 1978 Ordinary and rapid-hardening Portland cement
14. Cement and Concrete Association. Concrete practice (1976)
15. BS 1370: Low heat Portland cement, Part 2: 1974 metric units
16. *BRE Digest 174*: Concrete in sulphate-bearing soils and groundwaters. HMSO (1975)
17. DOE Advisory leaflet 7: Concreting in cold weather. HMSO (1975)
18. BS 882 Coarse and fine aggregates from natural sources, Part 2: 1973 Metric units
19. DOE Advisory leaflet 26: Making concrete on site. HMSO (1976)
20. D.C. Teychenné. Design of normal concrete mixes. HMSO (1977)
21. CP 114: 1969 Structural use of reinforced concrete in buildings
22. *BRE Digest 13*: Concrete mix proportioning and control. HMSO (1971)
23. BS 1881: Methods of testing concrete; Parts 1–5: 1970, Part 6: 1971
24. *BRE Digest 133*: Concrete pumping. HMSO (1971)
25. BS 4449: 1978 Hot rolled steel bars for the reinforcement of concrete; BS 4482: 1969 Hard drawn mild steel wire for the reinforcement of concrete
26. BS 4483: 1969 Steel fabric for the reinforcement of concrete
27. DOE Advisory leaflet 51: Part 1, Watertight basements. HMSO (1976)
28. CP 102: 1973 Protection of buildings against water from the ground
29. BS 4027: Sulphate-resisting Portland cement, Part 2: 1972 Metric units
30. BS 915: High alumina cement, Part 2: 1972 Metric units
31. BS 877: Foamed or expanded blastfurnace slag lightweight aggregate for concrete, Part 2: 1973 Metric units; BS 1165: 1966 Clinker aggregate for concrete; BS 3797: 1964 Lightweight aggregates for concrete, Part 2: 1976 Metric units
32. CP 110: The structural use of concrete, Part 1: 1972 Design, materials and workmanship
33. DOE Advisory leaflet 32: Handling concrete on housing sites. HMSO (1977)
34. DOE Advisory leaflet 43: Testing concrete. HMSO (1974)
35. Greater London Council. London Building (Constructional) Amending By-laws (1974)

4 WALLS

Walls to buildings can be constructed in various ways using a variety of materials. In order to appreciate the different constructional techniques and their relative merits, it is necessary to know the functions of walls in various locations. The principal methods used in the construction of both external and internal walls and partitions are examined in some detail.

FUNCTIONS OF WALLS

External walls should perform a number of functions

(1) support upper floors and roofs together with their superimposed loads;
(2) resist damp penetration;
(3) provide adequate thermal insulation;
(4) provide sufficient sound insulation;
(5) offer adequate resistance to fire;
(6) look attractive and satisfactorily accommodate windows and doors.

Strength of Walls

Walls must be of sufficient thickness to keep the stresses within the limits of the safe compressive stresses of the materials in the wall, for example bricks and mortar. Tensile stresses may be induced by unequal loadings on the wall and the bonding and jointing of the bricks or blocks must be able to withstand them. The thickness-to-height ratio must be sufficient to prevent buckling and there must be adequate lateral support to resist overturning. The materials in the wall must be durable and able to withstand soluble salts, atmospheric pollution and other adverse conditions.

Code of Practice 111[1] recommends permissible stresses for loadbearing walls, and from these stresses the thickness of walls can be determined in relation to the loads to be carried. To utilise the full capacity of high-strength bricks (70 N/mm^2 or more) a 1:3 cement−sand mix is needed. For lower strength bricks, mortars with an increased proportion of lime can be used without any great loss in brickwork strength. Suitable mortar mixes for various brick strengths are listed in table 4.1. The strength of a 255 mm cavity wall with loads spread over both leaves is 20 per cent less than that of a one-brick wall. Wall ties built into cavity walls are intended to share lateral forces and deflections between the two leaves.

Table 4.1   Suitable mortar mixes for various brick strengths

Brick strength	(N/mm^2)	Mortar mix (cement: lime: sand by volume)
Low	10	1:2:9
Medium	20 to 35	1:1:6
High (class B)	48 to 62	1:¼:3
Very high (class A)	70 or more	1:0:3

Source: *Building Research Station Digest 61*[2]

Both the Building Regulations[3] and the London Constructional By-Laws[4] prescribe ways of calculating slenderness ratios of walls in various situations. The slenderness ratio of a wall is the ratio of the effective height to the effective thickness. The effective height of a wall with lateral support at its top is three-quarters of the storey height. Schedule 7 in the Building Regulations[3] prescribes minimum thicknesses for walls of stone, brick and similar materials. Typical examples are 190 mm thickness for external and separating walls of any length not exceeding 3.5 m high, and for walls with a height between 3.5 m and 9 m and with a length not exceeding 9 m. For longer walls of similar height, the thickness shall be 290 mm from the base for the height of one storey and 190 mm for the rest of its height. (The London Constructional By-Laws contain similar provisions.) The thickness of walls for smaller buildings are however determined by considerations other than strength, such as weather resistance, thermal insulation and fire resistance.

Resistance to Dampness

Damp penetration is one of the most serious defects in buildings. Apart from causing deterioration of the structure, it can also result in damage to finishings and contents and can in severe cases adversely affect the health of occupants. Dampness may enter a building by a number of different routes and these are now examined.

Water introduced during construction. In building a traditional house several tonnes of water are introduced into the walls during bricklaying and plastering. The walls often remain damp until a summer season has passed. As the moisture dries out from inner and outer surfaces it is liable to leave deposits of soluble salts or translucent crystals. A porous wall with an impervious coating on one surface will cause drying out on the other surface. Typical moisture contents of some of the more common building materials are plaster: 0.2 to 1.0 per cent, lightweight concrete: up to 5 per cent; and timber: 10 to 20 per cent.

Penetration through roofs, parapets and chimneys. Tiled roofs may admit fine blown snow and fine rain, particularly in exposed situations. Both tiles and slates must be laid to an adequate pitch and be securely fixed. It is wise to provide a generous overhang at eaves. Parapets and chimneys can collect and deliver water to parts of the building below roof level, unless they have adequate damp-proof courses and flashings. Leakage through flat roofs is more difficult to trace and needs to be distinguished from condensation.[5]

Penetration through walls. Penetration occurs most commonly through walls exposed to the prevailing wet wind or where evaporation is retarded as in light wells. On occasions the fault stems from excessive wetting from a leaking gutter or downpipe. There must be a limit to the amount of rain that a solid wall can exclude. In the wetter parts of the country (for example SW, W and NW England rising to a maximum in NW Scotland) and on exposed sites, one-brick thick solid walls may permit penetration of water. In this connection Lacy has devised indices of exposure to driving rain.[7] The greatest penetration is likely to occur through the capillaries between the mortar joints and the walling units. The more impervious the mortar and the denser the bricks or blocks, the more serious is the penetration likely to be. Dense renderings can prevent moisture drying out more effectively than preventing its entry, and this tendency is accentuated in cracked renderings with moisture penetrating the cracks by capillary action, becoming trapped behind the rendering and subsequently drying out on the inner face of the wall.

Cavity walls when properly detailed and soundly constructed will not permit penetration of rain. Penetration when it occurs is usually the direct result of faulty detailing at openings or mortar droppings on wall ties. Finally, disintegration of brickwork may be caused by the action of sulphates or frost when the bricks are saturated.

Rising damp. In older buildings damp may rise up walls due to the lack of damp-proof courses. In newer buildings rising damp may occur through a defective damp-proof course, the bridging of the damp-proof course by a floor screed internally, or by an external rendering, path or earth outside the building, or mortar droppings in the cavity. Damp may also penetrate a solid floor in the absence of a damp-proof membrane. These sources are well illustrated in BRE Digest 27.[6]

Other causes. Dampness may result from leaks in a plumbing system, although this must not be confused with condensation on cold pipes. The amount of water vapour that air can contain is limited and when this is reached the air becomes saturated. The saturation point varies with the temperature — the higher the temperature of the air, the greater the weight of water vapour it can contain.[8]

When warm, damp weather follows a period of cold, the fabric of a building which has not been fully centrally heated may remain cold for some time. The warm, moist air will tend to condense on the cold wall and floor surfaces. Furthermore, the humidity inside a building is usually higher than outside. The occupants and many of their activities, such as cooking and washing, produce moisture vapour. Changing living habits and constructional methods, such as hard plasters and solid floors, accentuate the condensation problem. Water vapour from a kitchen or bathroom may circulate through a house and condense on colder surfaces of stairwells and unheated bedrooms. Remedial measures include the provision of background heating, thermal insulation and adequate ventilation.

Thermal Insulation

Thermal insulation serves a number of purposes

(1) to prevent excessive loss of heat from within a building;
(2) to prevent a large heat gain from the outside in hot weather;
(3) to prevent condensation;
(4) to reduce expansion and contraction of the structure.

The thermal transmittance or *U-value* of a wall, roof or floor of a building is a measure of its ability to conduct heat out of a building; the greater the *U*-value, the greater the heat loss through the structure. The total heat loss through the building fabric is found by multiplying *U*-values and areas of the externally exposed parts of the building, and then multiplying the result by the temperature difference between inside and outside. *U*-values are expressed in $W/m^2 {}^\circ C$ (watts per square metre for $1^\circ C$ difference between internal and external temperatures).

The Building Regulations[3] (section F) require a *U*-value for external walls of not more than 1.0, and 1.0 for floors and 0.6

for roofs. A cavity wall with a half-brick outer skin, unventilated cavity, 100 mm lightweight concrete block inner leaf and with 10 mm 'Perlite' plaster on the inside face, has a *U*-value of 0.95[10] and is within the statutory limit, whereas a one-brick plastered wall would not comply with the regulations (*U*-value: 2.10). The overall *U*-value of the wall can be reduced to about 0.5 by filling the cavity with urea—formaldehyde foam, mineral fibre batts or other suitable insulating material, subject to provisos on p. 53; and double-glazed windows. The calculated average *U*-value of perimeter walling, including openings, shall not exceed 1.8. DOE proposals in 1980 for dwellings were *U*-values of 0.6 for walls and exposed floors and 0.8 for roofs, and to limit the area of single glazing.

The *U*-value of a cavity wall can be calculated in the following manner.

	Resistance (m^2 °C/W)
External surface	0.053
Brickwork, 1700 kg/m^3, 102.5 mm thick,	
5 per cent moisture content	0.125
Cavity, unventilated	0.180
Aerated concrete blocks, 750 kg/m^3, 100 mm	
thick, 3 per cent moisture content	0.450
Lightweight plaster, 13 mm thick	0.060
Internal surface	0.120
Total resistance	0.988

$$U\text{-value} = \frac{1}{\text{total resistance}} = \frac{1}{0.988} = 1.01 \text{ W/m}^2\text{ °C}$$

The Building (First Amendment) Regulations 1978[3] extended thermal insulation requirements to heated buildings, other than dwellings, with a floor area greater than 30 m^2. The requirements are a *U*-value of 0.6 W/m^2 °C for external walls, roofs and exposed floors, except for factories and warehouses where the *U*-value is 0.7 W/m^2 °C. In addition there are restrictions on the amount of glazing.

Sound Insulation

Noise levels both inside and outside buildings have increased significantly in recent years. Sound can be either structure-borne by direct contact between various parts of the building, or airborne whereby vibration of the structure occurs and sound passes through the air from one part of the building to another.[13] The Building Regulations[3] require party walls and walls to habitable rooms to provide adequate resistance to the transmission of airborne sound, and give examples of deemed-to-satisfy provisions.

The performance of the external envelope of walls and roof in reducing the level of noise transmitted from outside to inside a building, is influenced largely by the mass of the envelope, its continuity and the extra insulation afforded by double-leaf construction. The sound reduction of the brick or block portions of an external wall is in the range 45 to 50 dB (decibels), whereas the reduction with closed single windows is in the order of 20 to 25 dB. Double glazing increases the reduction factor to much nearer that of the brick or block wall. With open windows, the basic noise reduction will be as little as 10 dB for the area of opening.[11]

For party walls between dwellings, a one-brick wall, plastered on both sides, and weighing not less than 415 kg/m^2 (including plaster) is generally acceptable. Wall linings on battens or studs can improve sound insulation, but the airspace should always be as wide as possible. Linings over shallow airspaces have been found of little value for insulation of low frequencies.[12] Internal partitions of plasterboard panels with a honeycomb core or plasterboard on each side of a 50 or 75 mm stud frame, give a sound reduction of about 28 to 30 dB, which is inadequate for use between habitable rooms. The reduction factor can be increased to the more acceptable level of 35 to 40 dB by the use of solid brick or blockwork, plastered both sides and weighing 75 to 150 kg/m^2.[12] Methods of sound-insulating floors are described in chapter 6.

Fire Resistance

Under the Building Regulations[3] the structure on the ground and upper floors of houses of not more than three storeys in height shall have a minimum period of fire resistance of half-an-hour, increasing to one hour for a basement. Load-bearing walls of brick or concrete blocks, 100 mm thick, have a fire resistance of two hours, whilst a 50 mm non-loadbearing wall has a resistance of half-an-hour. In general, loadbearing external walls of normal materials satisfying the conditions of strength, stability and weather resistance will usually provide sufficient resistance to fire. Furthermore, flats or maisonettes with a floor 4.5 m or more above ground level, must be provided with suitable means of escape.[3] Methods of upgrading existing partitions are described in BRE Digest 230.[46] Ends of cavities shall be effectively sealed to provide fire-stops, as described in BRE Digests 214 and 215.[47]

Other Defects in Brickwork

Defects in brickwork sometimes result from the use of unsuitable materials or an unsatisfactory combination of materials. The following examples will serve to illustrate some of these defects.

Sulphate attack on mortar. This comprises a chemical action between sulphate salts in clay bricks and the aluminate constituent of Portland cement in the mortar.[14] A new compound is formed with resultant expansion. The most vulnerable situations are parapet, boundary and retaining walls and chimney stacks, where deterioration and expansion of the mortar and cracking and spalling of the brickwork may occur. It is advisable to use bricks with a low sulphate content in these situations and to pay particular attention to damp-proofing arrangements.

Unsuitable mortars. Pitting and cracking of mortar may occur where the pointing mortar is excessively strong or there are unslaked particles of quicklime in the mortar.

Frost action. Damage to brickwork by frost is usually restricted to new work or walling which remains wet throughout the winter. Strong pointing mortar, porous bedding mortar and soft or weak bricks are particularly vulnerable. Precautions to be taken when bricklaying in cold weather are described in DOE Advisory Leaflet 8.[42]

Crystallisation of salts in brickwork. The crystallisation of water-soluble salts may result in a white powder or feathery crystals on the surface of the bricks (efflorescence), surface disintegration of the brickwork or decay of individual bricks in severe cases. The salts may originate from the bricks themselves, soil in contact with the bricks or sea water. Surface efflorescence is normally removed by periodic dry-wire brushing of the surface of the brickwork.[18] In more severe cases it may be necessary to cut out and replace individual defective bricks or to apply a rendering to more extensive areas. Nevertheless, efflorescence is usually a temporary springtime occurrence on new brickwork and is relatively harmless.[23]

Movements with changes in moisture content of brickwork. Shrinkage movements may cause stepped cracks following the horizontal and vertical joints or vertical joints through both bricks and mortar. This type of defect is most likely to occur with sandlime and concrete bricks, and is remedied by cutting out defective bricks and joints and making good. Preventive measures include using comparatively dry bricks, protecting work from rain and snow during erection and using a gauged mortar not stronger than 1:2:9.[48]

Corrosion of iron and steel in the brickwork. The corrosion may be caused by the action of moisture, acids, sulphates or chlorides on the metals, resulting in expansion and opening of brick joints, cracking of bricks and staining of brickwork by rust. Preventative measures include priming the metal, applying a coat of bituminous paint and then encasing it with cement mortar 25 mm thick.

Movement. Long runs of walling should be provided with 10 mm wide vertical expansion joints about 12 mm apart.[32]

BRICKS AND BLOCKS

Brick Types

Bricks are still the most popular form of walling unit for domestic construction. Their limited size and variety of colours and textures make them an attractive proposition. There is a wide range of bricks available, varying in the materials used, method of manufacture and form of brick, and these are now considered.

BS 3921[15] classifies *clay bricks and blocks* in three different ways

Varieties and functions. (a) Common: suitable for general building work but generally of poor appearance; (b) facing: specially made or selected to give an attractive appearance; (c) engineering: dense and strong semi-vitreous to defined limits for absorption and strength.

Qualities. (a) Internal: suitable for internal use only; (b) ordinary: normally sufficiently durable for external use; (c) special: durable in situations of extreme exposure.

Types. Solid: not more than 25 per cent small holes or 20 per cent in frogs, of volume of brick or block:; (b) perforated: small holes exceeding 25 per cent; (c) hollow: larger holes exceeding 25 per cent; (d) cellular: holes closed at one end exceeding 20 per cent.[18]

The usual brick size is 215 x 102.5 x 65 mm. Allowing for a 10 mm mortar joint, this gives an overall walling unit size of 225 x 112.5 x 75 mm. Some modular bricks are being produced, ranging from 288 or 190 mm long x 90 mm wide x 90 or 65 mm high.

BS 3921[15] prescribes minimum requirements as to dimensional tolerances, strength and absorption, finish and liability to efflorescence. In the case of special quality bricks and blocks, requirements are also prescribed for soluble salts content and frost resistance. Bricks and blocks of ordinary quality shall be well-fired and reasonably free from deep or extensive cracks and from damage to edges and corners, from pebbles and expansive particles of lime. They shall also, when a cut surface is examined, show a reasonably uniform texture.

Engineering bricks are classified into two groups

Class	Average compressive strength in N/mm² not less than	Average absorption (boiling or vacuum) per cent weight not greater than
A	69.0	4.5
B	48.5	7.0

Manufacture of Clay Bricks

These bricks are made from clays composed mainly of silica and alumina, with small amounts of lime, iron, manganese and other substances. The different types of clay produce a wide variety of colours and textures. Most clay bricks are kiln-burnt, either in a *Hoffman* kiln in which the fire passes through a series of chambers whilst others are used for heating and cooling, or a *tunnel* kiln where the bricks pass through a long chamber with the firing zone in the centre.

Hand-made bricks are rather irregular in shape and size with uneven arrises; they are used solely for facing work and weather to attractive shades. The majority of clay bricks are either machine-pressed or wirecut. With *machine-pressed* bricks the clay is fed into steel moulds and shaped under heavy pressure with a single or double frog, whereas with *wirecut* bricks, the clay is extruded from a pugmill in a continuous band and then cut into bricks by wires attached to a frame. The Building Research Establishment has designed vertically perforated 'V' bricks to provide the equivalent of a cavity wall in a single leaf with a reduction in labour, mortar and weight. Standard return bricks ('L' bricks) are available for bonding quoins and stopped ends. Solid and perforated 'Calculon' clay bricks are designed for highly stressed brickwork and are particularly well suited for crosswall construction.

Other Types of Brick

There are other types of brick available apart from clay. *Calcium silicate bricks* (sandlime and flintlime) are made from lime and sand or siliccous gravel moulded under heavy pressure and then subjected to steam pressure in an autoclave. BS 187[16] prescribes six acceptable classes of brick (2 to 7) with minimum compressive strengths ranging from 14.0 to 48.5 N/mm^2. The lower strength bricks are only really suitable for internal walls, whereas the stronger classes of brick can be used for external work, including exposed positions and below damp-proof courses. The drying shrinkage of calcium silicate bricks calls for care in design, in storage and during building.[40] The requirements for *concrete bricks* and fixing bricks are detailed in BS 1180.[17] They are manufactured in six strength categories ranging from 7.0 to 40.0 N/mm^2 average compressive strength. *Fire bricks* are made from refractory clay with a high fusing point and they are laid in refractory mortar with tight joints.

Choice of Bricks

The location in which the bricks are to be used has an appreciable effect on the choice of brick. Special quality bricks only should be used in positions of severe exposure, such as parapet, free standing and retaining walls. If a building has a parapet, the choice of brick for the whole wall will probably be restricted to those suitable for severe exposure as it is rarely acceptable architecturally to use one facing brick up to roof level and another for the parapets. Strength is not necessarily an index of durability, nor does high water absorption necessarily indicate a brick of low standard. Bricks of ordinary quality are suitable for normal exposure (in an external wall between eaves and damp-proof course).[18]

The student should examine and handle bricks used in his own locality and note their particular characteristics. He should also be aware of the main properties of some of the more commonly used bricks.

Flettons. These are made from Oxford clay, primarily in Bedfordshire and Northamptonshire. These common bricks are produced in very large quantities and have single frogs, sharp arrises, and faces crossed with bands of dark and light pink. They have a hard exterior skin but are relatively soft inside. Because of their unattractive appearance they are used primarily for internal walls and the inner leaf of cavity walls.

Stocks. These are mainly yellow in colour and made from London clay. They are made in several grades and are good all-purpose bricks. They are suitable for facings, and have the unusual property of increasing in strength with age.

Staffordshire blues. These are bluish-grey engineering bricks made from clays containing iron oxides and are burnt at high temperatures. They are very heavy and dense, with a high crushing strength and low water absorption. These bricks are ideal for positions requiring great strength, such as bridge abutments and retaining walls, and for damp-proof courses.

Southwater reds. These came from Sussex and have similar properties but are red in colour.

Facing bricks. They may be machine-made or hand-made to a variety of colours and textures. Some are multi-coloured and embrace a range of colours, such as red and purple to brown. Sandfaced bricks have a rough textured surface obtained by applying coloured sand prior to burning.

Concrete Blocks

BS 2028, 1364[19] prescribes the requirements for precast concrete blocks made from cement and one of a number of different aggregates from gravel to lightweight materials. The British Standard recognises three types of block, as indicated below, made in various forms (solid, hollow, cellular and aerated concrete). They are made to various sizes, a common size being 440 x 215 mm, and to thicknesses ranging from 75 to 215 mm.

Type	Uses
A	General use including work below damp-proof course.
B	Internal walls and inner leaf of cavity walls; a dense aggregate is needed for the outer leaf of cavity walls below damp-proof course.
C	Primarily for internal non-loadbearing walls.

Clay blocks

These are hollow with smooth or keyed surfaces and have a high thermal insulation value.

MORTARS

Brick and blocks are bedded in and jointed with mortar. A good mortar spreads readily, remains plastic while bricks are being laid to provide a good bond between bricks and mortar, resists frost and acquires early strength, particularly in winter. Mortars should not be stronger than necessary, as an excessively strong mortar concentrates the effects of any differential movement in fewer and wider cracks.

There are a number of different types of mortars in use.[22]

Lime mortar. In the past lime mortars composed of one part of lime to three parts of sand were widely used, although they developed strength slowly. With the advent of cement their use diminished rapidly.

Cement mortar. A mix of 1:3 (cement to sand) is workable but is too strong for everyday use. It would be suitable for heavily loaded brickwork or in extremely wet situations. The sand should comply with BS 1200[20] and be clean and well graded.

Cement—lime mortar. This is sometimes described as compo-mortar and is the most useful general purpose mortar. The best properties of both cement and lime are utilised to produce a mortar which has good working, water-retention and bonding qualities, and also develops early strength without an excessively high mature strength. The lime should be non-hydraulic or semi-hydraulic complying with BS 890.[21] White limes (chalk lime) and magnesium limes (from dolomitic limestone) are both non-hydraulic limes and do not set under water. Greystone limes (obtained from beds of greyish chalk or limestone) are semi-hydraulic limes and harden under water in a few weeks.

Air-entrained (plasticised) mortar. Mortar plasticisers, which entrain air in the mix, provide an alternative to lime for improving the working qualities of lean cement—sand mixes. Hence a 1:6 cement—sand mortar gauged with plasticiser is a good alternative to a 1:1:6 cement—lime—sand mix.

Masonry cement mortar. This consists of a mixture of Portland cement with a very fine mineral filler and an air-entraining agent,[33] complying with the requirements of BS 5224.[49]

Mortars containing special cements. High alumina cement may be used where high early strength or resistance to chemical attack is required. Sulphate-resisting cement may be used where sulphate attack is possible.

Choice of Mortars

It is advisable to restrict the cement content to a minimum consistent with obtaining mortar of adequate strength. In winter, richer mixes are needed to obtain early strength and so resist the effects of frost. Weak mortars should always be used with bricks or blocks of high-drying shrinkage. Tables 4.2 and 4.3 (page 47) provide useful guidelines for the selection of mortars for different situations and using alternative types of brick or block. These tables have been extracted from BRE Digest 160,[22] and the guiding principle is that the mortar should contain no more cement than is necessary to give adequate strength in the brickwork.

Pointing and Jointing

Mortar joints may be finished in a number of ways after the bricks have been laid. Where the work is carried out while the mortar is still fresh, the process is called *jointing*. When the mortar is allowed to harden and then some is removed and replaced with fresh mortar, the process is termed *pointing*. The mortar used in pointing is generally weaker than that used in the joints, to ensure that the load is carried by the bed joint. The five main types of finish are illustrated in figure 4.1.

1. *Struck flush* (figure 4.1.1) gives maximum strength and weather resistance to brickwork but may appear irregular with uneven bricks.

2. *Curved recessed* (bucket handle) (figure 4.1.2) gives a superior finish to struck flush but with little difference in strength or weather resistance.

3. *Struck or weathered* (figure 4.1.3) produces interplay of light and shadow on brickwork but has less strength and weather resistance than previous two finishes.

4. *Overhung struck* (figure 4.1.4) produces variation of light and shade but weakens brickwork and provides a surface on which rainwater may lodge, and this may result in discolouration or frost damage.

5. *Square recessed* (figure 4.1.5) should only be used with very durable bricks as it offers less strength and weather resistance, but has good appearance.

Jointing is quicker and cheaper than pointing and, as the surface finish is part of the bedding mortar, there is less risk of the face joints failing through frost action or insufficient adhesion. On the other hand it may not look as attractive as pointing and it is difficult to keep the work clean and of uniform colour. With pointing, the preparatory work is important — the joints must be properly raked out and brushed to remove loose mortar and dust, and wetted to provide good adhesion.

BUILDING BRICK AND BLOCK WALLS

Bricks and blocks are bonded to give maximum strength and adequate distribution of loads over the wall. Unbonded walls contain continuous vertical joints with the accompanying risk

Table 4.2 Mortar mixes (proportions by volume)

	Mortar group	Cement : lime : sand *	Masonry-cement : sand	Cement : sand, with plasticiser
Increasing strength but decreasing ability to accommodate movements caused by settlement, shrinkage, etc.	i	1 : 0–¼ : 3	—	—
	ii	1 : ½ : 4—4½	1 : 2½—3½	1 : 3—4
	iii	1 : 1 : 5—6	1 : 4—5	1 : 5—6
	iv	1 : 2 : 8—9	1 : 5½—6½	1 : 7—8
	v	1 : 3 : 10—12	1 : 6½—7	1 : 8

Direction of changes in properties

←——— equivalent strengths within each group ———→

increasing frost resistance ———→

←——— improving bond and resistance to rain penetration

Where a range of sand contents is given, the larger quantity should be used for sand that is well graded and the smaller for coarse or uniformly fine sand.

Because damp sands bulk, the volume of damp sand used may need to be increased. For cement : lime : sand mixes, the error due to bulking is reduced if the mortar is prepared from lime : sand coarse stuff and cement in appropriate proportions; in these mixes 'lime' refers to non-hydraulic or semi-hydraulic lime and the proportions given are for lime putty. If hydrated lime is batched dry, the volume may be increased by up to 50 per cent to obtain adequate workability.

* The addition of an air-entraining agent might improve the frost-resistance of cement : lime : sand mortars.

Table 4.3 Selection of mortar groups

Type of brick:	Clay		Concrete and calcium silicate	
Early frost hazard[a]	no	yes	no	yes
Internal walls	(v)	(iii) or (iv)[b]	(v)[c]	(iii) or plast (iv)[b]
Inner leaf of cavity walls	(v)	(iii) or (iv)[b]	(v)[c]	(iii) or plast (iv)[b]
Backing to external solid walls	(iv)	(iii) or (iv)[b]	(iv)	(iii) or plast (iv)[b]
External walls; outer leaf of cavity walls:				
—above damp-proof course	(iv)[d]	(iii)[d]	(iv)	(iii)
—below damp-proof course	(iii)[e]	(iii)[b, e]	(iii)[e]	(iii)[e]
Parapet walls; domestic chimneys:				
—rendered	(iii)[f, g]	(iii)[f, g]	(iv)	(iii)
—not rendered	(ii)[h] or (iii)	(i)	(iii)	(iii)
External free-standing walls	(iii)	(iii)[b]	(iii)	(iii)
Sills; copings	(i)	(i)	(ii)	(ii)
Earth-retaining walls (back-filled with free-draining material)	(i)	(i)	(ii)[e]	(ii)[e]

a during construction, before mortar has hardened (say 7 days after laying) or before the wall is completed and protected against the entry of rain at the top

b if the bricks are to be laid wet, *see* 'Cold weather bricklaying'

c if not plastered, use group (iv)

d if to be rendered, use group (iii) mortar made with sulphate-resisting cement

e if sulphates are present in the groundwater, use sulphate-resisting cement

f parapet walls of clay units should not be rendered on both sides; if this is unavoidable, select mortar as though *not* rendered

g use sulphate-resisting cement

h with 'special' quality bricks, or with bricks that contain appreciable quantities of soluble sulphates

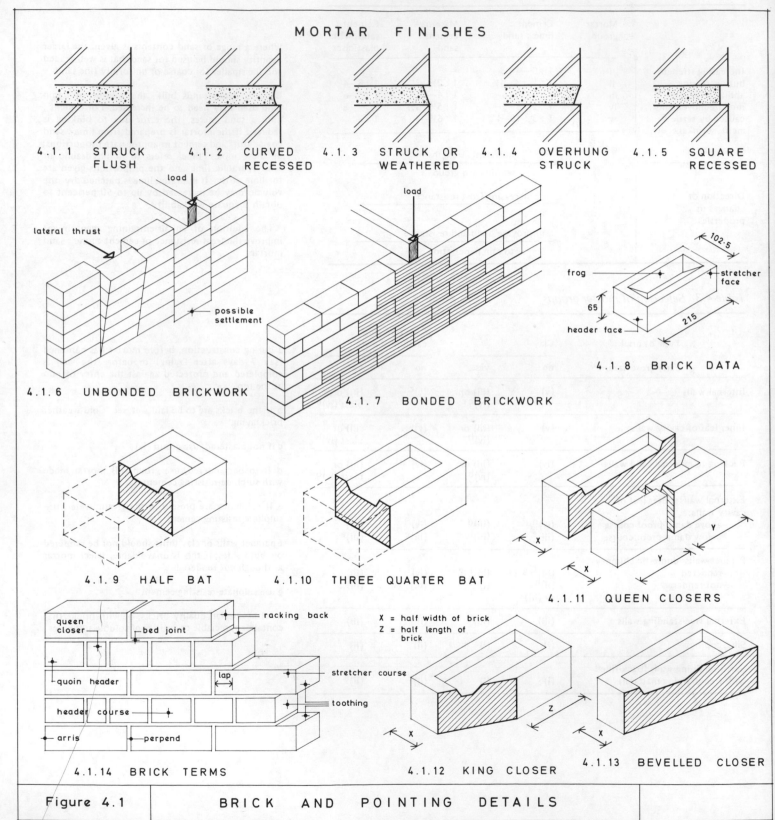

MORTAR FINISHES

4.1.1 STRUCK FLUSH

4.1.2 CURVED RECESSED

4.1.3 STRUCK OR WEATHERED

4.1.4 OVERHUNG STRUCK

4.1.5 SQUARE RECESSED

load

lateral thrust

possible settlement

4.1.6 UNBONDED BRICKWORK

load

4.1.7 BONDED BRICKWORK

102.5

frog

stretcher face

65

header face

215

4.1.8 BRICK DATA

4.1.9 HALF BAT

4.1.10 THREE QUARTER BAT

X   X   Y   Y

4.1.11 QUEEN CLOSERS

queen closer

bed joint

racking back

quoin header

lap

stretcher course

header course

toothing

arris

perpend

4.1.14 BRICK TERMS

X = half width of brick
Z = half length of brick

X   Z

4.1.12 KING CLOSER

X   Z

X

4.1.13 BEVELLED CLOSER

| Figure 4.1 | BRICK AND POINTING DETAILS | |

of failure as shown in figure 4.1.6. Bonded walls provide lateral stability and resistance to side thrust (figure 4.1.7), and the bond can be selected to give an attractive appearance to the wall face. The bond is achieved by bricks in one course overlapping those in the course below. The amount of lap is generally one-quarter of the length of a brick, but with half-brick walls in stretcher bond the lap is one-half of a brick.

A typical brick is illustrated in figure 4.1.8 with a frog or sinking on its top face which may be V or U-shaped. To obtain a bond it is necessary to introduce bats (parts of bricks — usually one-half or three-quarter bricks: figures 4.1.9 and 4.1.10) or closers. Closers may be queen closers of half a brick, cut in its length, or a quarter brick (figure 4.1.11), or be a tapered brick in the form of a king closer (figure 4.1.12) or bevelled closer (figure 4.1.13).

Some of the more commonly used brick terms are illustrated in figure 4.1.14. A stretcher course is a row of bricks with their stretcher or longer faces showing, while a header course is made up of ends of bricks. An edge of a brick is termed an *arris*, while an external angle of brickwork is called a *quoin*. *Perpends* are vertical brick joints, whilst vertical joints at right angles to the wall face are known as *cross joints*. The term *gauge* is used to indicate the number of courses to be built in a fixed height, for example four courses to 300 mm. *Racking back* is a stepped arrangement of bricks often formed during construction at corners and ends of walls, when one part of the brickwork is built to a greater height than that adjoining. *Toothing* refers to the provision for bonding at the end of brickwork where it is to be continued subsequently.

Types of Bond

The choice of bond is influenced by the situation, function and thickness of wall. For instance with external walls to buildings appearance is often important, whilst with manhole walls strength would be the main consideration. The number of bonds suitable for use with half-brick walls is very limited.
Flemish bond. This is the most commonly used bond for solid brick walls as it combines an attractive appearance with reasonable strength (figures 4.2.7 to 4.2.12). The term double Flemish is used where the bond is used on both faces. Bricks are laid as alternate headers and stretchers in the same course, the header in one course being in the centre of the stretcher in the course above and below. Single Flemish bond has Flemish bond on one face only with English bond as a backing.
English bond. This consists of stretchers throughout the length of one course and headers throughout the next course (figures 4.2.1 to 4.2.6). It is rather stronger than Flemish bond because

of the absence of internal straight joints, and is particularly well suited for use in manhole and retaining walls. It uses more facing bricks than Flemish bond.
Garden wall bonds. These are used to reduce the number of headers and the cost of facing bricks, yet at the same time to produce a wall of reasonable appearance and strength. English garden wall bond is made up of three courses of stretchers to each course of headers and Flemish garden wall bond consists of three stretchers to one header in each course (figure 4.3.7), and these are well suited for one-brick boundary walls.
Stretcher bond. This consists of stretchers throughout and the centre line of each stretcher is directly over the centre line of the cross joint in the course below (figure 4.3.6). This bond is used for half-brick walls, including the leaves of cavity walls. A half-bat is used to commence or finish alternate courses, and three-quarter bats may be necessary at junctions of crosswalls or where pilasters occur. On occasions snapped headers are introduced to imitate Flemish bond but the additional cost is rarely justified.
Rat-trap bond. This is used on one-brick walls to reduce the weight and cost of the wall (see figure 4.3.8). The bricks are laid on edge and a series of cavities are formed within the wall. If used externally, the outer face would normally be rendered.

Setting out Bonds

The method of setting out a brick bond on plan is illustrated in figures 4.3.1 to 4.3.5, and it needs to be undertaken systematically in logical stages.

1. Draw to scale the outline of the wall, normally starting at a right-angled corner, on alternate courses and draw in the quoin header in opposite directions on each course.

2. Insert queen closers next to the quoin header and continue them until they intersect the back line of the wall produced.

3. Draw in alternate headers and stretchers on the front faces in the case of Flemish bond, and headers on one face and stretchers on the other with English bond. The exposed header and stretcher faces of the quoin header determine the nature of the faces of the other bricks.

4. Continue the facework to the back faces. In English bond if the wall is one, two or three-bricks thick, stretchers or headers will appear on both wall faces in the same course, whereas if the wall is one-and-a-half or two-and-a-half-bricks thick and stretchers appear on one face, then headers will occur on the other side of the wall in the same course.

5. In thick walls, the interior gaps are filled in with headers.

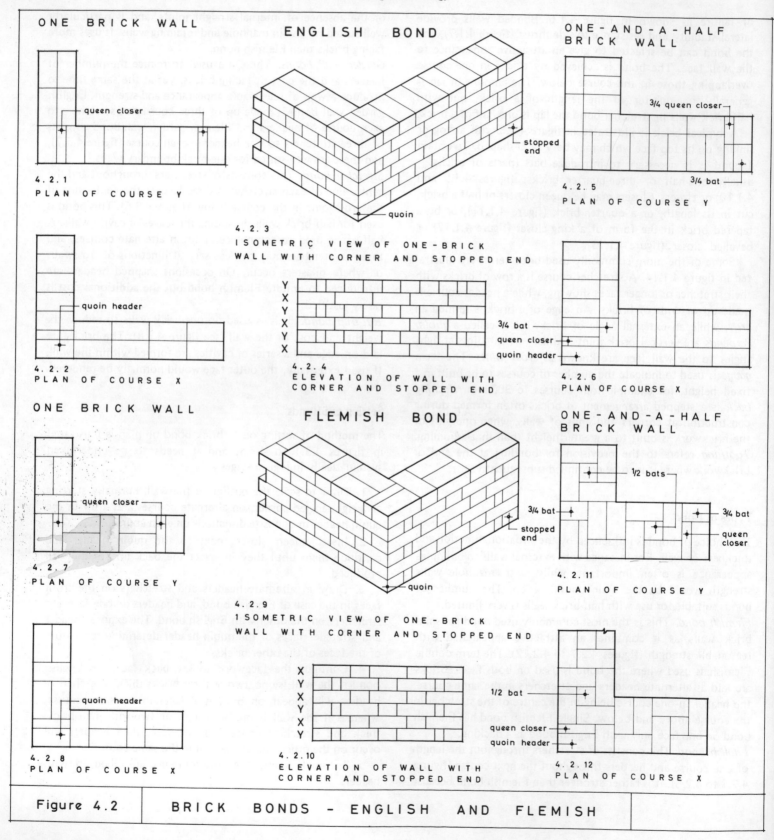

ONE BRICK WALL

4.2.1 PLAN OF COURSE Y

queen closer

4.2.2 PLAN OF COURSE X

quoin header

ONE BRICK WALL

4.2.7 PLAN OF COURSE Y

queen closer

4.2.8 PLAN OF COURSE X

quoin header

ENGLISH BOND

4.2.3 ISOMETRIC VIEW OF ONE-BRICK WALL WITH CORNER AND STOPPED END

stopped end

quoin

4.2.4 ELEVATION OF WALL WITH CORNER AND STOPPED END

Y X Y X Y X

FLEMISH BOND

4.2.9 ISOMETRIC VIEW OF ONE-BRICK WALL WITH CORNER AND STOPPED END

stopped end

quoin

4.2.10 ELEVATION OF WALL WITH CORNER AND STOPPED END

X Y X Y X Y

ONE-AND-A-HALF BRICK WALL

4.2.5 PLAN OF COURSE Y

3/4 queen closer

3/4 bat

ONE-AND-A-HALF BRICK WALL

4.2.6 PLAN OF COURSE X

3/4 bat
queen closer
quoin header

4.2.11 PLAN OF COURSE Y

1/2 bats

3/4 bat

3/4 bat
queen closer

4.2.12 PLAN OF COURSE X

1/2 bat

queen closer
quoin header

Figure 4.2   BRICK BONDS - ENGLISH AND FLEMISH

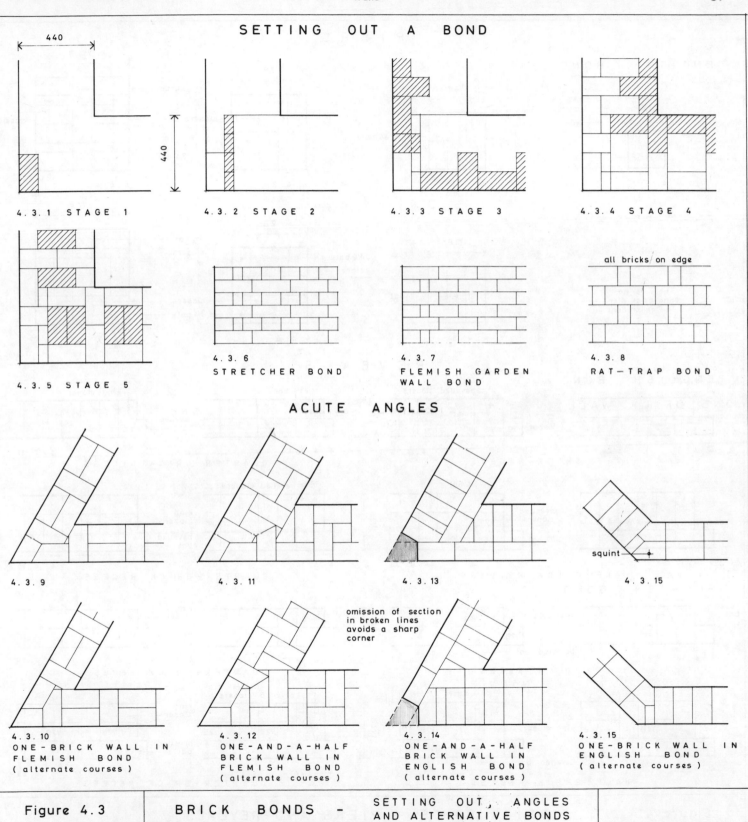

SETTING   OUT   A   BOND

440

440

4.3.1   STAGE   1

4.3.2   STAGE   2

4.3.3   STAGE   3

4.3.4   STAGE   4

4.3.5   STAGE   5

4.3.6
STRETCHER   BOND

4.3.7
FLEMISH   GARDEN
WALL   BOND

all bricks / on edge

4.3.8
RAT—TRAP   BOND

ACUTE   ANGLES

4.3.9

4.3.11

4.3.13

4.3.15

squint

4.3.10
ONE-BRICK   WALL   IN
FLEMISH   BOND
(alternate courses)

4.3.12
ONE-AND-A-HALF
BRICK   WALL   IN
FLEMISH   BOND
(alternate courses)

omission of section
in broken lines
avoids a sharp
corner

4.3.14
ONE-AND-A-HALF
BRICK   WALL   IN
ENGLISH   BOND
(alternate courses)

4.3.15
ONE-BRICK   WALL   IN
ENGLISH   BOND
(alternate courses)

| Figure 4.3 | BRICK BONDS - | SETTING OUT , ANGLES AND ALTERNATIVE BONDS | |

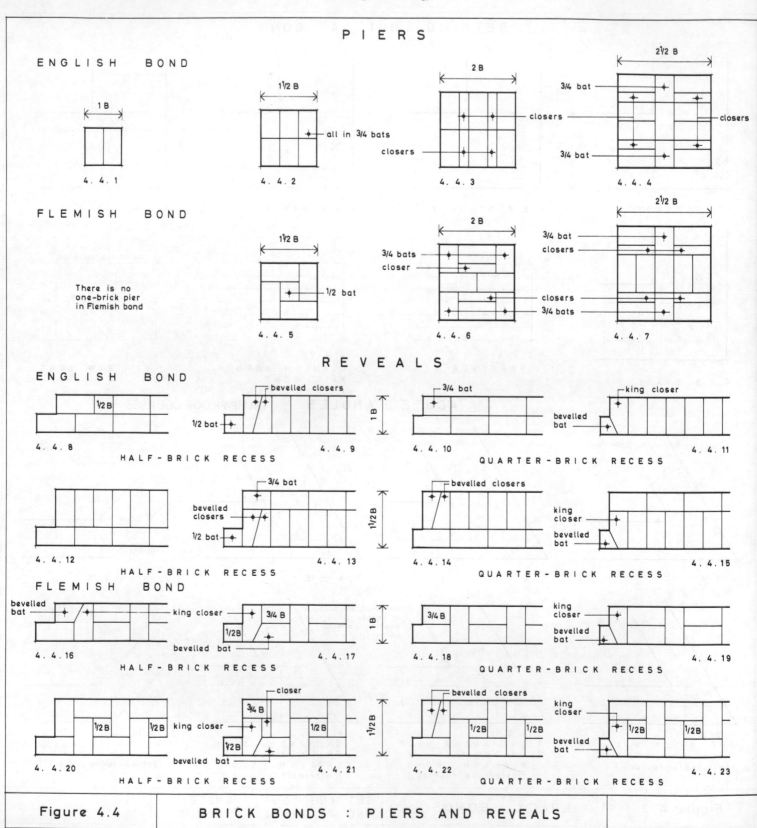

Figure 4.4   BRICK BONDS : PIERS AND REVEALS

Angles, Piers and Reveals

The bonding of angles, piers and reveals in English and Flemish bond is illustrated in figures 4.3 and 4.4. Squint corners usually incorporate a special squint brick as shown in figure 4.3.15. Acute angles introduce bats and closers around the corner, and figures 4.3.13 and 4.3.14 show a method of avoiding the sharp corner.

The bonding of square piers follows the normal rules of bonding described previously. Alternate courses are identical but turned through 90 degrees. One-brick piers in English bond (figure 4.4.1) consist of two headers laid side by side; there is no alternative arrangement in Flemish bond. Note how closers and/or bats are used in the larger piers to obtain bond (figures 4.4.2 to 4.4.7).

Recessed reveals may be formed in solid walls to accommodate doors and windows and provide added protection from the weather. Square reveals are identical to stopped ends which are illustrated in figure 4.2. Recessed reveals (figures 4.4.8 to 4.4.23) require the extensive use of closers and bats.

CAVITY WALLS

Solid brick walls of the thicknesses generally used in domestic construction are not always damp-proof. Moisture may penetrate the walls through the bricks, joints or through minute cracks which form between the bricks and joints as the mortar dries out. In addition, the thermal insulation value of solid walls is limited. These disadvantages can be overcome by using cavity or hollow walls, consisting of an air space between two leaves with upper floor loads supported by the inner leaf. Cavity walls often consist of a half-brick outer leaf, 50 mm wide cavity and inner leaf of 100 mm insulating, load-bearing concrete blocks, giving a total width of 252.50 mm. The head of the wall is built solid to spread the roof loads over both leaves and the cavity should be filled with fine concrete up to ground level to stiffen the base of the wall (figure 4.5.3).

The Building Regulations[3] (schedule 7) require the two leaves to each be not less than 90 mm thick and to be provided with adequate lateral support by roofs and floors. The cavity shall be not less than 50 mm nor more than 75 mm in width. The leaves shall be tied together with ties complying with BS 1243,[24] or suitable equivalent to give stability to the wall. The British Standard describes vertical twist, butterfly and double triangle (figure 4.5.4) types of wall ties made of mild steel coated with zinc, galvanised wire, copper, copper alloys or stainless steel.
Polypropylene wall ties provide another alternative; some failures of wall ties have occurred, mainly due to substandard galvanising or aggressive mortars such as the black ash type. The ties are spaced at distances apart not exceeding 900 mm horizontally and 450 mm vertically and there shall be, as near as practicable to openings, a tie every 300 mm of height if the leaves are not connected by a bonded jamb.[32]

Cavity walls offer distinct advantages with good water penetration prevention and thermal insulation properties, greater choice of bricks and blocks and less facing bricks required. It is essential to prevent mortar droppings falling into the cavity, ideally by bringing a wood lath up the cavity as the walls are built, as mortar droppings can form bridges across the cavity. A damp-proof course should be inserted over lintels to door and window openings (figure 4.5.1) and the ends of cavities at jambs to openings must be effectively sealed and damp-proofed (figures 4.5.2, 4.5.5 and 4.5.6). The satisfactory performance of cavity walls is very much dependent on close supervision and proper detailing. The cavity must not be ventilated or it will cease to act as a thermal insulator. The thermal insulation value of the cavity can be increased by filling it with urea-formaldehyde (UF) foam,[52] in the manner specified in BS 5618.[53] There are however risks with exposed porous outer skins and other problems, such as the shrinkage of the fill and emission of gases.[55]

DAMP-PROOF COURSES

Damp-proof courses are required in various locations in a building to prevent moisture penetration. The upward movement of moisture is prevented by damp-proof courses placed in walls at not less than 150 mm above ground level and in solid floors, either sandwiched in the concrete slab or underneath the floor screed. Barriers to resist the downward movement of moisture are inserted in parapets, chimneys and above lintels in cavity walls.[34] On occasions horizontal movement has to be resisted as where the outer leaf of a cavity wall is returned at an opening to close the cavity.[25] Typical details of damp-proof course arrangements around window and door openings are shown in figures 4.5.1 to 4.5.3 and 4.5.5 and 4.5.6.

Parapet walls have in the past often proved to be a source of moisture penetration into a building, either through the absence or incorrect placing of damp-proof courses, or through rendering parapet walls on both sides and the top; where the rendering cracks, rain enters and is trapped behind the rendering.[31] In low parapet walls, a damp-proof course should be inserted under the coping, and the roof covering extended up the inside face of the wall and tucked in under the damp-course (figure 4.6.1). With taller parapets two damp-proof courses are required; one under the coping and the other at or

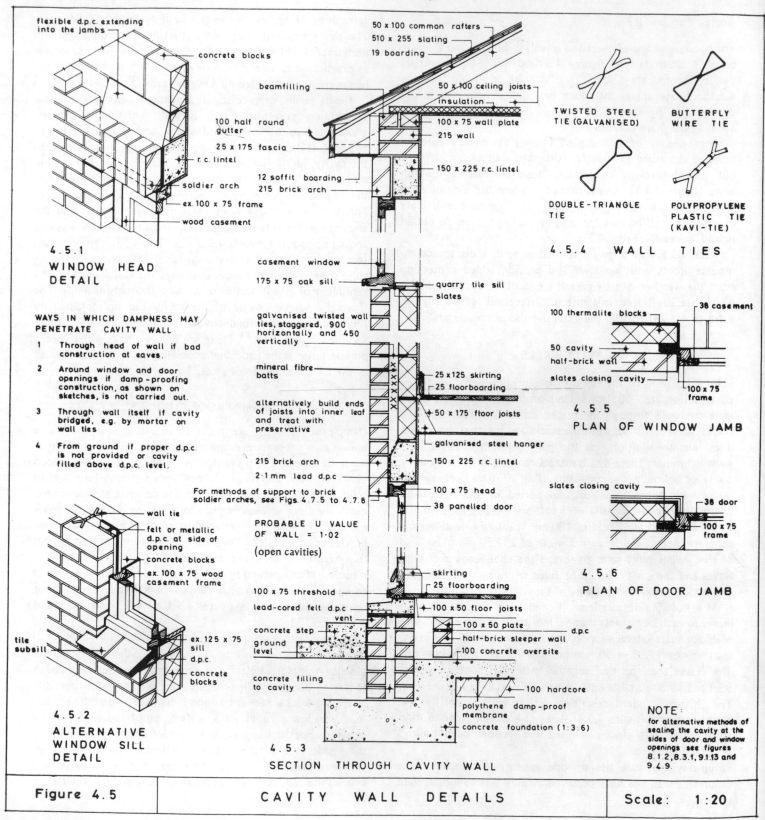

4.5.1
WINDOW HEAD DETAIL

flexible d.p.c. extending into the jambs
concrete blocks
r.c. lintel
soldier arch
ex. 100 x 75 frame
wood casement

50 x 100 common rafters
510 x 255 slating
19 boarding
beamfilling
100 half round gutter
25 x 175 fascia
12 soffit boarding
215 brick arch
50 x 100 ceiling joists
insulation
100 x 75 wall plate
215 wall
150 x 225 r.c. lintel

WAYS IN WHICH DAMPNESS MAY PENETRATE CAVITY WALL

1　Through head of wall if bad construction at eaves.

2　Around window and door openings if damp-proofing construction, as shown on sketches, is not carried out.

3　Through wall itself if cavity bridged, e.g. by mortar on wall ties

4　From ground if proper d.p.c. is not provided or cavity filled above d.p.c. level.

casement window
175 x 75 oak sill
galvanised twisted wall ties, staggered, 900 horizontally and 450 vertically
mineral fibre batts
alternatively build ends of joists into inner leaf and treat with preservative
215 brick arch
2·1 mm lead d.p.c
For methods of support to brick soldier arches, see Figs. 4.7.5 to 4.7.8

quarry tile sill
slates
25 x 125 skirting
25 floorboarding
50 x 175 floor joists
galvanised steel hanger
150 x 225 r.c. lintel
100 x 75 head
38 panelled door

PROBABLE U VALUE OF WALL = 1·02
(open cavities)

4.5.2
ALTERNATIVE WINDOW SILL DETAIL

wall tie
felt or metallic d.p.c. at side of opening
concrete blocks
ex. 100 x 75 wood casement frame
tile subsill
ex. 125 x 75 sill
d.p.c.
concrete blocks

100 x 75 threshold
lead-cored felt d.p.c
concrete step
ground level
concrete filling to cavity
vent
skirting
25 floorboarding
100 x 50 floor joists
100 x 50 plate
d.p.c
half-brick sleeper wall
100 concrete oversite
100 hardcore
polythene damp-proof membrane
concrete foundation (1:3:6)

4.5.3
SECTION THROUGH CAVITY WALL

4.5.4　WALL TIES

TWISTED STEEL TIE (GALVANISED)
BUTTERFLY WIRE TIE
DOUBLE-TRIANGLE TIE
POLYPROPYLENE PLASTIC TIE (KAVI-TIE)

100 thermalite blocks
38 casement
50 cavity
half-brick wall
slates closing cavity
100 x 75 frame

4.5.5
PLAN OF WINDOW JAMB

slates closing cavity
38 door
100 x 75 frame

4.5.6
PLAN OF DOOR JAMB

NOTE:
for alternative methods of sealing the cavity at the sides of door and window openings see figures 8.1.2, 8.3.1, 9.1.13 and 9.4.9.

| Figure 4.5 | CAVITY WALL DETAILS | Scale: 1:20 |

Table 4.4 Materials for damp-proof courses (Source: BRE Digest 77[25])

Material	Minimum weight kg/m²	Minimum thickness mm	Joint treatment to prevent water moving: upward	downward	Liability extrusion	Durability	Other considerations
Group A: Flexible							
Lead to BS 1178[26]	Code Nr 4		Lapped at least 100 mm	Welted	Not under pressures met in normal construction	Corrodes in contact with fresh lime or Portland cement mortar	To prevent corrosion apply bitumen or bitumen paint of heavy consistency to the corrosion-producing surface and both lead surfaces
Copper to BS 2870,[27] Grade A, annealed condition	Approx. 2.28*	0.25	Lapped at least 100 mm†	Welted†	Not under pressures met in normal construction	Highly resistant to corrosion	Where soluble salts are known to be present (eg sea salt in sand) similar treatment to lead is recommended. Copper may stain external surfaces, especially stone
Bitumen to BS 743:[28] A—hessian base B—fibre base C—asbestos base D—hessian base and lead E—fibre base and lead F—asbestos base and lead G—sheeting with hessian base	3.8 3.3 3.8 4.4 4.4 4.9 5.4	— — — — — — —	Lapped at least 100 mm	Lapped and sealed	Likely to extrude under heavy pressure but this is unlikely to affect water resistance	The hessian or felt may decay but this does not affect efficiency if the bitumen remains undisturbed. Types D, E and F are most suitable for buildings that are intended to have a very long life or where there is risk of movement	Materials should be unrolled with care. In cold weather the rolls should be warmed before use
Polythene, black, low density (0.915–0.925 g/ml)	Approx 0.5	0.46	Lapped for distance at least equal to width of dpc	Welted	Not under pressures met in normal construction	No evidence of deterioration in contact with other building materials	Welted joints must be held in compression. When used as cavity trays exposed on elevation these materials may need bedding in mastic for the full thickness of the outer leaf of walling, to prevent penetration by driving rain
Pitch/polymer	Approx 1.5	1.27	Lapped at least 100 mm	Lapped and sealed	Not under pressures met in normal construction		
Group B: Semi-rigid							
Mastic asphalt to BS 1097[29] or BS 1418[30]			No joint problems		Liable to extrude under pressures above 65 kN/m². Material of hardness appropriate to conditions should be used	No deterioration	To provide key for mortar below next course of brickwork, grit should be beaten into asphalt immediately after application and left proud of surface. Alternatively the surface should be scored whilst still warm
Group C: Rigid							
Epoxy resin/sand	—	6.0	No joint problems		Not extruded	No evidence of deterioration in contact with other building materials	Resin content should be about 15%. The appropriate hardener should be used
Brick to BS 3921:[15] max water absorption 4.5%	—	—	—	Not suitable	Not extruded	No deterioration	Use two courses, laid to break joint, bedded in 1:3 Portland cement:sand
Slates at least 230 mm long; should pass wetting and drying test and acid-immersion test of BS 3798[39]	—	5.0	—	Not suitable although they may sometimes be used to resist lateral movement of water	Not extruded	No deterioration	

* Where copper is to form in one piece a dpc and a projecting drip or flashing it should weigh at least 4.12 kg/m²
† Copper may be welded, in which case phosphorus-deoxidised copper to BS 1172 should be used

near roof level and lapped with the roof covering (figure 4.6.2).[25]

The most important characteristics of a damp-proof course are that it shall be impervious and durable, in order to last the life of the building without permitting moisture to penetrate. A reasonable degree of flexibility is important to prevent fracture with movement of the building. Other desirable properties include ease of application, ability to withstand loads, and availability in thin sheets at reasonable cost.

Building Research Establishment Digest 77[25] classifies materials for damp-proof courses into three categories

Flexible. Lead, copper, bitumen, polythene and pitch/polymer — these are suitable for most locations and they alone should be used over openings and for bridging cavities. Flexible materials are the most satisfactory for stepped damp-proof courses and they can be dressed to complex shapes without damage.

Semi-rigid. Mastic asphalt — particularly suitable for very thick walls and to withstand high water pressure.

Rigid. Epoxy resin/sand, brick and slate — these can withstand heavy loads but are more likely to be fractured by building movement than materials in the other groups.

Table 4.4 lists the requirements and principal characteristics of damp-proof course materials.

Damp-proofing Old Walls

Various methods are available for damp-proofing existing buildings which are subject to rising damp in the absence of damp-proof courses.[6] Probably the best method is to saw a slot in a mortar bed joint and to insert a damp-proof membrane in the slot. The membrane is normally inserted in 500 mm lengths and may be of slate, bitumen-felt, copper, lead or polythene. Suitable saws are described in DOE Advisory Leaflet 58.[35]

Another process is called electro-osmotic damp-proofing. Damp rises in walls from soils by capillarity, osmotic diffusion within the porous substance of the wall and by evaporation from its surface. Damp walls are negatively charged in relation to the underlying soil. The wall acts as an accumulator which, as long as evaporation losses are replenished from the soil, is virtually on permanent trickle charge. The electro-osmotic installation consists of 25 mm holes drilled from the outside with strip electrodes of high conductivity copper mortared into the drillings and looped into copper strips set into bed joints at damp-course level along the wall face. Earth electrodes are also installed consisting of 15 mm diameter copper or copper-clad steel rod electrodes driven from depths of 4 to 6 m at intervals of 10 to 13 m. The object is to provide a bridge between the wall at damp-course level and the soil, thus destroying the surface tension and preventing rising moisture.

Yet another process is to inject a siliconate/rubber latex fluid into the lowest accessible mortar bed joint through drill holes 10 mm in diameter at 50 mm intervals. The siliconate acts as a 'carrier', stabilising the rubber within the wall and allowing the area to harden off, and so to form an effective damp-proof membrane.

SCAFFOLDING

Most brickwork is erected from scaffolding made up of steel or aluminium alloy tubes of 38 mm nominal bore and a variety of fittings as illustrated in DOE Advisory Leaflet 36.[36] Upright members are called standards; horizontal members along the length of the scaffold are termed ledgers; and the short cross members are called putlogs when one end is supported by the wall, and transoms when both ends rest on ledgers. A putlog scaffold (figure 4.6.3) has a single row of standards set about 1.25 m or 1.30 m from the wall and is partly supported by the building. An independent tied scaffold (figure 4.6.6) has a double row of standards about one metre apart, and does not rely on the building for support.

The spacing of standards depends on the load to be carried and varies from about 1.80 m to 2.40 m for bricklayers. Every scaffold should be securely tied into the building at intervals of approximately 3.60 m vertically and 6.00 m horizontally. Putlogs cannot be relied upon to act as ties. Alternative methods of fixing to window openings are shown in figures 4.6.3 and 4.6.6, as also are bracing arrangements. Substantial working platforms are required of 32 to 50 mm thick boards with butted joints and guard rails and toe boards for safety purposes, all fixed in accordance with the Construction (Working Places) Regulations, 1966.

ARCHES AND LINTELS

Openings in walls are spanned by arches or lintels, or a combination of both. An *arch* is an arrangement of wedge-shaped blocks mutually supporting one another. The superimposed load is transmitted through the blocks and exerts a pressure, both downwards and outwards, upon its supports. A *lintel* is a solid horizontal member or beam spanning an opening to carry the load and transmit it to the wall on either side.

Arches

The simpler form of brick arch can be classified into three main types.

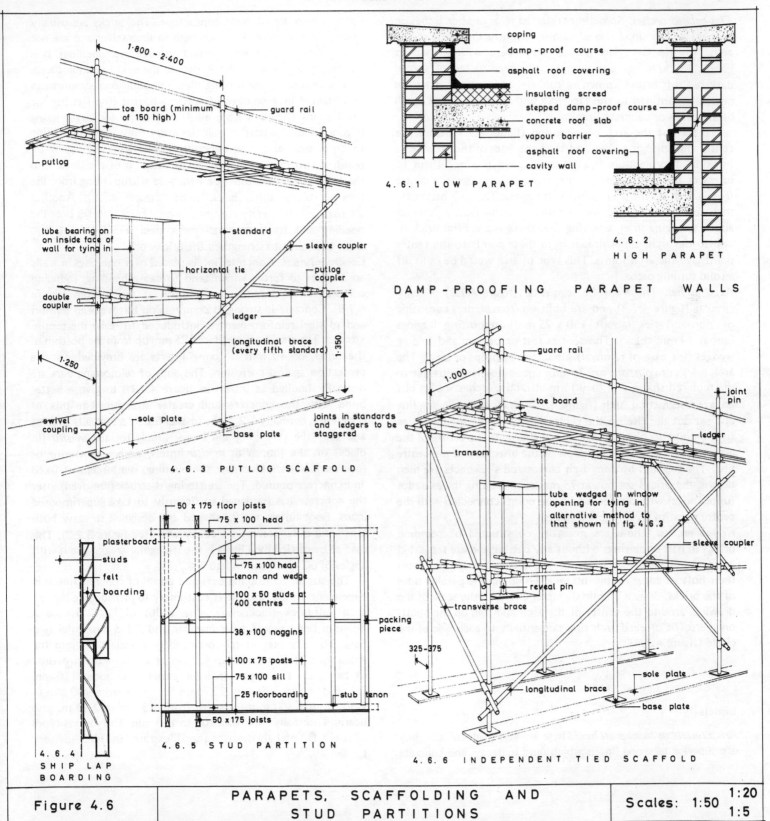

1·800 – 2·400

toe board (minimum of 150 high)

guard rail

putlog

tube bearing on on inside face of wall for tying in

standard

sleeve coupler

horizontal tie

putlog coupler

double coupler

ledger

longitudinal brace (every fifth standard)

1·250

1·350

swivel coupling

sole plate

base plate

joints in standards and ledgers to be staggered

4.6.3 PUTLOG SCAFFOLD

coping
damp-proof course
asphalt roof covering
insulating screed
stepped damp-proof course
concrete roof slab
vapour barrier
asphalt roof covering
cavity wall

4.6.1 LOW PARAPET

4.6.2 HIGH PARAPET

DAMP-PROOFING PARAPET WALLS

guard rail

1·000

joint pin

toe board

ledger

transom

tube wedged in window opening for tying in. alternative method to that shown in fig. 4.6.3

sleeve coupler

reveal pin

transverse brace

325–375

longitudinal brace

sole plate

base plate

4.6.6 INDEPENDENT TIED SCAFFOLD

50 x 175 floor joists
75 x 100 head
plasterboard
studs
felt
boarding
75 x 100 head
tenon and wedge
100 x 50 studs at 400 centres
38 x 100 noggins
packing piece
100 x 75 posts
75 x 100 sill
25 floorboarding
stub tenon
50 x 175 joists

4.6.5 STUD PARTITION

4.6.4
SHIP LAP
BOARDING

| Figure 4.6 | PARAPETS, SCAFFOLDING AND STUD PARTITIONS | Scales: 1:50 | 1:20 1:5 |

Flat-gauged arches. Sometimes referred to as camber arches on account of the small rise or camber given to the soffit of the arch. In the Georgian period they were widely used in good-class buildings and were usually constructed of special dark red soft bricks known as 'rubbers' and 'cutters'. They are normally confined to the facework and are backed with rough brick arches or concrete lintels.

To set out the arch on elevation (figure 4.7.1), draw the centre and springing lines and then the lines of the skewbacks (maximum projection $\frac{1}{10}$ of span) and produce them to intersect the centre line of the arch. This locates the centre from which all the arch joints will radiate. The arch bricks are then set out along the flat top of the arch (usually 300 mm above springing line), ensuring that there is a central brick or key to the arch. Lines drawn from these points to the centre point give the arch joints. This type of arch would be built off a solid turning piece.

Axed arches. These can be segmental (figure 4.7.2) or semi-circular (figure 4.7.3) and are built up from centres consisting of ribs and ties (about 150 x 25 mm) supporting laggings (about 19 mm thick). The centres rest on bearers and folding wedges (for ease of removal) supported by props or posts. The arch bricks or *voussoirs* are axed or cut as closely as possible to the required shape, and they form attractive arches. To set out an axed segmental arch (figure 4.7.2) draw the springing line and set out the rise on the centre line. Join the springing line to the point so obtained and bisect this line. The centre of the arch curve is at the intersection of the bisector and the centre line. The top and bottom arch curves and skewbacks are then drawn; the bricks set out at 75 mm centres along the extrados (upper edge) with dividers and these points connected with the centre point to give the brick joints.

Rough arches. These are normally constructed of common bricks in half-brick rings without any cutting and are separated by wedge-shaped joints. Hence the arches are set out from their bottom edges in this instance. To obtain the parallel sides of the bricks, draw a 75 mm diameter circle, to the scale of the drawing, around the centre of the arch and join up the points on the soffit of each arch ring tangentially to each side of the circle (figure 4.7.4).

Lintels

Brick lintels or soldier arches. These are very popular, and they are superior to wood lintels which tend to decay, and concrete lintels which are of poor appearance. The bricks are usually placed on end, offer little strength in themselves and are not true arches. When bricks are laid on edge over openings, it is customary to make up the 40 mm to the next brick joint with two courses of nibless roofing tiles. The additional support can be obtained in a number of ways: wrought iron flat bar for small spans (figure 4.7.5); mild steel angle for larger spans (figure 4.7.6); 'butterfly' wall ties every third joint connecting to an *in situ* reinforced concrete lintel (figure 4.7.7); or by building two mild steel bars across the top of the lintel into the brick jambs on either side with wire stirrups hung from the bars at every third brick joint (figure 4.7.8). Another alternative is to merely continue normal brick courses over the opening and to insert small mesh steel reinforcement into three bed joints to strengthen the brickwork.

Concrete lintels. Concrete lintels placed over openings in walls are subject to forces which tend to cause bending, inducing compression at the top and tension at the bottom (figure 4.7.9). Concrete is strong in compression but weak in tension and so steel reinforcement is introduced to resist the tensile stresses. The steel bars are placed 25 mm up from the bottom of the lintel, or 40 mm if fixing inserts are provided, to give protection against corrosion. The ends of reinforcing bars are normally hooked as shown in figure 4.7.10 to give a better bond with the concrete and greater strength. The mix of concrete is normally 1:2:4; the smaller lintels are usually precast and only the very long and heavy lintels formed *in situ* (in place) on the job. With *in situ* lintels, timber shuttering or formwork is assembled over the opening, reinforcement fixed and concrete poured. The shuttering is struck (removed) after the concrete has hardened sufficiently to take superimposed loads. Boot lintels are 'L' shaped and designed to carry both inner and outer leaves of a cavity wall (see figure 8.2.2). The load to be carried by a lintel is the triangular area above it with angles of $60°$ as shown in figure 4.7.11.

The size of reinforcing bars and depth of concrete lintels is dependent on the span. It is customary to provide one bar for each half-brick thickness in the width of lintel. Hence a 105 mm lintel would have one bar and a 215 mm lintel two bars. The diameter of reinforcing bars is usually 12 mm for spans up to 1.60 m, 16 mm for longer spans or the provision of two bars. The depth of lintels varies from about 150 mm for spans up to 700 mm, 225 mm for spans from 700 mm to 1.30 m, and 300 mm for spans from 1.30 m to 2.20 m. End bearings normally vary from 100 to 150 mm. The various types of concrete lintel are described and illustrated in DOE Advisory Leaflet 49.[41]

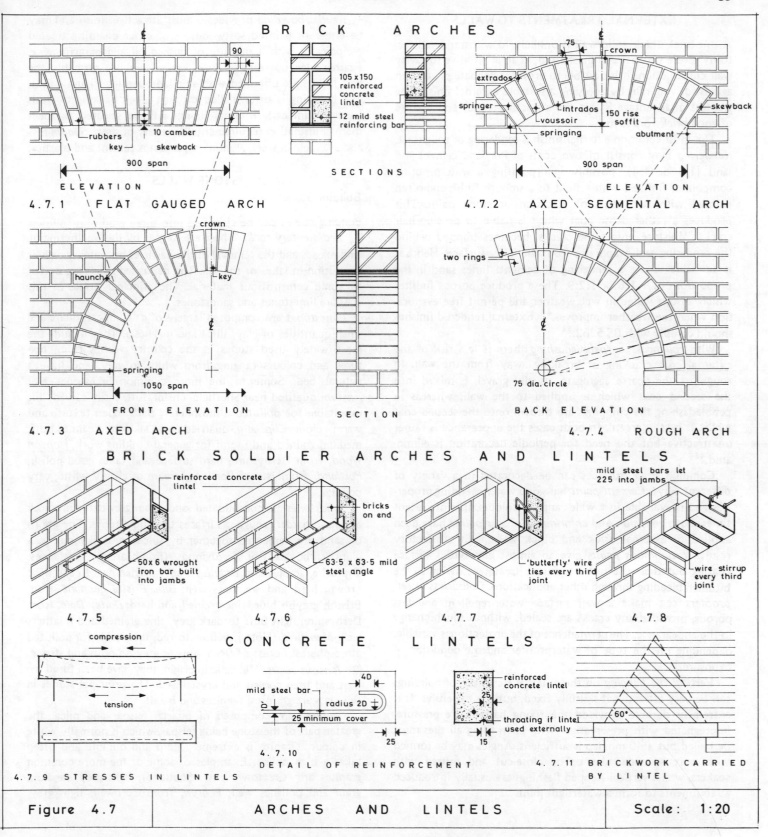

BRICK ARCHES

90

105 x 150 reinforced concrete lintel

12 mild steel reinforcing bar

rubbers
key
skewback
10 camber

900 span

ELEVATION

SECTIONS

4.7.1 FLAT GAUGED ARCH

75
crown
extrados
springer
intrados
voussoir
springing
skewback
150 rise
soffit
abutment

900 span

ELEVATION

4.7.2 AXED SEGMENTAL ARCH

crown
haunch
key
springing

1050 span

FRONT ELEVATION

SECTION

4.7.3 AXED ARCH

two rings

75 dia. circle

BACK ELEVATION

4.7.4 ROUGH ARCH

BRICK SOLDIER ARCHES AND LINTELS

reinforced concrete lintel

50 x 6 wrought iron bar built into jambs

4.7.5

bricks on end

63.5 x 63.5 mild steel angle

4.7.6

'butterfly' wire ties every third joint

4.7.7

mild steel bars let 225 into jambs

wire stirrup every third joint

4.7.8

CONCRETE LINTELS

compression

tension

4.7.9 STRESSES IN LINTELS

4D
mild steel bar
radius 2D
25 minimum cover
25

4.7.10 DETAIL OF REINFORCEMENT

reinforced concrete lintel

25

throating if lintel used externally

15

4.7.11 BRICKWORK CARRIED BY LINTEL

60°

Figure 4.7 ARCHES AND LINTELS Scale: 1:20

EXTERNAL TREATMENTS TO WALLS

A one-brick thick solid wall is unlikely to withstand satisfactorily severe weather conditions and, as an alternative to cavity wall construction, it is possible to provide adequate protection against driving rain by applying a suitable external finish. The greatest disadvantage is that the attractive colour and texture of good brickwork is lost.

The most common external finish is rendering or roughcast. *Rendering* may consist of two coats of Portland cement and sand (1:3 or 1:4), possibly incorporating a waterproofing compound, finished with a float to a smooth finish and often painted with two coats of emulsion or stone paint. This produces a rather dense coat which is liable to develop hair cracks. Moisture enters the cracks, becomes trapped behind the rendering and evaporates from the inner surface. Hence it is advisable to use a rendering of cement: lime: sand in the proportions of 1:1:6 or 1:2:9. These produce porous finishes which absorb water in wet weather and permit free evaporation when the weather improves.[44] External rendered finishes should comply with BS 5262.[51]

With *roughcast* or *pebbledashing* there is less risk of the external coat cracking or breaking away from the wall. In roughcast the coarse aggregate, usually gravel, is mixed into the second coat which is applied to the wall, whereas in pebbledashing the chippings are thrown onto the second coat whilst it is still 'green'. In both cases the appearance is rather unattractive but the need for periodic decoration is eliminated.[43]

Common brick surfaces can be decorated in a variety of different ways. *Cement paint* has good water-shedding properties and is available in a wide range of colours. In more recent times, both *emulsion* and *chlorinated rubber paints* have been applied to external brick and block walls with satisfactory results. *Bituminous paints* give an almost impervious surface coating but their use restricts future treatment owing to the bitumen bleeding through other applications. *Colourless waterproofers* can make a wall surface water-repellent and less porous, provided any cracks are sealed, without much change in the appearance. The permanence of the protection is variable, depending on the type of waterproofer and the condition of exposure.[45]

Vertical tile hanging on the external elevations of buildings can be very attractive if skilfully fixed, but it is expensive. It is extremely durable provided the fixing battens are pressure-impregnated with preservative. With plain tiling all tiles must be nailed but a 40 mm lap is sufficient. Angles may be formed with purpose-made tiles or be close-cut and mitred with soakers, whilst vertical stepped flashings are usually introduced at abutments to form a watertight joint.

Weatherboarding provides a most attractive finish and may be either of painted softwood, or of cedar boarding treated periodically with a mixture of linseed oil and paraffin wax. Feather-edge boarding is used extensively but probably the best results are obtained with ship-lap boarding (figure 4.6.4). The boarding is often nailed through a felt backing to studs (vertical timbers). Internal linings will be selected to meet thermal insulation requirements. A variety of *sheet claddings* is also available in steel, aluminium, asbestos cement and plastics.

STONE WALLS

Building Stones

Building stones can be classified into three groups as follows: (1) *Sedimentary rocks*. These are formed by the weathering of land masses and the removal of the particles by water and their deposition in lakes or the sea. The deposited grains are united by some cementitious material. The principal stones in this class are limestones and sandstones.

Limestones are composed largely of calcium carbonate and small quantities of clay, silica and magnesia. They include the most widely used stones in the country with varying textures and colours ranging from white to grey, and have a natural bed. Some of the more common limestones are: *Anston*, quarried near Sheffield, cream, soft, easily carved, but unsuitable for polluted atmospheres; *Bath*, open texture and warm colour; *Clipsham*, quarried bear Oakham, cream to buff, medium-grained and useful for general building work; *Hopton Wood*, from Derbyshire, hard surface and takes good polish; *Portland*, from Isle of Portland, cream or chalky white, very popular.

Sandstones are consolidated sands from grey to red in colour and have harder, rougher surfaces than limestones. The grains of sand are cemented together by siliceous, calcareous (calcium carbonate) or ferruginous (iron-bearing) substances. Typical examples are: *Bramley Fall*, quarried near Leeds, light brown, hard and weathers well; *Blue Bristol Pennant*, from Bristol, greyish blue, fine grained and hard; *Darley Dale*, from Derbyshire, light buff to dark grey, fine grained and weathers well; *Mansfield* (Notts), yellow to red, unsuitable for polluted atmospheres; *Forest of Dean*, blue or grey, smooth and strong. (2) *Igneous rocks*. Volcanic in origin, they have been fused by heat and then cooled and crystallised. The principal stones in this class are granites, syenites and basalt.

Granites are composed of quartz, felspar and mica, the greater part of the stone being felspar, which is normally white in colour. Granite is extremely hard and durable and often takes a high polish. Examples of some of the more common granites are: *Creetown* (S.W. Scotland), light grey, medium grain and polishes well; *Penryn*, from Cornwall, light grey, rather coarse grain, takes good polish, good all-purpose stone; *Peterhead* (N.E. Scotland), red, takes good polish, used extensively in building, civil engineering and monumental work; *Rubislaw*, from N.E. Scotland, grey in colour, fine grain, polishes well, used extensively in building work; *Shap* (Westmorland), pink to brown, fine grained.

(3) *Metamorphic rocks*. These may be sedimentary or igneous in origin and have been affected by heat or earth movement to such an extent as to assume a new structure and/or form new materials. The principal stones in this class are marbles, serpentines and slates.

Marble is a limestone which has been subjected to intense heat and pressure deep down in the earth's crust. The calcium carbonate is crystallised into calcite and a rough cleavage develops which gives the stone a veined appearance. True marbles are virtually non-existent in the British Isles, as Purbeck marble from Dorset is really a very compact limestone which has not undergone metamorphosis. Marbles are imported from Norway, Sweden, Belgium, Italy and France. They are available in a variety of colours and are mainly used as thin internal linings, owing to their high cost.[37]

Useful guidance on the choice and use of stone is given in CP 298.[54]

Quarrying, Conversion, Working and Bedding of Stone

Sedimentary rocks are generally quarried with the aid of picks, bars and wedges and are split along their natural bed, while igneous rocks split readily in various directions. On quarrying the stone contains some moisture, known as 'quarry sap', which aids conversion (reducing to required sizes) by saws or feathers and wedges. Stone may be worked by hand or machine to a variety of finishes, some of which are illustrated.

Plain face. Consists of working the stone to an accurately finished face with a saw or chisel.

Hammer dressed. This is worked to rough hammered finish to give appearance of strength (figure 4.8.1).

Boasted. Parallel but not continuous chisel marks are formed across face of stone with a boaster (figure 4.8.2).

Tooled. This is a good-class finish with continuous chisel cuts (figure 4.8.3).

Vermiculated. Draughted margin with haphazard formation to interior; suitable for quoins (figure 4.8.4).

Furrowed. Draughted margin and base of furrows at same level, for quoins, string courses and similar features (figure 4.8.5).

Combed. A steel comb produces an irregular pattern on face of stone; also known as 'dragged' (figure 4.8.6).

Rubbed. A smooth face obtained by rubbing similar stones with sand and water.

Polished: Consists of polishing stone with putty powder.

Stones should be bedded with a mortar which matches the stone. It is usually made up of mason's putty, consisting of one part of lime putty to three parts of stone dust, and joints are often pointed with Portland cement and stone dust (1:3). In walling, stones should be laid with their natural beds horizontal, arch stones with their natural beds normal to the curve of the arch and vertical in the case of cornices and coping stones.

Walling Types

Stone was once a popular building material, but owing to the high cost, its use has diminished and it is now mainly employed as a facing to prestige buildings with a backing of brickwork, concrete or other material. The stone facework is made up of accurately dressed stones with fine joints and the surface finish can vary from a plain face to one of the finishes illustrated in figures 4.8.1 to 4.8.6. This type of stonework is called *ashlar* and is illustrated in figure 4.8.12. The stone facing can be bonded into the backing material or alternatively be fixed by non-ferrous metal angles, dowels, cramps or corbels built into slots formed in the back of the stonework. It is advisable to paint the back of the stone with bituminous paint where it adjoins brickwork or concrete to prevent cement stains appearing on the front face.

Stone members often have to be joined by special fittings or fastenings where the mortar joint offers insufficient adhesion. Coping stones can be connected by slate cramps (figure 4.8.13) about 150 x 62 x 50 mm set in chases in the adjoining stones in Portland cement, or by metal cramps (figure 4.8.14) of copper or bronze, about 250 – 300 x 25–50 x 6–15 mm set in sinkings in Portland cement, lead or asphalt, with the bedding material covering the cramp. Joggle joints (figure 4.8.15) help to bind adjoining stones with the jointing material filling the grooves. Dowels (figure 4.8.16) can be used for joining sections of vertical members like columns and mullions. They may be of slate or metal.

In stone-producing districts, waste stone cut to rectangular blocks about 150 mm thick (bed joint) are sometimes used to form the outer leaf of cavity walls but they are more expensive than brick. Another alternative quite widely adopted in the past, is to use rubble walling composed of roughly dressed stone built to one of a number of patterns with wide joints. *Uncoursed random rubble* (figure 4.8.7) consists of roughly dressed stones taken at random and not laid to any pre-determined pattern. It is used in boundary walls where it may be laid dry (without mortar joints), as well as in cottages and farm buildings. With *random rubble built to courses* (figure 4.8.8), the stone walling is levelled up at intervals in the height of the

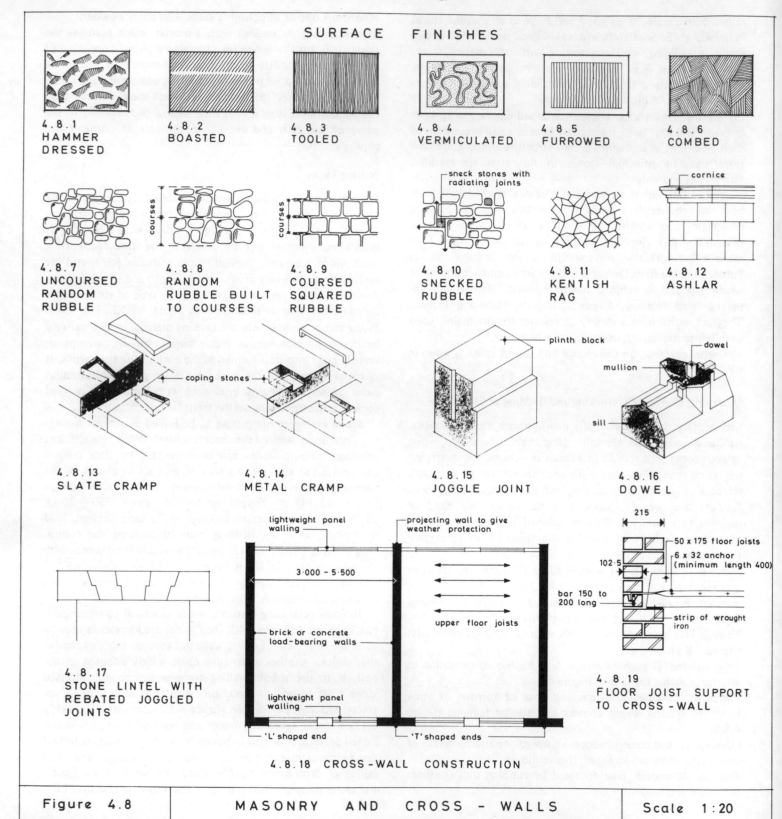

SURFACE FINISHES

4.8.1 HAMMER DRESSED

4.8.2 BOASTED

4.8.3 TOOLED

4.8.4 VERMICULATED

4.8.5 FURROWED

4.8.6 COMBED

4.8.7 UNCOURSED RANDOM RUBBLE

4.8.8 RANDOM RUBBLE BUILT TO COURSES

4.8.9 COURSED SQUARED RUBBLE

4.8.10 SNECKED RUBBLE — sneck stones with radiating joints

4.8.11 KENTISH RAG

4.8.12 ASHLAR — cornice

4.8.13 SLATE CRAMP — coping stones

4.8.14 METAL CRAMP

4.8.15 JOGGLE JOINT — plinth block

4.8.16 DOWEL — dowel, mullion, sill

4.8.17 STONE LINTEL WITH REBATED JOGGLED JOINTS

4.8.18 CROSS-WALL CONSTRUCTION
- lightweight panel walling
- projecting wall to give weather protection
- 3.000 – 5.500
- upper floor joists
- brick or concrete load-bearing walls
- lightweight panel walling
- 'L' shaped end
- 'T' shaped ends

4.8.19 FLOOR JOIST SUPPORT TO CROSS-WALL
- 215
- 50 x 175 floor joists
- 102.5
- 6 x 32 anchor (minimum length 400)
- bar 150 to 200 long
- strip of wrought iron

| Figure 4.8 | MASONRY AND CROSS - WALLS | Scale 1:20 |

wall, such as at damp-proof course, window sill, window head and eaves. *Coursed squared rubble* (figure 4.8.9) consists of walling stones dressed to a roughly rectangular shape and every course is one stone height, although the heights of the individual courses can vary. Indeed all the walling arrangements can be made up of either random rubble or squared rubble, but it is customary to build quoins, window and door jambs and chimneys in dressed stonework. *Snecked rubble* has small sneck stones, about 100 x 100 mm, spaced at approximately one metre intervals to break the joint. Joints radiate from the sneck stone in the manner illustrated in figure 4.8.10. *Kentish rag* (figure 4.8.11) uses roughly dressed stone as a backing to brickwork or concrete. It has an attractive appearance and is used quite extensively in panels to the front elevations of modern houses.

Rubble walling is often constructed of selected facing stones on both faces with the intervening voids in the heart of the wall filled with stone chippings or 'spalls' grouted up with liquid lime mortar. Bonders or through stones (about one to the square metre) will tie the two faces together. Stone walls should be at least one-third thicker than the corresponding brick wall, as stone is apt to be rather porous. The use of preservative impregnated battens on the inside face to take plasterboard or other finishings, assists in combatting condensation. A useful finish to a window head is a stone lintel with rebated joggled joints, as illustrated in figure 4.8.17.

Reconstructed stone is widely used as a substitute for natural stone in order to reduce cost. Natural stone, sand and cement form the basic materials and reinforcement can be introduced during casting, in addition to any required sinkings. Special facing mixes permit a wide range of finishes. It eliminates the defects of natural stone but problems may arise through crazing (formation of fine hair cracks).

CROSSWALL CONSTRUCTION

In crosswall construction, all the vertical loads of the roof and upper floors are transmitted to the ground by transverse party and flank walls, as distinct from the more orthodox method whereby all enclosing and some internal walls carry the loads. The relative costs are influenced by a large number of factors. The front and rear walls no longer carry loads and this permits flexibility in the choice of claddings (figure 4.8.18). On the other hand the uniform spacing of crosswalls, normally between 3 and 5½ m, places constraints on layout. Some stiffening of the crosswalls by longitudinal members such as upper floor joists is advisable. One method is to stagger the joists on each side of a party wall and to tie the floor to the

wall with metal anchors (figure 4.8.19) at intervals not exceeding 1.20 m. Anchor straps have one end screwed to the floor joist and the other end mortared into a joint in the crosswall and bent over a steel bar built into the wall. An alternative is to use mild steel joist hangers. The part of the load of a traditional pitched roof which is normally supported by the front and rear walls, is transferred to the crosswalls by beams. To further stabilise the walls, it is good practice to return the ends of the crosswalls in the manner shown in figure 4.8.18.

INTERNAL PARTITIONS

There is a wide range of materials available for the construction of thin dividing walls in buildings. The choice will be influenced by a number of factors: thickness, weight, sound insulation, cost, decorative treatment and possibly fire resistance.

Brick

Half-brick walls may be used on the ground floors of domestic buildings, but they need some form of foundation on account of their weight. They have good loadbearing qualities and have a notional fire resistance period of two hours. Brick-on-edge partitions could be used to enclose small areas such as cupboards.

Clay Block

Hollow clay blocks containing cavities are available with smooth or keyed surfaces as specified in BS 3921.[15] A common size is 290 x 215 mm with thicknesses ranging from 62.5 to 150 mm. The blocks are usually laid to lap half a block and special blocks are available for stopped ends and junctions. Some blocks have interlocking or tongued joints which increase rigidity.

Concrete Block

Concrete blocks can be either loadbearing or non-loadbearing and are made from a wide range of lightweight aggregates, including sintered pulverised fly ash from power stations, foamed slag, expanded clays and shales, clinker, expanded vermiculite and aerated concrete, as prescribed in BS 2028, 1364.[19] The loadbearing blocks range from 75 to 215 mm in thickness and non-loadbearing are 60 and 75 mm thick with a commonly used block size of 440 x 215 mm. Suitable mortars include cement: lime: sand 1:2:9 or 1:1:6; masonry cement: sand 1:5, and cement: sand 1:7 with a plasticiser. The blocks should be allowed to dry out before plastering and in long lengths of partition vertical joints filled with mastic or bridged by metal strips should be provided at 6 m intervals. Concrete block partitions can be tied to brick walls by leaving indents or recesses in the brickwork, to receive the ends of blocks on alternate courses, or by building expanded metal strips into the bed joints of the brickwork and blockwork.

Timber

Timber stud partitions (figure 4.6.5) are still used to a significant extent in the upper floors of domestic buildings. They are generally constructed of a 100 x 75 mm head and sill with vertical members or studs, ranging from 75 x 38 mm to 100 x 50 mm, framed between them at about 400 mm centres. Nogging pieces are often inserted between the studs to stiffen the partition. Plasterboard is generally nailed to either side of the partition with rose-headed galvanised nails and finished with a 3 to 5 mm skim coat of plaster. The sound insulating properties of a stud partition can be increased by staggering the studs and mounting the partition on a pad of resilient material. Stud partitions on upper floors are usually supported by floor joists. Where loads are to be carried, trussed partitions incorporating two sets of braces may be used.

Plasterboard

There are several types of dry partitioning incorporating plasterboard faces which can be used to form lightweight partitions with good sound insulating properties. One quite popular variety consists of two gypsum wallboards separated by a core of cellular construction in accordance with BS4022,[38] to thicknesses of 38 to 100 mm depending on the area of the partition. The surfaces may be prepared for immediate decoration, or to receive gypsum plaster or plastic sheeting. Another alternative is to use hot-dipped galvanised steel track and studs, faced each side with gypsum wallboard as a non-loadbearing relocatable partition. Laminated partitions are also available made up of layers of wallboard and plank to an overall thickness of 50 or 65 mm.

Other Types

There is a variety of relocatable steel partitions on the market which although expensive may be ideally suited for offices. It is also possible to use panels of wood wool or compressed strawboard which are comparatively easy to cut and erect. Glass bricks can be used where it is required to transmit light through a partition. Non-loadbearing partitions should comply with BS 5234.[50]

REFERENCES

1. CP 111: 1970 Structural recommendations for loadbearing walls
2. *BRE Digest 61*: Strength of brickwork, blockwork and concrete walls. HMSO (1970)
3. The Building Regulations 1976. SI 1676 HMSO (1976) and the Building (First Amendment) Regulations 1978 SI 723 HMSO (1978)
4. Greater London Council, London Building Acts 1930–1939, Constructional by-laws (1972)
5. DOE Advisory leaflet 47: Dampness in buildings. HMSO (1975)
6. *BRE Digest 27*: Rising damp in walls. HMSO (1969)
7. R. E. Lacy, An index of exposure to driving rain: *BRE Digest 127*. HMSO (1971)
8. *BRE Digest 110*: Condensation. HMSO (1972)
9. DOE Advisory leaflet 61: Condensation. HMSO (1976)
10. *BRE Digest 108*: Standard *U*-values. HMSO (1975)
11. *BRE Digest 128*: Insulation against external noise Part 1. HMSO (1971)
12. *BRE Digest 102*: Sound insulation of traditional buildings Part 1. HMSO (1971)
13. *BRE Digest 143*: Sound insulation: basic principles. HMSO (1976)
14. *BRE Digest 89*: Sulphate attack on brickwork. HMSO (1971)
15. BS 3921: 1974 Clay bricks and blocks
16. BS 187: Calcium silicate (sandlime and flintlime) bricks, Part 2: 1970 Metric units
17. BS 1180: 1972 Concrete bricks and fixing bricks
18. *BRE Digests 164 and 165*: Clay brickwork. HMSO (1979)
19. BS 2028, 1364: 1968 Precast concrete blocks
20. BS 1200: 1976 Sands for mortar for plain and reinforced brickwork; blockwalling and masonry
21. BS 890: 1972 Building limes
22. *BRE Digest 160*: Mortars for bricklaying. HMSO (1973)
23. DOE Advisory leaflet 75: Efflorescence and stains on brickwork. HMSO (1972)
24. BS 1243: 1978 Metal ties for cavity wall construction
25. *BRE Digest 77*: Damp-proof courses. HMSO (1971)
26. BS 1178: 1969 Milled lead sheet and strip for building purposes
27. BS 2870: 1968 Rolled copper and copper alloys, sheet strip and foil
28. BS 743: 1970 Materials for damp-proof courses
29. BS 988, 1076, 1097, 1451: 1973 Mastic asphalt for building (limestone aggregate)

30. BS 1162, 1418, 1410: 1973 Mastic asphalt for building (natural rock asphalt aggregate)
31. DOE. Principles of modern building Vol. 1. HMSO (1975)
32. CP 121: Walling. Part 1: 1973
33. DOE Advisory leaflet 16: Mortars for brickwork and blockwork. HMSO (1976)
34. DOE Advisory leaflet 23: Damp-proof courses. HMSO (1976)
35. DOE Advisory leaflet 58: Inserting a damp-proof course in an existing building. HMSO (1974)
36. DOE Advisory leaflet 36: Metal scaffolding. HMSO (1975)
37. A. Everett. *Mitchell's Building Construction: Materials*. Batsford (1978)
38. BS 4022: 1970 Prefabricated gypsum wallboard panels
39. BS 3798: 1964 Coping units
40. *BRE Digest 157*: Calcium silicate brickwork. HMSO (1973) and DOE Advisory leaflet 65: Calcium silicate bricks and how to use them. HMSO (1976)
41. DOE Advisory leaflet 49: Simple concrete lintels. HMSO (1976)
42. DOE Advisory leaflet 8: Bricklaying in cold weather. HMSO (1975)
43. DOE Advisory leaflet 27: Rendering outside walls. HMSO (1976)
44. *BRE Digest 196*: External rendered finishes. HMSO (1976)
45. *BRE Digest 125*: Colourless treatments for masonry. HMSO (1971)
46. *BRE Digest 230*: Fire performance of walls and linings. HMSO (1979)
47. *BRE Digests 214 and 215*: Parts 1 and 2: Cavity barriers and fire stops. HMSO (1978)
48. *BRE Digest 200*: Repairing brickwork. HMSO (1977)
49. BS 5224: 1976 Masonry cement
50. BS 5234: 1975 Code of practice for internal non-load-bearing partitions
51. BS 5262: 1976 Code of practice for external rendered finishes
52. BS 5617: 1979 Urea-formaldehyde (UF) foam for thermal insulation of cavity walls
53. BS 5618: 1978 Code of practice for thermal insulation of cavity walls
54. CP 298: 1972 Natural stone cladding (non-loadbearing)
55. *BRE Digest 236*: Cavity insulation. HMSO (1980)

5 FIREPLACES, FLUES AND CHIMNEYS

Before examining the constructional requirements and techniques associated with fireplaces, flues and chimneys, it is helpful to consider their purpose, the changing form of appliances and the general objectives, against the background of statutory requirements.

GENERAL PRINCIPLES OF DESIGN

The primary objective of an open fire is to heat the room in which it is placed, but it often fulfils a secondary objective of providing a focal point in the living room or lounge of a dwelling. Indeed so much importance may be attached to the secondary objective that it may become the deciding factor in providing an open fire in a new centrally heated dwelling, when heating efficiency considerations alone (it is probably less than 45 per cent efficient) would preclude its use. Hence the appearance of the fireplace is important and there is a wide range of materials from which to choose.

Many of the older types of open fire were inefficient and much of the obtainable heat was lost as a result of incomplete combustion. Furthermore, they tended to emit considerable quantities of dark smoke. The efficiency of modern open fires has been increased substantially by means of improved appliances, flues, air supply and fuels. A number of variants of the traditional open fire have been produced to secure greater efficiency and the more important of these will be described and illustrated later.

The principal requirements of a fireplace and flue are threefold

(1) to secure maximum heat for the benefit of occupants;
(2) to take adequate precautions against spread of fire;
(3) to ensure effective removal of smoke and avoidance of downdraught.

All fires require sufficient air for combustion purposes, and as air is drawn into the flue from the room, further air is needed to replace it. Apart from blockage of the flue by soot or debris, there are three sets of conditions which can prevent the chimney operating satisfactorily, namely

(1) insufficient air entering the room to replace that passing up the chimney;
(2) adverse flow conditions resulting from the poor design of passages through which the smoke passes (throat, gathering and flue);
(3) downdraught caused by the build-up of pressure at the chimney top; this is influenced by the form of the building itself, neighbouring buildings, trees and the topography of the site.[1]

The Building Regulations[2] prescribe structural requirements covering fireplace recesses, constructional hearths, chimneys and flues, and flue outlets. The Regulations classify appliances into two classes; class 1 appliances include most domestic solid fuel and oil-burning appliances, while class 2 appliances include most gas appliances. The detailed requirements of The Building Regulations will be considered later.

An open fire in a living room or dining room may incorporate a *boiler unit* in place of the usual firebrick at the back of the fireplace, so that the fire may heat water as well the room in which it is placed.

Convector fires are an efficient type of solid-fuel fire and they can also be fitted with a back boiler for domestic hot water supply. They emit convected warm air into the room in addition to radiant heat, and can either be inset in a normal fireplace recess (figure 5.3.5) or be freestanding, and they have a heat efficiency of 45 to 65 per cent. *Room heaters* are generally cleaner and more efficient than open fires, and can emit radiant heat. They can be quite attractive in appearance and generally contain a transparent or translucent door(s) which can be left open if desired (figures 5.2.5, 5.2.6 and 5.2.7). They have a heat efficiency of 45 to 65 per cent when the doors are open, and up to 75 per cent when they are closed.

The Clean Air Act 1956 made provision for the establishment of smoke control areas in which authorised smokeless fuels only can be used. The Building Regulations prescribe that in new buildings no appliances shall be installed which discharge the products of combustion into the atmosphere, unless they are designed to burn as fuel either gas, coke or anthracite — that is, smokeless fuels.

HEARTHS

Every fireplace must have a constructional hearth extending both under and in front of the opening. The Building Regulations[2] and the London Constructional By-laws[3] prescribe that the hearth shall project not less than 500 mm in front of the jambs, extend not less than 150 mm beyond each side of the opening between the jambs and be not less than 125 mm thick. These requirements are illustrated in figures 5.1.1 and 5.1.2. Where the hearth adjoins a floor of combustible material, it shall be laid so that the upper surface of the hearth is not lower than the surface of the floor. No combustible material, other than timber fillets supporting the edges of a hearth, where it adjoins a floor, shall be placed under a constructional hearth, serving a class 1 appliance, within a distance of 250 mm, measured vertically from the upper side of the hearth, unless such material is separated from the underside of the hearth by an air space of not less than 50 mm.

Hearths are almost invariably constructed of concrete and, when formed on a solid concrete ground or upper floor, become part of the normal floor construction. When constructed in association with a suspended timber ground floor, the hearth is supported on fender walls and hardcore in the manner shown in figures 5.1.1, 5.1.2 and 5.1.3. With upper timber floors, the hearth is normally supported by the brick chimney breast at the back, and by asbestos cement sheeting or temporary timber formwork as illustrated in figures 5.1.4 and 5.1.6.

Suspended concrete hearths may be formed of precast concrete, reinforced with steel fabric reinforcement, and built in as the work proceeds. They are more frequently cast *in situ* on timber formwork, which must be removed, or on asbestos cement or metal flat sheeting which can be left in position. Some hearths in very old buildings are supported by brick trimmer arches.

Timber upper floors need to be trimmed around the hearth and chimney breast in the manner shown in figures 5.1.4 and 5.1.6. The trimmer and trimming joists are each 25 mm thicker than the normal floor joists and the joints of the intersecting timbers must be well framed together.

FIREPLACE CONSTRUCTION

Fireplace Recesses

Where the structure accommodating the fireplace opening and flue projects into a room, as in figures 5.1.5 and 5.2.1, the projection is termed a *chimney breast*. The fireplace opening is the recess housing the firegrate (figure 5.1.3); opening sizes in fireplace surrounds vary from 360 to 460 mm wide by 560 mm high.[4] The depth of the recess is usually about 338 mm (one-and-a-half bricks deep).

The Building Regulations[2] (clause L3(2)), prescribe minimum dimensions for the backs and jambs of fireplace recesses. The jambs must be at least 200 mm thick. The back of the recess if a solid wall shall be not less than 200 mm thick, or if a cavity wall each leaf shall be not less than 100 mm thick (figure 5.3.1), extending for the full height of the recess. In the London Constructional By-laws[3] the operative height is 300 mm above the top of the opening. Where the recess is situated in an external wall and no combustible external cladding is carried across the back of the recess, the back wall of the recess may be reduced to 100 mm thick. Nevertheless in practice it would be desirable to provide a thickness of 200 mm to achieve a reasonable standard of thermal insulation. Where a wall, other than one separating dwellings, serves as the back of two recesses built on opposite sides of the wall, it may be reduced to 100 mm thick (see figure 5.3.3). Economy in construction, in addition to greater efficiency in heating, can be obtained if fireplaces are grouped together preferably on an internal wall.

To improve the air supply to a fireplace the air may be ducted up through the front hearth as shown in figure 5.3.5 or through an ash container pit as in figure 5.3.4. In the latter case the Building Regulations (clause L4(4)) prescribe that the sides and bottom of the pit shall be constructed of non-combustible material not less than 50 mm thick and that no combustible material shall be built into a wall or beside the pit within 225 mm of the inner surface of the pit.

Throats and Lintels

The throat is the part of the flue immediately above the fireplace opening (contraction at bottom of flue).[9] It is restricted in size, often about 100 x 250 mm in cross-section, to accelerate the flow of the flue gases and to ensure adequate draught. The throat may be provided in a precast concrete throat unit of the form illustrated in figures 5.1.1 and 5.1.3 or be formed between a reinforced concrete lintel and the top of the fireback, as in figures 5.2.2 and 5.2.4. On occasions the brickwork is corbelled over to reduce the size of the opening

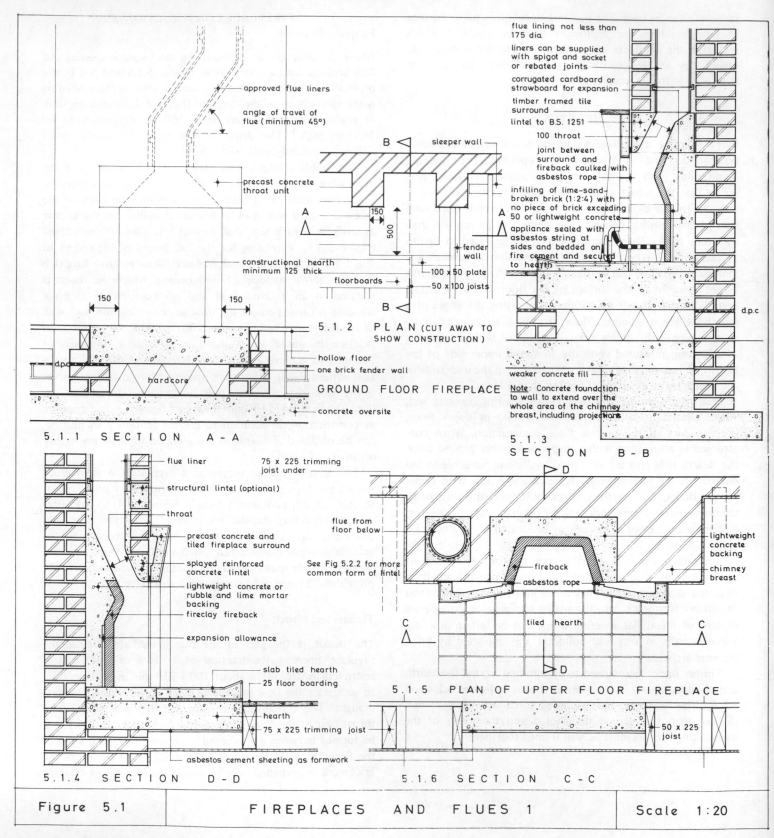

approved flue liners

angle of travel of flue (minimum 45°)

precast concrete throat unit

B ▷ sleeper wall

constructional hearth minimum 125 thick

A

500

150

fender wall

100 x 50 plate

floorboards

50 x 100 joists

150   150

B

5.1.2 P L A N (CUT AWAY TO SHOW CONSTRUCTION)

d.p.c

hardcore

hollow floor

one brick fender wall

GROUND FLOOR FIREPLACE

concrete oversite

5.1.1 S E C T I O N   A - A

flue lining not less than 175 dia.

liners can be supplied with spigot and socket or rebated joints

corrugated cardboard or strawboard for expansion

timber framed tile surround

lintel to B.S. 1251

100 throat

joint between surround and fireback caulked with asbestos rope

infilling of lime-sand-broken brick (1:2:4) with no piece of brick exceeding 50 or lightweight concrete

appliance sealed with asbestos string at sides and bedded on fire cement and secured to hearth

d.p.c

weaker concrete fill

Note: Concrete foundation to wall to extend over the whole area of the chimney breast, including projections

5.1.3
S E C T I O N   B - B

flue liner

structural lintel (optional)

throat

precast concrete and tiled fireplace surround

splayed reinforced concrete lintel

lightweight concrete or rubble and lime mortar backing

fireclay fireback

expansion allowance

slab tiled hearth

25 floor boarding

hearth

75 x 225 trimming joist

asbestos cement sheeting as formwork

75 x 225 trimming joist under

flue from floor below

See Fig 5.2.2 for more common form of lintel

D

C   C

tiled hearth

D

fireback

asbestos rope

lightweight concrete backing

chimney breast

5.1.5 P L A N O F U P P E R F L O O R F I R E P L A C E

50 x 225 joist

5.1.4 S E C T I O N   D - D

5.1.6 S E C T I O N   C - C

| Figure 5.1 | FIREPLACES   AND   FLUES 1 | Scale 1:20 |

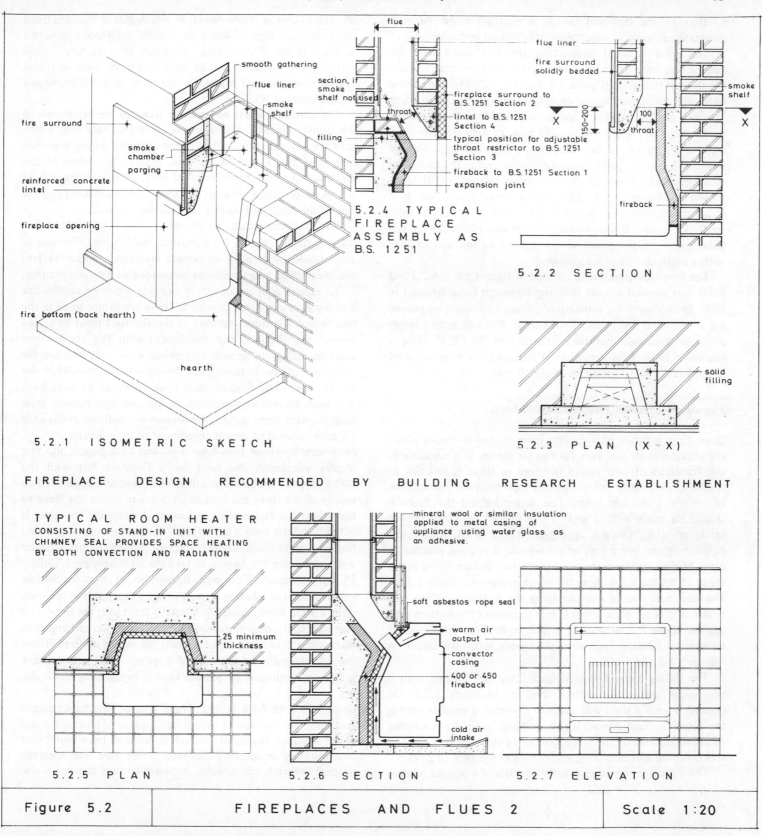

smooth gathering

flue liner

section, if smoke shelf not used

smoke shelf

fire surround

smoke chamber

parging

reinforced concrete lintel

fireplace opening

fire bottom (back hearth)

hearth

throat

filling

flue

flue liner

fireplace surround to B.S. 1251 Section 2

lintel to B.S. 1251 Section 4

typical position for adjustable throat restrictor to B.S. 1251 Section 3

fireback to B.S. 1251 Section 1

expansion joint

flue liner

fire surround solidly bedded

smoke shelf

150-200

100

throat

X   X

fireback

5.2.1 ISOMETRIC SKETCH

5.2.4 TYPICAL FIREPLACE ASSEMBLY AS B.S. 1251

5.2.2 SECTION

solid filling

5.2.3 PLAN (X-X)

FIREPLACE DESIGN RECOMMENDED BY BUILDING RESEARCH ESTABLISHMENT

TYPICAL ROOM HEATER
CONSISTING OF STAND-IN UNIT WITH
CHIMNEY SEAL PROVIDES SPACE HEATING
BY BOTH CONVECTION AND RADIATION

25 minimum thickness

mineral wool or similar insulation applied to metal casing of uppliance using water glass as an adhesive.

soft asbestos rope seal

warm air output

convector casing

400 or 450 fireback

cold air intake

5.2.5 PLAN

5.2.6 SECTION

5.2.7 ELEVATION

| Figure 5.2 | FIREPLACES AND FLUES 2 | Scale 1:20 |

to that of the flue and this is termed *gathering*. Another alternative is to fit an adjustable metal throat restrictor.

The splayed lintel supporting the brickwork over the fireplace opening has a rounded internal bottom edge to assist the flow of flue gases (figures 5.1.4 and 5.2.2). The ends of the lintel will be of rectangular cross-section to facilitate building into the brickwork. They are usually 125 mm thick and the depth may vary from 145 to 225 mm. Lintels are reinforced with a 8 mm or 12 mm diameter mild steel bar. To give flexibility, it is good practice to provide an additional rectangular lintel several courses above the splayed lintel as shown in figure 5.1.4. The higher lintel then acts as a structural lintel and permits the installation of a taller appliance should it later be required. It also allows the splayed lintel to be positioned off the centre line of the wall to tie in with a particular fireplace surround.

The fireplace assembly illustrated in figures 5.2.1, 5.2.2 and 5.2.3 was devised by the Building Research Establishment in 1955 to overcome the problem of smoky chimneys, by providing a restricted throat and smoke shelf. More recently a rather different approach has been adopted in BS 1251[4] using a shallower lintel with provision, if required, for a smoke shelf or adjustable throat restrictor (figure 5.2.4).

Fireplace Surrounds, Firebacks and Ashpits

Open fires without back boilers have shaped firebacks which are obtainable in one, two, four or six pieces. It is advisable to use firebacks of two pieces or more as there is less risk of cracking, preferably using asbestos tape between the sections to ensure a durable joint. The filling behind the fireback should be made with a weak 1:2:4 mix of lime:sand:broken brick or a lightweight aggregate mixed with lime. After positioning the lower half of a fireback, it is good practice to place behind it thin corrugated cardboard before filling in the space at the back, to allow for some expansion (figure 5.1.3). The top part should not overhang the bottom section; it is better if it is set back about 2 mm to protect the bottom edge against flame.[5] The way in which a fireback is shaped to transmit maximum heat into the room is shown clearly in figures 5.1.3 and 5.1.4.

The filling behind the fireback may be finished with a horizontal surface or smoke shelf as in figure 5.2.1. On occasions such a shelf will assist in preventing smoke entering the room through downdraught, although generally it is better to finish the top of the filling to a splay as in figure 5.1.4, and this prevents soot collecting and enhances the flow of gases.

The fireplace surround usually consists of a precast ceramic tile slab in one or more pieces, in which glazed tiles are fixed on a precast high alumina or similar reinforced concrete backing about 50 mm thick, produced in a factory. Alternatively, the surround can be built *in situ* of brick or stone. The surround should preferably conform to the dimensional requirements of BS 1251.[4]

The surround must be securely fixed to the brickwork of the chimney breast by two clips or eyelets on each side, which are normally cast into the back of the surround during manufacture, and the plaster is finished against the edges of the surround. An expansion joint of soft asbestos rope should always be provided between the back of the surround and the face of the lintel and leading edges of the fireback.

The matching hearth for the surround may be made of tiles laid *in situ* directly on the constructional hearth. Tiles laid *in situ* should be bedded in cement:lime:sand mortar (1:1:6) and preslabbed hearths should be bedded in a similar manner.

To ensure proper control of burning, *inset or continuous burning fires* are often provided. These are firmly fixed to the hearth and the sides and base of the firefront must be sealed against the fireback and the hearth with fire cement and asbestos string, to ensure that primary air only reaches the underside of the bottom grate through the air control in the ashpit cover. The firebars must have clearance at both back and sides to permit expansion of the cast iron bottom grate fittings. With some designs the firebars are notched at the ends for easy removal to permit adjustment of their length.

Deep ashpit. These fires have a sunken back hearth and this usually eliminates the need for a firefront fret with the cast iron grate set below hearth level. A specially controlled air supply direct from the outside air is taken under the floor to the ashpit for combustion purposes. A deep ash container is fitted to hold a week's ash refuse.[10] In the case of a suspended timber ground floor, a hole is formed through the brick fender wall supporting the hearth, to take the air supply pipe, usually 75 mm diameter. This pipe should be of noncombustible material such as asbestos cement or light gauge galvanised iron. With a solid floor, ducts are normally built into the floor. It is best to install a dual system of pipe ducts laid at right angles to each other from the outside walls to avoid suction. These should terminate on the face of external walls with suitable grilles or ventilators to prevent vermin penetrating the building.

Open fires with back boilers. These kind of fires have become quite popular because of their dual function of both space and water heating. The usual fireback is replaced by a boiler unit incorporating a specially designed boiler flue and specially shaped firebrick side cheeks. A boiler control damper is also

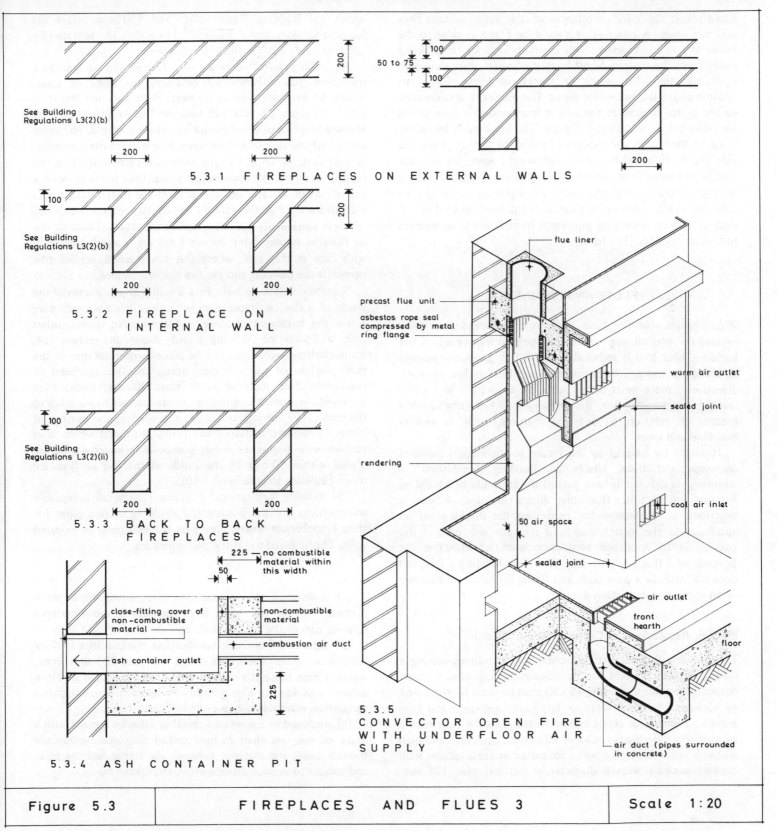

See Building
Regulations L3(2)(b)

200 200

5.3.1 FIREPLACES ON EXTERNAL WALLS

100

See Building
Regulations L3(2)(b)

200 200

**5.3.2 FIREPLACE ON
INTERNAL WALL**

100

See Building
Regulations L3(2)(ii)

200 200

**5.3.3 BACK TO BACK
FIREPLACES**

225 — no combustible
material within
this width

50

close-fitting cover of
non-combustible
material

non-combustible
material

combustion air duct

ash container outlet

225

5.3.4 ASH CONTAINER PIT

200

50 to 75

100

100

200

flue liner

precast flue unit

asbestos rope seal
compressed by metal
ring flange

warm air outlet

sealed joint

rendering

50 air space

cool air inlet

sealed joint

air outlet

front
hearth

floor

**5.3.5
CONVECTOR OPEN FIRE
WITH UNDERFLOOR AIR
SUPPLY**

air duct (pipes surrounded
in concrete)

Figure 5.3	FIREPLACES AND FLUES 3	Scale 1:20

fitted. Most approved appliances of this type contain fires 400 mm wide. A relief valve should be fitted as close to the boiler as possible together with an emptying cock for draining water from the system. Short lengths of metal pipes should be built into the surrounding brickwork to form 'sleeves' to accommodate the hot-water pipes. The pipework is connected to the boiler as soon as the unit is fixed and it is then tested for leaks before installing the grate. The grate must be rigidly fixed to the hearth and the space behind the appliance and the side cheeks filled with a weak lightweight aggregate or brick rubble and weak lime mortar. The voids around pipes passing through sleeves in the brickwork are sealed by caulking with asbestos string. Such an appliance might heat up to $4 \, m^2$ of heating surface, including pipework, in addition to an indirect hot-water cylinder (136 litres).

FLUES AND CHIMNEYS

Where practicable fireplaces, flues and chimneys should be located on internal walls as the flues help to warm parts of the building, heat loss is reduced and the flue gases are warmer resulting in more efficient operation. For a flue to work effectively, there must be an adequate air supply to the fire and the outlet to the flue (the chimney stack) must be suitably located in relation to external features, such as nearby buildings and trees.

It might be helpful to the reader to distinguish between *chimneys* and *flues*. Under the Building Regulations[2] a 'chimney' is defined as 'any part of the structure of a building forming any part of a flue other than a flue pipe'. A 'flue' is described as 'a passage for conveying the discharge of an appliance to the external air and includes any part of the passage in an appliance ventilation duct which serves the purpose of a flue'. A 'flue pipe' is 'a pipe forming a flue, but does not include a pipe built as a lining into either a chimney or an appliance ventilation duct'.

Building Regulations for Correct Flue Installation

The Building Regulations[2] prescribe that a chimney serving a class 1 appliance (solid fuel or oil-burning appliance with an output rating not exceeding 45 kW) shall be lined with rebated or socketed linings of clay or kiln-burnt aggregate and high alumina cement, or clay pipes and fittings, or be constructed of concrete flue blocks, also of kiln-burnt aggregate and high alumina cement. A flue when measured in cross-section shall contain a circle with a diameter of not less than 175 mm,

under the Building Regulations[2] and 150 mm under the London Constructional By-laws.[3] Flues shall be separated by solid materials not less than 100 mm thick.

As far as practicable bends in a flue should be kept to a minimum. Where bends are necessary the angle of travel should be kept as steep as possible, preferably not less than 60°.[11] It is good practice to take the flue as high as possible above a fireplace opening before introducing a bend. The inner surface of the flue should be smooth to prevent the accumulation of soot, which in its turn reduces the size of the flue and retards the flow of combustion gases. Flue liners provide a smooth inner surface and also protect the surrounding brickwork from the harmful effect of acid gases and condensation, which is particularly prevalent with boiler flues. The majority of flues in houses built before 1965 are 'parged' internally with lime mortar and, where this has cracked, smoke may penetrate the building and the fire risk is increased.

No timbers shall be built into a wall within 200 mm of the inside of a flue or fireplace recess, unless it is a wooden plug where the minimum distance is 150 mm. No timber, other than a floorboard, skirting board, dado rail, picture rail, mantelshelf or architrave, can be placed within 38 mm of the outer surface of the chimney, unless the flue surround of non-combustible material is at least 200 mm thick. Flue surrounds in domestic buildings rarely exceed half-a-brick in thickness (102.5 mm), and care is needed in the placing of trimming timbers in floors and roofs. No metal which is in contact with timber or other combustible material shall be placed within 50 mm of the inside of the flue or fireplace recess (Building Regulations[2] L10).

The Building Regulations[2] (L9) also incorporate acceptable arrangements for the placing and shielding of flue pipes for class 1 appliances where they pass through roofs or external walls. The 'deemed-to-satisfy' provisions are

(1) a distance of not less than three times the external diameter of the flue pipe from any combustible material forming part of the roof or wall; *or*

(2) separated from any combustible material in a roof by solid non-combustible material not less than 200 mm thick, where a pipe passes through a roof; or not less than 200 mm below or beside the pipe or 300 mm above it where it passes through an external wall; *or*

(3) enclosed in a sleeve of metal or asbestos cement with a space of not less than 25 mm packed with noncombustible thermal insulating material between the sleeve and the pipe, and subject to various other specific requirements.

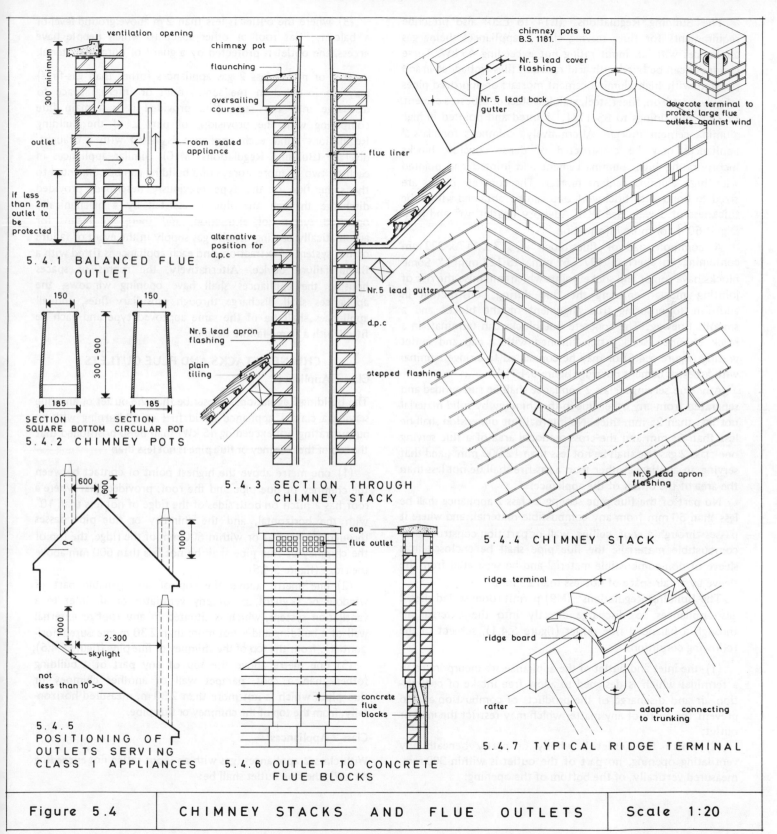

5.4.1 BALANCED FLUE OUTLET

ventilation opening

300 minimum

outlet

if less than 2m outlet to be protected

room sealed appliance

5.4.2 CHIMNEY POTS

150

150

300 - 900

185

185

SECTION SQUARE BOTTOM POT

SECTION CIRCULAR POT

chimney pot

flaunching

oversailing courses

flue liner

alternative position for d.p.c

Nr.5 lead gutter

d.p.c

Nr.5 lead apron flashing

plain tiling

5.4.3 SECTION THROUGH CHIMNEY STACK

chimney pots to B.S. 1181.

Nr. 5 lead cover flashing

Nr. 5 lead back gutter

dovecote terminal to protect large flue outlets against wind

stepped flashing

Nr.5 lead apron flashing

5.4.4 CHIMNEY STACK

5.4.5 POSITIONING OF OUTLETS SERVING CLASS 1 APPLIANCES

600

600

1000

∞

1000

2·300

skylight

not less than 10°∞

cap

flue outlet

concrete flue blocks

5.4.6 OUTLET TO CONCRETE FLUE BLOCKS

ridge terminal

ridge board

rafter

adaptor connecting to trunking

5.4.7 TYPICAL RIDGE TERMINAL

| Figure 5.4 | CHIMNEY STACKS AND FLUE OUTLETS | Scale 1:20 |

The Building Regulations[2] (L14 to L20) also prescribe requirements for flues serving class 2 appliances, being gas appliances with an input rating not exceeding 45 kW. These chimneys can be lined with acid-resistant tiles embedded in and pointed with high alumina cement mortar; or approved pipes of clay, cast iron, sheet steel, stainless steel or asbestos cement; or clay flue linings to BS 1181,[7] jointed and pointed in high alumina cement mortar. Alternatively, chimneys for class 2 appliances may be constructed of dense concrete blocks incorporating high alumina cement and jointed and pointed with high alumina cement mortar. The concrete blocks are sized to bond with brickwork and can be obtained within the thickness of a wall. See DOE Advisory Leaflet 50[8] and BRE Digest 60[12] for more information.

A common size of flue block is 317 x 117 x 222 mm containing an aperture 197 x 67 mm (13 123 mm²).[6] Some blocks have spigot and socket joints which minimise the risk of jointing mortar entering the flue during construction in addition to ensuring true alignment of the blocks and a smooth flueway. The concrete flue blocks can terminate in a brick chimney stack often finished with a cap and outlet grating (figure 5.4.6) or may be connected to a ridge terminal with asbestos cement ducting (figure 5.4.7).[13]

Any flue serving a class 2 appliance shall be surrounded and separated from any other flue in the chimney by solid material not less than 25 mm thick. No internal flue dimension shall be less than 63 mm and the cross-sectional area of a flue serving one class 2 gas fire shall be not less than 12 000 mm², and that serving an appliance other than a gas-fire shall be not less than the area of the outlet of the appliance.

No part of the flue pipe serving a class 2 appliance shall be less than 50 mm from any combustible material, and where it passes through a roof, ceiling, wall or partition constructed of combustible materials, the flue pipe shall be enclosed in a sleeve of non-combustible material and be separated from the sleeve by an air space of not less than 25 mm.

The Building Regulations[2] (M9) permit room-sealed class 2 gas appliances to discharge directly into the external air through a balanced flue outlet (figure 5.4.1), subject to the following conditions

(1) the inlet and outlet of the appliance are incorporated in a terminal which is designed to allow free intake of combustion air and discharge of the products of combustion and to prevent the entry of any matter which may restrict the inlet or outlet;

(2) where the outlet is wholly or partly beneath any ventilating opening, no part of the outlet is within 300 mm, measured vertically, of the bottom of the opening;

(3) where the outlet is less than 2 m above ground level or a balcony, flat roof or other space to which people have access, the outlet is protected by a guard of durable material.

Two or more class 2 gas appliances (other than gas-fires) may be installed in the same room or internal space to discharge into the same flue provided it is a main flue complying with the provisions of part L of the Building Regulations[2] and each appliance is fitted with a draught-diverter (Building Regulations: M10). Similar appliances in each of two or more storeys of a building can be connected to the same flue of the type previously described, provided discharge through the flue is assisted by a mechanically operated system of extraction, and there are means for automatically cutting off the gas supply in the event of failure of the system of extraction, and each appliance is fitted with a flame failure device. Alternatively, the rooms or spaces housing the appliances shall have opening windows, the appliances shall discharge through subsidiary flues, and all appliances shall be of the same approved type and each be fitted with a flame-failure device.

CHIMNEY STACKS AND FLUE OUTLETS

Class 1 Appliances

The Building Regulations[2] prescribe that the outlet of any flue serving a class 1 appliance (solid fuel or oil-burning with an output rating not exceeding 45 kW) shall be so positioned that the top of the chimney or flue pipe is not less than

(1) one metre above the highest point of contact between the chimney or flue pipe and the roof; provided that where a roof has a pitch on both sides of the ridge of not less than 10° with the horizontal, and the chimney or flue pipe passes through the roof at or within 600 mm of the ridge, the top of the chimney or flue pipe shall be not less than 600 mm above the ridge (figure 5.4.5);

(2) one metre above the top of an openable part of window or skylight, or of any ventilator or air inlet to a ventilation system, which is situated in any roof or external wall of a building and is not more than 2.30 m, measured horizontally, from the top of the chimney or flue pipe (figure 5.4.5);

(3) one metre above the top of any part of a building (other than a roof, parapet wall or another chimney or flue pipe) which is not more than 2.30 m, measured horizontally, from the top of the chimney or flue pipe.

Class 2 Appliances

With class 2 appliances (gas with an input rating not exceeding 45 kW), the flue outlet shall be:

(1) fitted with a flue terminal designed to allow free discharge, to minimise downdraught and to prevent the entry of any matter which might restrict the flue (figure 5.4.7);

(2) so situated externally that a current of air may pass freely across it at all times;

(3) so situated in relation to any opening (as previously described) that no part of the outlet is less than 600 mm from the opening.

Hence with flues serving class 2 appliances, the outlets do not have to project above the ridge.

Chimneys and Chimney Stacks

When determining the height of a chimney or flue pipe, the pot or other flue terminal is disregarded. Furthermore, the height of a chimney stack should not exceed six times the least width to ensure adequate stability. Chimney pots as illustrated in figure 5.4.2 provide a neat and effective finish to the outlet of a flue. The taper of a pot reduces the entry of rain and improves draught and flow of gases. It is best to use square-based pots with square flues. Pots may be 300, 450, 600, 750 or 900 mm high.[7]

Chimney pots should be built into the stack not less than 150 mm, or one-quarter the length of the pot, whichever is the greater, to ensure ample support. The pots are flaunched around in cement and sand (1:3) to throw water off the stack and give additional support to the pots (figures 5.4.3 and 5.4.4). Pots should extend above the highest point of the flaunching for at least 50 mm.

The head of a chimney stack may be finished in different ways. One method is to build brick oversailing courses (each projecting about 28 mm) and to flaunch around the pots as shown in figures 5.4.3 and 5.4.4. Another method is to use a precast concrete capping which is weathered and throated on all sides, overhangs the stack by about 38 to 50 mm in all directions, and is generally bedded on a damp-proof course. Another damp-proof course is inserted lower down the stack where it passes through the roof (figure 5.4.3).

Rafters and ceiling joists are trimmed around the chimney and these timbers must be kept at least 38 mm clear of the outside face of the brick stack. A watertight joint has to be provided around the chimney where the roof covering of tiles or slates adjoins the brickwork. A gutter and flashing are normally used at the top edge, an apron flashing at the bottom edge and soakers and stepped flashings at the sides, as illustrated in figures 5.4.3 and 5.4.4. Lead is often used in these positions but various other materials are available, including zinc, aluminium, copper and nuralite. The stepped

flashings overhang the right-angled soakers, and the top edges of the flashings are wedged into brick joints and pointed. At the top edge the gutter sheeting is dressed 150 mm up the brickwork, across the gutter board, and up the roof slope over a tilting fillet, whilst a cover flashing with its top edge built into the brickwork is dressed down over the gutter upstand. The constructional aspects will be considered in greater detail in chapter 7. SI 1370 (1975) reduced the minimum height to width ratio of 4½:1.

CURING SMOKY CHIMNEYS

As described in the early part of this chapter smoky chimneys, whereby the chimney does not draw satisfactorily causing smoke to be blown back into the room through a fireplace opening, can result from a number of factors. These include a blocked flue, insufficiency of air supply, poor design of throat, unsatisfactory gathering of flue, and overshadowing of flue outlet by tall trees or buildings.

A normal open fire needs between 110 and 170 m^3 of air per hour and, in the absence of open windows or doors, this must find its way into the room through cracks. Some occupants aggravate the problem by sealing many of the cracks with draught excluders. In extreme cases an underfloor air supply to the hearth will overcome the difficulty, whilst a throat restrictor may be introduced to reduce the amount of air passing up the chimney.

When inspecting a smoky chimney, the following procedures should be adopted.

(1) Ensure that the chimney is swept; excessive quantities of soot may be the result of using unsuitable fuel. Check on the size of throat, shape and position of lintel, and similar matters.

(2) Open the window(s) and door(s) of the room in which the fire is located; if smoking ceases then the fire is starved of air. There is probably a need for ventilators or underfloor ducts.

(3) Determine the effect of reducing the height of the fireplace opening with a strip of sheet metal. Where an improvement results from a restriction of the opening by up to 100 mm, then the fitting of a permanent canopy will probably be the answer.[8]

(4) On occasions the streamlining of entrances and restriction of the throat with a piece of bent sheet metal may produce a considerable improvement. In these circumstances a variable throat restrictor could be built into the fireplace.

(5) Determine the effect of increasing the height of the chimney stack with pieces of sheet metal pipe of varying

lengths, with a throat restrictor in position in the fireplace. Where good results are secured with a certain length of pipe, it will be advisable to fit a longer pot, increase the height of the stack, or possibly to fit a cowl or louvred pot.[1]

REFERENCES

1. DOE Advisory leaflet 44: Smoky chimneys. HMSO (1972)
2. The Building Regulations 1976, SI 1676 HMSO (1976)
3. Greater London Council, London Building Acts 1930–1939 and Constructional By-laws (1972)
4. BS 1251: 1970 Open fireplace components
5. DOE Advisory leaflet 30: Installing solid fuel appliances; Open fires and convectors. HMSO (1977)
6. Marley Buildings. Marley gas flue system (1978)
7. BS 1181: 1971 Clay flue linings and flue terminals
8. DOE Advisory leaflet 50: Chimneys for domestic boilers. HMSO (1975)
9. BS 3589: 1963 Glossary of general building terms
10. CP 403: 1974 Installation of domestic heating and cooking appliances burning solid fuel
11. CP 131: 1974 Chimneys and flues for domestic appliances burning solid fuel
12. *BRE Digest 60*: Domestic chimneys for independent boilers. HMSO (1973)
13. BS 5440: Code of practice for flues and air supply for gas appliances, Part 1: 1978 Flues; Part 2: 1976 Air supply

6 FLOORS

This chapter examines the various types of floor and the constructional techniques and materials used in their provision. The techniques employed must be considered in relation to the functional requirements, in order to achieve a satisfactory form of construction. The form of construction selected may also influence the provision of other elements. For instance, the thickness of floors will affect the height of walling, and the choice between a solid or suspended ground floor can affect the location of services and the size of the heating installation.

FUNCTIONS OF FLOORS

Floors need to satisfy a number of functional requirements which are now outlined.

Ground Floors

(1) To withstand the loads that will be imposed upon them; with domestic buildings they are normally confined to persons and furniture, but in other classes of building, such as factories, warehouses and libraries, the floors may be subjected to much heavier loads and must be of sufficient strength to carry them.

(2) To prevent the growth of vegetable matter inside the building, by the provision of the concrete oversite.

(3) To prevent damp penetrating the building by inserting a damp-proof membrane in or below the floor; suspended ground floors also require underfloor ventilation to prevent stagnant, moist air accumulating below them.

(4) To meet certain prescribed insulation standards; for instance, with ground floors of centrally heated buildings it is advisable to incorporate a layer of insulating material to reduce the heat loss into the ground below.

(5) To be reasonably durable and so reduce the amount of maintenance or replacement work to a minimum.

(6) To provide an acceptable surface finish which will meet the needs of the users with regard to appearance, comfort, safety, cleanliness and associated matters.

Upper Floors

The functional requirements of upper floors differ considerably from those of ground floors. Their requirements are as follows.

(1) To support their own weight, ceilings and super-imposed loads.

(2) To restrict the passage of fire; this is particularly important in high-rise buildings, and in buildings where parts are in different ownerships, where there are many occupants or where large quantities of combustible goods are stored.

(3) To restrict the transmission of sound from one floor to another, particularly where this may seriously interfere with the activities undertaken.

(4) To possess an adequate standard of durability.

(5) To bridge the specific span economically and be capable of fairly quick erection.

(6) To accommodate services readily.

(7) To provide an acceptable surface finish in the manner described for ground floors.

SOLID GROUND FLOORS

With domestic buildings a choice often has to be made between solid and suspended ground floors. Solid floors are likely to prove cheaper on fairly level sites, may reduce the quantity of walling, eliminate the need for underfloor ventilation and hence reduce heat loss through the floor, can avoid the risk of dry rot and offer a greater selection of floor finishes. For these reasons the majority of ground floors in modern dwellings are of solid construction, comprising a concrete slab laid on a bed of hardcore. The concrete slab usually supports a cement and sand screed on which the floor finish is bedded.

Hardcore

A bed of hardcore may be required on a building site to fill hollows and to raise the finished level of an oversite concrete slab after removal of turf and other vegetation.[1] On wet sites it may be used to provide a firm working surface and to prevent contamination of the lower part of the wet concrete during placing and compaction. It can reduce the amount of rising ground moisture but it cannot eliminate the need for a waterproof membrane.[2]

The best filling materials are those with particles that are hard and durable and chemically inert and will not attack concrete or brickwork mortar, and that can readily be placed in a compact and dense condition, requiring only limited consolidation. Materials with fairly large particles, such as hard rock wastes, gravels and coarse sands, drain easily and consolidate quickly. If the material is well graded, with a mixture of coarser and finer grains, it will form a dense fill.[1] The concrete oversite slab can be protected against excess sulphates by a layer of bitumen felt or plastics sheeting placed on the hardcore; alternatively, or in addition, the concrete can contain a more resistant type of cement.[3]

Some of the more commonly used materials for hardcore beds are now considered.

Brick or tile rubble. This is clean, hard and chemically inert but it may contain little or no fine material and so may not consolidate readily.

Clinker. Ideally this should be hard-burnt with the material fused and sintered into hard lumps.

Colliery shale. Ideally this should be well-burnt brick-red shale, but some may contain sulphates.

Gravel. If well graded, it meets most requirements but is expensive. A cheaper alternative is to use coarse screened gravel in excess of 40 mm for the bottom layer with a consolidated upper layer of well-graded material.

Quarry waste. This is clean, hard and safe to use, but it may be unevenly graded and consequently difficult to consolidate.

Pulverised-fuel ash (PFA). This is recovered from flue gases produced by boilers with pulverised coal, as at power stations. It is lightweight, and when conditioned with the correct amount of water and compacted, has self-hardening properties.

Concrete Oversite

The oversite bed of concrete should not be less than 100 mm thick, although it is often 150 mm thick. The mix of concrete should be at least 1:3:6 with a maximum size of coarse aggregate of 38 mm, but a mix of 1:2:4 is to be preferred incorporating coarse aggregate with a maximum size of 19 mm. Figure 6.1.1 shows a typical concrete ground floor slab. It should be noted that the edges of the slab are not built into the surrounding walls to allow the two elements with their differing loads to move independently of one another.

The Building Regulations[4] (C3) prescribe that 'any floor which is next to the ground shall be so constructed as to prevent any part of the floor being adversely affected by moisture or water vapour from the ground'. Oversite concrete and floor screeds as commonly laid cannot by themselves keep back all ground moisture. In the absence of a satisfactory waterproof membrane, the moisture rises to the surface where it evaporates or accumulates under a less pervious material. Some floor finishes are particularly susceptible to damp, such as magnesium oxychloride, PVA emulsion/cement, rubber, flexible PVC, linoleum, cork-carpet or tiles, and wood flooring.

Damp-proof Membrane

To provide satisfactory protection for applied finishes, a damp-proof layer must be

(1) impermeable to water (liquid or vapour);
(2) continuous with the damp-proof course in adjoining walls;[2]
(3) tough enough to remain undamaged when laying the screed or finish.

The best location for the damp-proof membrane is below the floor screed and linking with the horizontal damp-proof course in adjoining walls, as illustrated in figure 6.1.1. Suitable materials are listed in BRE Digest 54[2] and they include mastic asphalt to BS 1097 or 1418; bitumen sheets to BS 743 with properly sealed joints; hot-applied pitch or bitumen laid on a primed surface to give an average thickness of 3 mm; two or three coats of cold-applied bitumen solutions or coal tar pitch/rubber emulsion or bitumen/rubber emulsion; and polythene film sheets at least 0.12 mm thick with properly sealed joints. Polythene film may not always offer sufficient protection for the more vulnerable finishes listed earlier, and it seems advisable to use material of double the minimum thickness (1000 gauge) laid on a 12 mm bed of sand to avoid the film being punctured by any irregularities in the concrete below.

Solid Floors with Timber Finish

The Building Regulations[4] (C5) prescribe the following damp-proofing requirements for solid floors incorporating timber.

(1) The concrete incorporates a damp-proof sandwich membrane consisting of a continuous layer of hot applied soft bitumen or coal tar pitch not less than 3 mm thick, or not less than three coats of bitumen solution, bitumen/rubber emulsion or tar/rubber emulsion; *or*

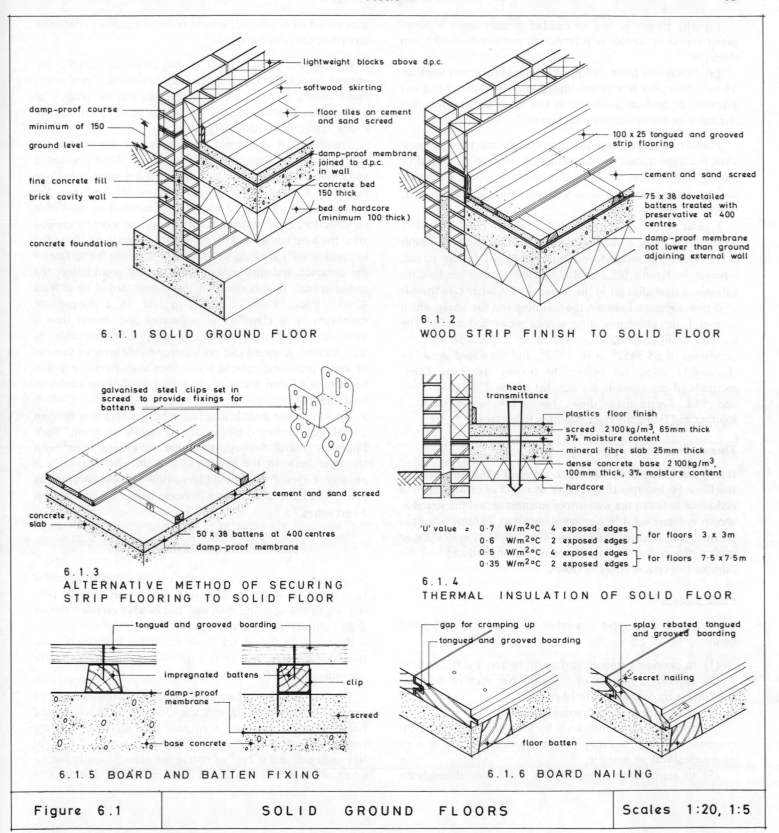

6.1.1 SOLID GROUND FLOOR

lightweight blocks above d.p.c.

softwood skirting

floor tiles on cement and sand screed

damp-proof membrane joined to d.p.c. in wall

concrete bed 150 thick

bed of hardcore (minimum 100 thick)

damp-proof course

minimum of 150

ground level

fine concrete fill

brick cavity wall

concrete foundation

6.1.2 WOOD STRIP FINISH TO SOLID FLOOR

100 x 25 tongued and grooved strip flooring

cement and sand screed

75 x 38 dovetailed battens treated with preservative at 400 centres

damp-proof membrane not lower than ground adjoining external wall

6.1.3 ALTERNATIVE METHOD OF SECURING STRIP FLOORING TO SOLID FLOOR

galvanised steel clips set in screed to provide fixings for battens

cement and sand screed

concrete slab

50 x 38 battens at 400 centres

damp-proof membrane

6.1.4 THERMAL INSULATION OF SOLID FLOOR

heat transmittance

plastics floor finish

screed 2 100 kg/m^3, 65mm thick 3% moisture content

mineral fibre slab 25mm thick

dense concrete base 2 100 kg/m^3, 100mm thick, 3% moisture content

hardcore

'U' value =	0·7	W/m^2°C	4 exposed edges	for floors 3 x 3m
	0·6	W/m^2°C	2 exposed edges	
	0·5	W/m^2°C	4 exposed edges	for floors 7·5 x 7·5m
	0·35	W/m^2°C	2 exposed edges	

6.1.5 BOARD AND BATTEN FIXING

tongued and grooved boarding

impregnated battens

damp-proof membrane

base concrete

clip

screed

6.1.6 BOARD NAILING

gap for cramping up

tongued and grooved boarding

splay rebated tongued and grooved boarding

secret nailing

floor batten

| Figure 6.1 | SOLID GROUND FLOORS | Scales 1:20, 1:5 |

(2) The timber is laid or bedded directly upon a damp-proof course of asphalt or pitchmastic not less than 12.5 mm thick; *or*

(3) Where the floor incorporates wood blocks not less than 16 mm thick, the blocks are dipped in an adhesive of hot soft bitumen or coal tar pitch and so laid upon the concrete that the adhesive forms a continuous layer.

Furthermore, the damp-proof membrane must not be lower than the highest level of the surface of the outside ground or paving, and it must be continuous with, or joined and sealed to, the damp-proof course in any adjoining wall, floor, pier, column or chimney.

Figures 6.1.2, 6.1.3 and 6.1.5 show two alternative methods of fixing a boarded floor to a concrete slab. Figures 6.1.2 and 6.1.6 show dovetailed battens set in the screed, whereas in figures 6.1.3 and 6.1.5 the battens are held by galvanised steel clips set in the screed. In the latter case there is a 38 mm air space between the boarding and the screed which is useful for accommodating service pipes and cables. The battens or fillets should be treated in accordance with the provisions of BS 3452[6] or BS 4072[7], and the screed should be thoroughly dried out before the boards are fixed. Other methods of preservation are detailed in BSs 1282, 3453, 144 and 913. Figure 6.1.6 shows two alternative methods of jointing and fixing the floor boarding to battens.

Thermal Insulation

It may be considered desirable to reduce the heat loss through the floor by incorporating a layer of material of high thermal resistance between the waterproof membrane and the screed as shown in figure 6.1.4.[47] A mineral fibre slab, 25 mm thick, has a thermal resistance of 0.66 to 0.71 m^2 °C/W. The *U*-value of such a floor increases as the area of floor reduces and the number of exposed edges increases.[8]

Floor Screeds

Floor screeds may serve a number of functions as described below

(1) to provide a smooth surface to receive the floor finish;
(2) to bring a number of floor finishes each of different thicknesses up to the same finished level;
(3) to provide falls for drainage purposes;
(4) to give thermal insulation by incorporating lightweight concrete, although this may not be the most efficient or most economical way of doing it;
(5) to accommodate service pipes and cables, although the

thin screed over pipes is liable to crack and access to defective services is costly and difficult.[9]

Cement and sand screeds in the ratios of 1:3–4½ (by weight) are suitable for thicknesses up to 40 mm. Mixes with a lower cement content will be subject to less shrinkage. For thicker screeds, fine concrete (1:1½:3) using 10 mm maximum coarse aggregate is satisfactory.

The thickness of a screed is influenced by the state of the base at the time the screed is to be placed. When a screed is laid on an *in situ* concrete base before it has set (within three hours of placing), complete bonding is obtained and the thickness need only be 12 mm. This is described as *monolithic* construction. With *separate* construction, the screed is applied after the base has set and hardened. Maximum bond is secured by mechanically hacking, cleaning and damping the surface of the concrete, and then applying wet cement grout before the screed is laid. In this situation the screed should be at least 40 mm thick. Where a screed is laid on a damp-proof membrane it is classified as *unbonded* and should have a thickness of at least 50 mm, or, if it contains heating cables, at least 65 mm. A screed laid on a compressible layer of thermal or sound insulating material is described as *floating* and should be at least 65 mm thick, or, if it contains heating cables, at least 75 mm.[19]

Screeds should desirably be laid in bays, where they contain underfloor warming cables or are to receive an *in situ* floor finish. In general the bay size should not exceed 15 m^2 with the ratio between the lengths of bay sides as near 1:1½ as possible. Edges of bays should be vertical with closely abutting joints, and expansion joints will rarely be needed at less than 30 m centres.[9]

SUSPENDED GROUND FLOORS

Suspended ground floors generally consist of boarding nailed to floor joists which in their turn are laid on timber plates running in the opposite direction and bedded on brick sleeper walls built off the concrete oversite. This form of construction is illustrated in figures 6.2.2 and 6.2.4. The Building Regulations[4] (C4) prescribe certain mimimum requirements for suspended timber floors.

(1) The ground surface is to be covered with a layer of concrete not less than 100 mm thick, consisting of cement and fine and coarse aggregate conforming to BS 882,[10] in the proportions of 50 kg of cement to not more than 0.1 m^3 of fine aggregate and 0.2 m^3 of coarse aggregate, properly laid on a bed of hardcore consisting of clean clinker, broken brick or

similar inert material largely free from water soluble sulphates or other harmful substances.

(2) The concrete is to be finished with a trowel or spade finish so that the top surface is not below the highest level of the surface of the ground or paving adjoining external walls.

(3) There is a space above the upper surface of the concrete of not less than 75 mm to the underside of a wall plate, and of not less than 125 mm to the underside of any suspended timbers, and this space is to be free of debris and have through ventilation.

(4) There are damp-proof courses to prevent ground moisture reaching any timbers.

All of these requirements are illustrated in figure 6.2.2. A better understanding of suspended floor construction can be obtained by examining the isometric sketch in figure 6.2.4.

The sleeper walls are built half-a-brick thick and honeycombed, being similar to Flemish bond with spaces replacing the headers, to permit a free flow of air under the floors from the air bricks built into the external walls. These walls are usually either 150 or 225 mm high, although the Building Regulations[4] allow one course in height (75 mm). The sleeper walls are built across each room at a spacing of from 1200 to 1800 mm centres, with the end walls positioned about 50 to 100 mm from the loadbearing walls. Sufficient airbricks should be provided in the external walls to give a minimum open area of 3000 mm^2 per metre of length of external wall.[11] They are often provided at about 1.50 m centres. The airbricks should be positioned as high as possible on opposite walls, and care should be taken to avoid unventilated airpockets such as may occur near corners, projections or bay windows. A typical clay airbrick conforming to BS 493[12] is illustrated in figure 6.2.5, and a duct may be formed through the cavity wall behind the airbrick with slates in the manner shown in figure 6.2.4. Another method is to use hollow tile ducts as illustrated in figure 6.2.6, while yet another alternative is to install purpose-made asbestos cement ducts.

A damp-proof course must be provided on top of all sleeper walls to prevent moisture reaching any timber. The damp-proof course should ideally be flexible to allow for any movement without fracturing and must be lapped at any joints. A timber wall plate, often 100 x 50 mm, is bedded in mortar above the damp-proof course and this assists in spreading the load from each joist over the wall below. The floor joists are often 50 x 100 mm, spaced at about 400 mm centres and skew-nailed to the wall plate. The actual spacing of the joists is influenced by the thickness of the floor-boarding and the load on it, whereas the size of joists is affected by the loading and the spacing of joists and sleeper walls. The sizes of joists can be determined quite easily from the table in schedule 6 of the Building Regulations.[4] The ends of joists are often cut on the splay and are kept back a short distance from the walls to prevent any moisture being absorbed by the vulnerable open grain at the ends of joists. A suitable distance would be 20 mm at the top of the joist and 30 mm at the bottom. In like manner, the end joists should be kept 50 mm from the loadbearing walls as shown in figure 6.2.1.

The normal arrangement at a ground floor hearth is illustrated in figure 6.2.1. A one-brick thick fender wall encloses and supports the filling material under the front hearth and also carries a wall plate which supports the ends of joists. It is important to provide adequate ventilation where a suspended floor joins a solid floor, and particularly to avoid any stagnant corners. Ventilating pipes or ducts should be laid below the solid floor, connected to airbricks in the external wall, as shown in figure 6.2.3. The damp-proof membrane under the solid floor should be linked with the damp-proof course in the division wall by a vertical damp-course.

A suspended timber floor may be covered with boards or strips. The majority of timber floors are finished with softwood boards between 100 and 150 mm wide, with a nominal thickness of 25 mm. The boards are laid at right angles to the joists with each board nailed to each joist with two floor brads, with the brads being 40 mm longer than the thickness of the boards and the heads of brads well punched down below the top surface of the boards. Floor boards are normally joined together with tongued and grooved joints of the type shown in figure 6.1.6, although in cheap work, plain or square-edged boards may be used. The use of plain-edge boards is not recommended, as with ventilated suspended floors draughts can penetrate through the joints, and with timber finished solid floors spilt liquid or washing-down water could penetrate the joints, would only dry out very slowly and might result in an outbreak of dry rot. With timber-covered solid floors it is advisable to pressure-impregnate the battens and to apply a brush coat of preservative to the underside of the boards as an additional safeguard against decay. Where it is considered desirable to avoid punch holes for brads, boards may be secret nailed as shown in figure 6.1.6, using splay rebated, tongued and grooved joints, to reduce the risk of splitting the tongue when nailing. This method of jointing and nailing is particularly suited for hardwood floorboarding and narrow strip boarding where appearance is especially important.

Floor boards should preferably be rift sawn, when they are cut as near radially from the log as practicable on conversion.

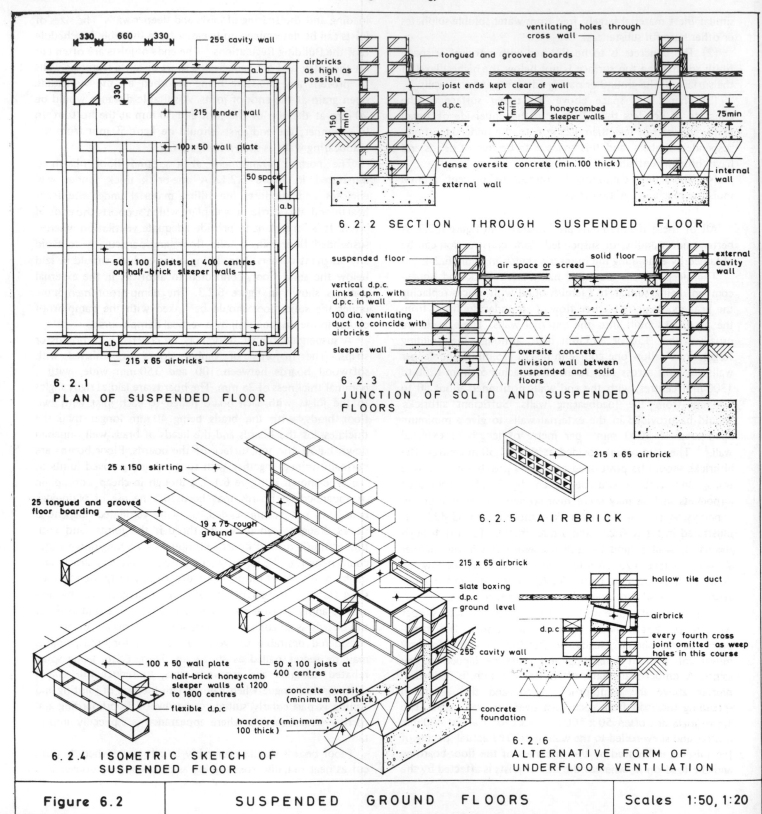

330   660   330
255 cavity wall
330
airbricks as high as possible
215 fender wall
100 x 50 wall plate
50 space
50 x 100 joists at 400 centres on half-brick sleeper walls
a.b
215 x 65 airbricks

6.2.1 PLAN OF SUSPENDED FLOOR

ventilating holes through cross wall
tongued and grooved boards
joist ends kept clear of wall
d.p.c
honeycombed sleeper walls
150 min
125 min
75 min
dense oversite concrete (min. 100 thick)
external wall
internal wall

6.2.2 SECTION THROUGH SUSPENDED FLOOR

suspended floor
air space or screed
solid floor
external cavity wall
vertical d.p.c. links d.p.m. with d.p.c. in wall
100 dia. ventilating duct to coincide with airbricks
sleeper wall
oversite concrete
division wall between suspended and solid floors

6.2.3 JUNCTION OF SOLID AND SUSPENDED FLOORS

215 x 65 airbrick

6.2.5 AIRBRICK

25 x 150 skirting
25 tongued and grooved floor boarding
19 x 75 rough ground
215 x 65 airbrick
slate boxing
d.p.c
ground level
255 cavity wall
100 x 50 wall plate
50 x 100 joists at 400 centres
half-brick honeycomb sleeper walls at 1200 to 1800 centres
flexible d.p.c
concrete oversite (minimum 100 thick)
hardcore (minimum 100 thick)
concrete foundation

6.2.4 ISOMETRIC SKETCH OF SUSPENDED FLOOR

hollow tile duct
airbrick
d.p.c
every fourth cross joint omitted as weep holes in this course

6.2.6 ALTERNATIVE FORM OF UNDERFLOOR VENTILATION

| Figure 6.2 | SUSPENDED GROUND FLOORS | Scales 1:50, 1:20 |

This reduces later warping of the boards to a minimum. Heading or end joints of floorboards are usually splayed and should be made over joists. They should be staggered in adjoining boards for greater strength. The edges of floorboards should be kept 13 mm away from surrounding walls to allow for movement and reduce the risk of damp penetration. The gap is closed by a skirting at the base of the wall or partition, which also masks the gap between the bottom edge of the wall plaster and the floor and provides adequate resistance to kicks. Skirtings are considered in more detail in chapter 10.

UPPER FLOORS

Timber Floor Construction

Most upper floors in domestic buildings are constructed of timber floorboarding supported by timber joists which, in their turn, bear upon the loadbearing walls below. Hence the main difference between suspended floors at ground level and those on upper floors is the longer spans of joists in the latter case, with resultant deeper joists. The sizes of joists vary with the dead load, (including ceiling below), the span of the joists and their spacing. Schedule 6 of the Building Regulations[4] shows joist sizes varying from 38 x 75 mm up to 75 x 225 mm for dead loads ranging from 25 to 125 kg/m^2, joist spacings of 400 to 600 mm and spans ranging from 1 to 6 m. A useful rule-of-thumb method for calculating the depth of floor joists is: $\frac{1}{25}$ span in mm + 50 = depth in mm. For example if the span of the joists is 3 m, then their depth = 3000/25 + 50 = 170 mm (nearest standard size is 175 mm). In fact, 50 x 175 mm floor joists at 400 mm centres are quite common in normal domestic construction.

The ends of upper floor joists are usually supported by the inner leaves of external cavity walls or loadbearing internal walls or partitions, as illustrated in figure 6.3.6. Joists should run the shortest span wherever practicable to economise in timber. The ends of the joists may be supported in a number of different ways.

(1) Ends of joists are treated with preservative and built into walls, preferably with a space all round. The joists may require packing and cutting of brickwork to keep their tops level. Care must be taken to prevent joists penetrating into the cavity and thus providing a bridge for moisture and probable deterioration of the timber. It is good practice to taper the ends of joists as in figure 6.3.1.

(2) Ends of joists may rest on a wall plate bedded on top of an internal partition (figure 6.3.4). This provides a sound, uniform bearing with maximum distribution of loads. It is not suitable for external walls, where brickwork will be superimposed upon the plate and damp from the cavity might cause decay of the non-ventilated timber. Another alternative with external walls is to support the joists on wall plates located clear of the wall on wrought iron or mild steel corbels built into the wall at about 750 mm centres (figure 6.3.3). In the latter situation a cornice is needed to mask the projecting plate in the rooms below.

(3) Ends of joists can rest on a galvanised mild steel or wrought iron bearing bar built into an external wall to assist in spreading the load, although this is not usually vital in domestic construction.

(4) Ends of joists are supported on galvanised steel hangers which are built into the wall, and the joists rest on the bottoms of hangers and are nailed from the sides (figure 6.3.2). Another form of steel joist hanger is shown in figure 6.3.10; this is welded. This is the best method of providing support to joists from external walls as it avoids building timber into the walls with its resultant risks and the hangers can usually be adjusted on the site to suit the particular conditions.

Where joists from either side meet on a loadbearing internal wall, they are usually placed side by side and nailed to each other and the wall plate (figure 6.3.4), with passings of about 150 mm. Where the span of joists exceeds 3 m, they should be strutted at about 2 m centres to stiffen the joists and prevent them twisting, and also to strengthen the floor. The most common form is herringbone strutting made up of 40 x 40 mm or 40 x 50 mm timbers with their ends cut on the splay and nailed to each other where they intersect and to the upper part of one joist and the lower part of the next (figure 6.3.5). The space between the first and last joists and the adjoining wall is wedged to take the thrust of the strutting. Another alternative is to use solid strutting, usually about 32 mm thick, which may be staggered to allow nailing of the ends to the joists, or laid in a continuous line and skew-nailed to joists. This method is not so effective as herringbone strutting, as the strutting may become loose due to rounding of joist sides or shrinkage in timbers.

The internal walls on the ground and first floors of a dwelling do not always coincide, and lightweight block partitions may be required on the first floor where there are no loadbearing walls beneath. Where the partition on the upper floor is parallel to the floor joists, two joists can be bolted together to form a base on which the partition can be built. Another alternative is to fix solid bridging pieces 50 mm thick at about 400 mm centres between a pair of joists, and the bridging pieces support floorboarding and a wall plate on

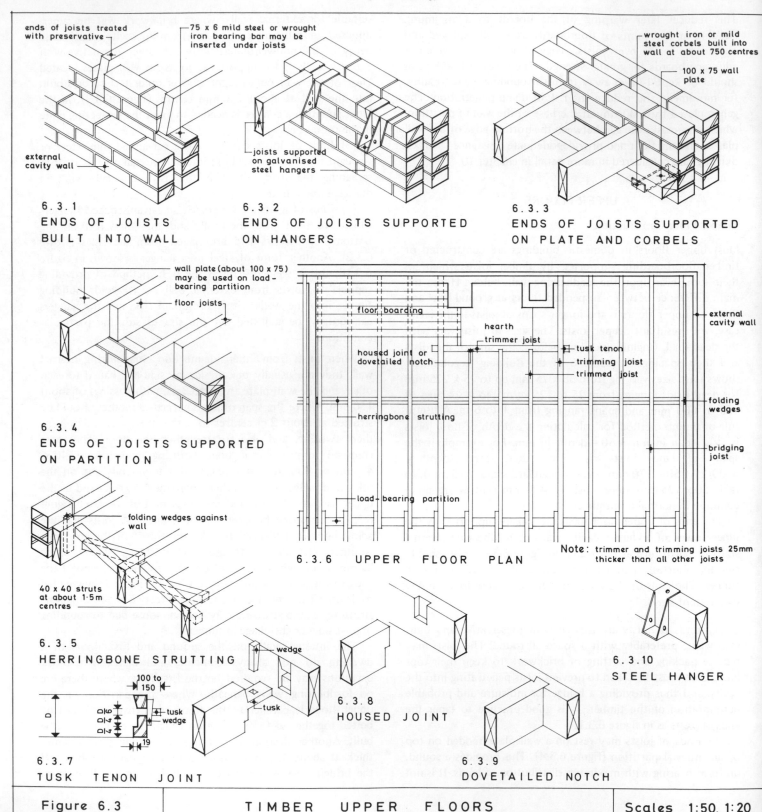

ends of joists treated with preservative

75 x 6 mild steel or wrought iron bearing bar may be inserted under joists

external cavity wall

6.3.1
ENDS OF JOISTS BUILT INTO WALL

joists supported on galvanised steel hangers

6.3.2
ENDS OF JOISTS SUPPORTED ON HANGERS

wrought iron or mild steel corbels built into wall at about 750 centres

100 x 75 wall plate

6.3.3
ENDS OF JOISTS SUPPORTED ON PLATE AND CORBELS

wall plate (about 100 x 75) may be used on load-bearing partition

floor joists

6.3.4
ENDS OF JOISTS SUPPORTED ON PARTITION

floor boarding

hearth

trimmer joist

housed joint or dovetailed notch

tusk tenon

trimming joist

trimmed joist

herringbone strutting

load-bearing partition

external cavity wall

folding wedges

bridging joist

6.3.6 UPPER FLOOR PLAN

Note: trimmer and trimming joists 25mm thicker than all other joists

folding wedges against wall

40 x 40 struts at about 1·5 m centres

6.3.5
HERRINGBONE STRUTTING

100 to 150

D

tusk
wedge

19

6.3.7
TUSK TENON JOINT

wedge

tusk

6.3.8
HOUSED JOINT

6.3.9
DOVETAILED NOTCH

6.3.10
STEEL HANGER

| Figure 6.3 | TIMBER UPPER FLOORS | Scales 1:50, 1:20 |

which the partition is built. Where the partition runs at right angles to the floor joists, solid bridging pieces can be fixed between the joists on the line of the partition, with boarding and a wall plate above them.

Construction around Openings

Figure 6.3.6 shows the method of trimming floor joists around a fireplace opening. A trimming joist is placed about 50 mm from each outside edge of the chimney breast, and a trimmer is provided adjacent to the front edge of the hearth and is supported at each end by the trimming joists. Both trimmers and trimming joists are 25 mm thicker than the normal floor or bridging joists as they have to withstand heavier loads and it is not possible to increase their depth. Strong and soundly constructed joints are required between the trimming members, and tusk tenon joints (figure 6.3.7) are often advocated for joints between trimmers and trimming joists and either dovetail notches (figure 6.3.9) or housed joints (figure 6.3.8) between trimmed joists and trimmers. In a tusk tenon joint the tenon passes through the trimming joist and is morticed out for a key or tusk, which when driven into position brings the shoulders up tight. The recommended dimensions of this joint are shown in figure 6.3.7. The housing or dovetail notch will project into the trimmer for at least one-half the depth of the trimmer. An alternative approach is to hang galvanised steel brackets or hangers (figure 6.3.10) over trimming joists to receive trimmers and over trimmers to receive trimmed joists, which are both nailed to prevent movement. The same principles apply in trimming floor joists around stairwells and rafters and ceiling joists around chimney stacks and rooflights.

Sound Insulation

Wood-joist floors are not sufficiently heavy and stiff to restrain vibration of the walls and the maximum net sound insulation is accordingly controlled by the thickness of the walls, even though the floor may have a higher sound insulation. Hence insulation values for wood floors can only be given in conjunction with the wall system,[13] and the recommended forms of construction aim at achieving grade 1 insulation for both airborne and impact sound. The passage of noise through walls is known as *flanking transmission*.[14] Sound transmission through floors is particularly important in the construction of flats, but may also have significance in ordinary dwellings where first-floor rooms are used as study bedrooms.

The only really satisfactory method of soundproofing wood-joist upper floors adjoining thin walls is to provide a ceiling of expanded metal lath and three-coat plaster, loaded directly with a *pugging* of dry sand 50 mm thick or other loose material weighing not less than 80 kg/m^2, together with a properly constructed floating floor, as illustrated in figure 6.4.1. The sand must not be omitted from the narrow spaces between the end joists and the walls; it should be as dry as possible when it is placed in the floor and should not contain deliquescent salts. An alternative to metal lath and plaster would be to use strong board ceilings of thick plasterboard or asbestos board that are firmly bonded to the walls, with solid, airtight joints between boards and a combined weight of ceiling and pugging of not less than 120 kg/m^2.[13]

With thick walls, it is possible to use a lighter form of floor construction (figure 6.4.2), incorporating a floating floor but reducing the pugging to 15 kg/m^2. The ceiling may be of plasterboard with a single-coat plaster finish, but a heavier ceiling is to be preferred. Wire netting stapled to the joists is sometimes inserted above the ceiling to keep the pugging in position and so ensure that the floor has a half-hour fire resistance. Another way of achieving the same fire resistance is to increase the plaster finish on the plasterboard ceiling to a thickness of 13 mm. A commonly recommended pugging material is high density slag wool (about 200 kg/m^3), laid to a thickness of about 75 mm.

The floating floor consists of woodboard or strip, plywood, or chipboard flooring, nailed to battens to form a *raft*, which rests on resilient quilt draped over the joists. A Building Research Establishment Digest[13] recommends that the raft should not be nailed to the joists and should be isolated from the surrounding walls, either by turning up the quilt at the edges or by the less satisfactory alternative of leaving a gap round the edges, which can be covered by a skirting. Skirtings should be fixed only to walls and not to floors. The flooring should be at least 20 mm thick, preferably tongued and grooved, and battens should be 50 x 40 to 50 mm in size, laid parallel to the joists. A common method of constructing the raft is to place the battens on the quilt along the top of each joist and to nail the boards to the battens (figure 6.4.1). An alternative method is to prefabricate the raft in separate panels for the length of the room and up to one metre wide, with the battens across the panels positioned between the joists (figure 6.4.2).

With concrete floors between flats a number of methods can be used to achieve a reasonable standard of sound insulation. A common method is to separate the floor screed from the concrete slab by a layer of wire netting (20 to 50 mm mesh chicken wire) on waterproof paper or plastics sheeting, laid on a resilient layer of glass wool or mineral wool to a thickness of 13 mm. The floor screed is then described as a

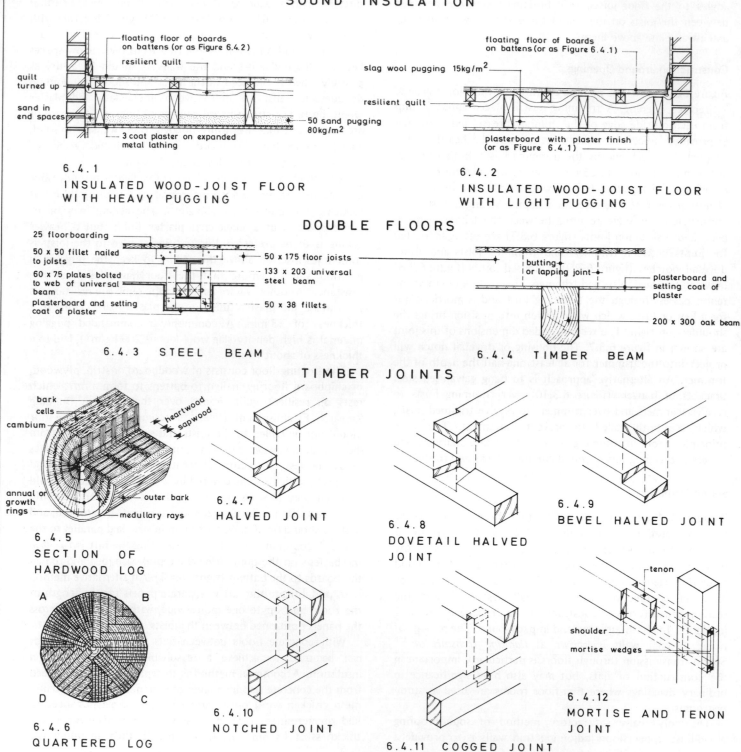

SOUND INSULATION

floating floor of boards on battens (or as Figure 6.4.2)

resilient quilt

quilt turned up

sand in end spaces

50 sand pugging 80kg/m²

3 coat plaster on expanded metal lathing

6.4.1 INSULATED WOOD-JOIST FLOOR WITH HEAVY PUGGING

floating floor of boards on battens (or as Figure 6.4.1)

slag wool pugging 15kg/m²

resilient quilt

plasterboard with plaster finish (or as Figure 6.4.1)

6.4.2 INSULATED WOOD-JOIST FLOOR WITH LIGHT PUGGING

DOUBLE FLOORS

25 floorboarding

50 x 50 fillet nailed to joists

60 x 75 plates bolted to web of universal beam

plasterboard and setting coat of plaster

50 x 175 floor joists

133 x 203 universal steel beam

50 x 38 fillets

6.4.3 STEEL BEAM

butting or lapping joint

plasterboard and setting coat of plaster

200 x 300 oak beam

6.4.4 TIMBER BEAM

TIMBER JOINTS

bark

cells

cambium

heartwood

sapwood

annual or growth rings

outer bark

medullary rays

6.4.5 SECTION OF HARDWOOD LOG

A   B

D   C

6.4.6 QUARTERED LOG

6.4.7 HALVED JOINT

6.4.8 DOVETAIL HALVED JOINT

6.4.9 BEVEL HALVED JOINT

6.4.10 NOTCHED JOINT

6.4.11 COGGED JOINT

tenon

shoulder

mortise wedges

6.4.12 MORTISE AND TENON JOINT

Figure 6.4	TIMBER CONVERSION, SOUND INSULATION OF WOOD-JOIST FLOORS, DOUBLE TIMBER FLOORS AND JOINTING TIMBERS	Scale 1:20

floating screed. In like manner, wood flooring on battens should be separated from the concrete slab by a similar resilient layer. A suspended ceiling may be used as a barrier against airborne sound consisting of battens wired to the underside of the slab which, in their turn, support an absorbent quilt and a heavy ceiling, such as plaster on expanded metal lathing.

Double Timber Floors

The longest span which can be effectively bridged with a normal joist floor is 6.02 m, using 75 x 225 mm joists at 400 mm centres with a dead load not exceeding 25 kg/m².[4] Spans in excess of this are broken down into convenient lengths by the use of cross-beams forming what is commonly known as a *double floor*. The majority of double floors consist of timber joists supported on steel beams in the manner shown in figure 6.4.3. The steel beams bridge the shortest span of the room and are generally placed at about 3 m centres with 50 x 175 mm timber floor joists running between them. The ends of steel beams are often supported on precast concrete padstones built into the walls, and should be protected against corrosion where they adjoin a cavity in an external wall. The ends of the timber joists are notched around the steel beam and supported on a plate, often about 50 x 75 or 75 x 75 mm, bolted to the beam at about 750 to 900 mm centres. Bracketing or cradling of 50 x 38 mm timber fillets around the bottom of the steel beam provides a framework to receive the plasterboard, finished with a setting coat of plaster. Timber bearers or fillets, often about 50 x 50 mm, are nailed to the upper part of each joist and bridge the gap over the steel beam to provide a bearing for the floorboards. The size of the steel beam will depend on the span, the load to be carried and the shape and thickness of the flange and web of the beam. In normal domestic work the depth of the beam is often about $\frac{1}{20}$ of the span and some idea of the range of sizes can be obtained from steelwork tables or from BS 4.[15]

An alternative to the steel beam is an exposed timber beam, which might be of oak or other suitable hardwood, probably with chamfered bottom edges (figure 6.4.4). Timber beams can provide an attractive feature to the ceiling of the room below.

Fire-resisting Floors

The selection of the most suitable form of fire-resisting floor in a particular situation will be influenced by such factors as cost, availability, rigidity, ease and speed of erection, sound and thermal insulating properties, permissible span and load,

and ease of variation if required. A selection of floor types is now considered.

Filler joist floors. These represent the earliest method of combining steel and concrete in floor construction. Concrete panels are formed between the main steel beams and are supported by light steel beams or filler joists, often at about 750 mm centres, encased in the concrete. The light steel beams may rest on the tops of the supporting beams or on a shelf angle riveted to the webs of the main beams. The concrete in the floor requires shuttering, but this can be supported from the beams and filler joists. Details of a typical filler joist floor are illustrated in figure 6.5.1.

To overcome objections to the standard filler joist floor arising from its weight, need for shuttering and sound transmission, hollow clay tiles or tubes are sometimes placed between filler joists in place of concrete. The tiles vary in depth from 100 to 175 mm and the filler joists are usually at about 900 mm centres. Another alternative is to use sheets of expanded metal instead of shuttering to retain the concrete.

Reinforced concrete floors. Concrete is strong in compression but weak in tension, whereas steel is strong in tension. In reinforced concrete the steel makes good the inadequacies of the concrete and the concrete protects the steel from the effects of fire. The steel reinforcement is generally in the form of round mild steel bars ranging from 6 to 40 mm in diameter. These bars are positioned where the tensile stresses are greatest. For instance the bars will be placed near the bottom of a suspended slab, but will be fixed near the top where the slab passes over or rests on supports. All bars must have adequate cover of concrete, normally not less than 19 mm in slabs and 25 mm in beams, or the diameter of the bar in either case, whichever is the greater.

Larger floors are split into bays which are supported at their edges on beams or intermediate walls. Reinforcing bars lap each other where extended and hooked ends are introduced where the bond stress (friction between concrete and steel) is high. Shear bars are incorporated to strengthen concrete which is subject to shear stresses.

For short spans of up to 5 m, a reinforced concrete floor can span between walls as in figure 6.5.2, but for larger spans it is advisable to introduce secondary beams. Reinforced concrete floors carry heavier loads than timber floors of equal thickness, provide good lateral rigidity, and offer good fire resistance and insulation against airborne sound. On the other hand, they need support from formwork and props, as illustrated in figure 6.5.2, for about seven days.

Hollow pots. These form a popular variety of patent floor as they are economical, of lightweight material, offer good

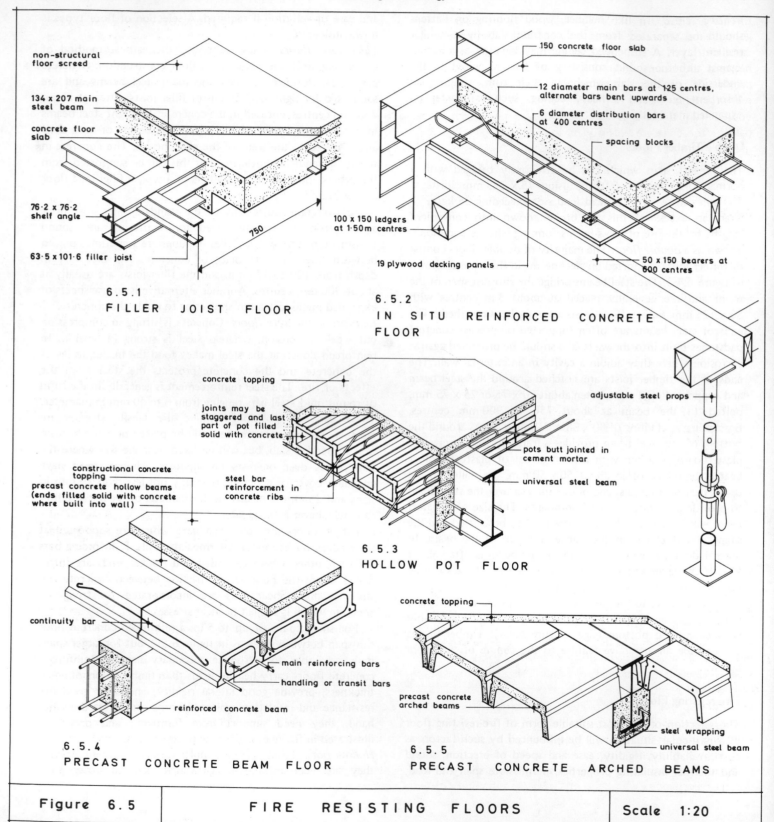

non-structural floor screed

134 x 207 main steel beam

concrete floor slab

76·2 x 76·2 shelf angle

63·5 x 101·6 filler joist

750

6.5.1
FILLER JOIST FLOOR

150 concrete floor slab

12 diameter main bars at 125 centres, alternate bars bent upwards

6 diameter distribution bars at 400 centres

spacing blocks

100 x 150 ledgers at 1·50m centres

19 plywood decking panels

50 x 150 bearers at 600 centres

adjustable steel props

6.5.2
IN SITU REINFORCED CONCRETE FLOOR

concrete topping

joints may be staggered and last part of pot filled solid with concrete

steel bar reinforcement in concrete ribs

pots butt jointed in cement mortar

universal steel beam

6.5.3
HOLLOW POT FLOOR

constructional concrete topping

precast concrete hollow beams (ends filled solid with concrete where built into wall)

continuity bar

main reinforcing bars

handling or transit bars

reinforced concrete beam

6.5.4
PRECAST CONCRETE BEAM FLOOR

concrete topping

precast concrete arched beams

steel wrapping

universal steel beam

6.5.5
PRECAST CONCRETE ARCHED BEAMS

| Figure 6.5 | FIRE RESISTING FLOORS | Scale 1:20 |

thermal and sound insulation, economise in concrete and have good fire resistance. The floor consists of *in situ* reinforced concrete tee beams with hollow clay or terra cotta blocks cast in between the beams or ribs, as illustrated in figure 6.5.3, to provide a flat ceiling. The blocks are laid on shuttering and the reinforcing bars placed in position. The concrete filling is then poured from the top, together with the topping which forms part of the concrete beams. The thickness of the floor and quantity and size of reinforcement varies with the span and superimposed load. In some cases the concrete beams or ribs extend the full depth of the floor, while in others clay slip tiles are inserted at the bottom of ribs and this reduces the possibility of pattern staining, gives better fire resistance and even suction for plaster.

Precast concrete beams. These also provide a useful form of floor. Figure 6.5.4 shows hollow precast beams laid side by side. Each beam is reinforced at the bottom corners and the gaps between beams are filled with jointing material. When the floor is supported by a main beam, continuity bars are inserted in the top of the slab or between the precast beams. The hollow beams have rather thin walls and a constructional concrete topping is laid over the tops of beams to spread superimposed loads. Figure 6.5.5 shows another type of floor using precast concrete arched beams. Precast concrete beam floors require no formwork and can be quickly erected.

Methods of increasing the fire resistance of existing timber floors are described in BRE Digest 208.[48]

TIMBER

This section is concerned with the properties, types, conversion processes, diseases and methods of jointing timber. Timber, of course, is probably the most important flooring material.

Properties of Timber

Timber is composed of cells which conduct sap (water fluid), give strength and provide storage space for food materials such as salts, starches, sugars and resins. A layer or band of cellular tissue is deposited round the trunk or stem and branches each year, and these are termed *annual rings*. In many timbers the wood formed in the early part of the growing season (spring or early wood) is more open, porous, and lighter in colour than the timber produced in the summer (summer or late wood). There is also a series of horizontal tubes or open cells radiating from the pith at the centre of the stem to the outside of the tree, known as *medullary rays*. Some down-flowing sap passes through medullary rays to interior layers which become durable *heartwood*, while the other layers remain as lighter

sapwood. Bark consists of cells and woody fibres and forms the outside protective layer of the tree. The wood from trees of rapid growth will have wide annual rings and is described as *coarse-grained*, while slow-growing trees produce narrow growth rings and the wood is termed *close-grained*. Many of these terms are illustrated in figure 6.4.5.

Timber Types

Botanically trees are grouped into two classifications:
Broad-leaved trees or hardwoods. These are generally hard, tough and dark-coloured with acrid, aromatic or even poisonous secretions, although not all hardwoods are hard. Typical examples of hardwoods are oak, teak, mahogany, walnut and elm.
Needle-leaved trees or softwoods. These are from coniferous trees which have cone-shaped seed vessels and narrow, needle-shaped leaves. They are usually elastic and easy to work, with resinous or sweet secretions. Some softwoods, like pitch pine, are quite hard. Typical examples of softwoods are European redwood, yellow pine, Douglas fir, whitewood and Canadian spruce.

Some of the more commonly used timbers are detailed in table 6.1 but readers are referred to the listed publications.[16,17,18]

Conversion of Timber

Felling of trees is best undertaken when there is minimum moisture in the tree, usually between mid-October and mid-January. An axe or cross-cut saw is generally used for felling, although sometimes they are uprooted with winches. The felled tree trunks are converted into logs by sawing off all branches and the logs are converted to timbers of marketable sizes (scantlings) using an axe or saw. Four methods of quartering a log are illustrated in figure 6.4.6. The quartering method is valuable when cutting oak and similar boards parallel to the medullary rays to show the figuring or grain to best advantage. Method A is considered the best but is also the most expensive, whereas B and C are less wasteful of timber and hence more economical. Method D is the most economical and is sometimes called the *slash* method. Methods A and B are sometimes described as *rift* or *radial* sawing. To reduce waste still further, the log can be sawn into boards with all cuts parallel ('through and through') but some boards will twist on shrinking. Another method of conversion is to cut the timbers parallel to the annual rings, leaving a 'boxed heart' in the centre of the log. Veneers can be obtained by mounting the log horizontally along its longitudinal axis and cutting off a fine circumferential shaving with a long knife edge.

Table 6.1   Timbers in common use

Variety	Hardwood or softwood	Average weight (kg/m³)	Origin	Main properties	Principal uses
Beech	hardwood	720	Europe, Asia Minor and Japan	Usually straight-grained with fine texture, stronger than oak, works easily and stains and polishes well	Veneers, kitchen equipment, cabinet-making and block and strip floors
Cedar (Western red)	softwood	360	British Columbia and Western United States	Straight-grained, coarse and soft, corrodes iron	Flush doors, weatherboarding and ceiling finishes
Elm	hardwood	560	English and Continental varieties	Twisty grain and warps unless carefully seasoned, tough and durable under wet conditions	Coffins, weatherboarding, piling, furniture and panelling
Fir (Douglas)	softwood	530	Pacific coast of United States and Canada	Straight-grained with pronounced figure, readily worked and stains well	Carpentry timbers, doors, flooring and panelling
Greenheart	hardwood	990 to 1090	Guyana	Very hard, heavy and strong, difficult to work, resists fungus attack and marine borers	Heavy construction, flooring, piling, dock and harbour work
Jarrah	hardwood	900	Western Australia	Hard, heavy, coarse timber, generally straight-grained, dull red in colour and resists decay and fire	Heavy construction, flooring and stair treads
Mahogany (African)	hardwood	560	West Africa, Nigeria and Ghana	Strong and durable; not very easy to work, cheapest variety of mahogany	Veneers, joinery, and interior decorative work
Mahogany (Honduras)	hardwood	540	Central America	Dark red with good figuring, strong but works well	High-class joinery, veneers and panelling

Seasoning of Timber

In 'green' timber large quantities of free water are present in the cell cavities and the cell walls are also saturated. Seasoning consists of drying out the free water and some of the water from the cell walls, which on withdrawal causes the timber to shrink, with the object of reducing the moisture content to a level consistent with the humidity of the air in which the timber will be placed. There are two principal methods of seasoning timber

Air seasoning. This is where the 'green' timber is stacked with laths or 'stickers' between the timbers to allow the passage of air and assist in the evaporation of moisture from the timber. A suitable roof is needed to protect the timber from sun and rain. Air seasoning is unlikely to reduce the moisture content below 17 per cent even under ideal conditions and may take up to two years. Timber used internally in centrally heated buildings should not have a moisture content exceeding ten

per cent to prevent shrinkage and warping.

Kiln seasoning. This is normally carried out in a forced draught compartment kiln, in which the air is heated by steam pipes and humidified by water sprays or steam jets. The temperature, degree of humidity and rate of air flow are all controlled from outside the kiln.

Recommended moisture contents, at time of building in, range from 18 per cent for carpentry timbers to 9 per cent for flooring over embedded heating.[44]

Defects in Timber

The strength and usefulness of timber can be affected by a wide variety of defects, some of which can occur during natural growth, others during seasoning or manufacture, while others result from attack by fungi or insects. The principal defects are listed and briefly described, with some definitions extracted from BS 565.[20]

Table 6.1 (cont.)

Variety	Hardwood or softwood	Average weight (kg/m³)	Origin	Main properties	Principal uses
Oak (European)	hardwood	740 to 770	European countries	Strong and durable with colour ranging from light to dark brown and has pronounced silver grain when quarter sawn	Furniture, internal and external joinery, gates and posts
Pitch pine	softwood	660	United States	Very hard, tough and resinous with many hard-wood characteristics, works and finishes well	Heavy construction, piling and flooring
Pine (Scots)	softwood	530	Europe and W. and N. Asia	Straight-grained, resinous and strong, may be excessively knotty, finishes well, yellow in colour	Carpentry timbers, flooring and joinery
Sycamore	hardwood	610	Great Britain and temperate Europe	Straight-grained, fine texture, good figure but not very durable, white in colour	Joinery, cabinet making and ladders
Teak	hardwood	640 to 720	Burma, India and Thailand	Very strong, durable and resistant to decay, moisture and fire, brown in colour	Garden furniture, doors and windows, flooring and panelling
Walnut (European)	hardwood	610 to 770	Europe, Turkey, Iran, N. India and China	Good figure, hard, easy to work and polishes well	Veneers, cabinet work and panelling
Whitewood (European)	softwood	430	Northern Europe	White to light yellow, not very strong or durable, works easily but obstructed by knots, finishes well	Interior joinery, flooring and scaffold poles

A. Defects arising from natural causes

(1) *Knots.* These are portions of branches enclosed in the wood by the natural growth of the tree; they affect the strength of the timber as they cause a deviation of the grain and may leave a hole. There are several types of knot. A *sound* knot is one free from decay, solid across its face and at least as hard as the surrounding wood. A *dead* knot has its fibres interwoven with those of the surrounding wood to an extent of less than one-quarter of the cross-sectional perimeter; a *loose* knot is a dead knot not held firmly in place. *Rind gall* is a surface wound that has been enclosed by the growth of the tree.

(2) *Shakes.* These consist of a separation of fibres along the grain due to stresses developing in the standing tree, or during felling or seasoning. A *cross* shake occurs in cross-grained timber following the grain; a *heart* shake is a radial shake originating at the heart; a *ring* or *cup* shake follows a growth ring; and a *star* shake consists of a number of heart shakes resembling a star.

(3) *Bark pocket.* Bark in a pocket associated with a knot which has been partially or wholly enclosed by the growth of the tree (inbark or ingrown bark).

(4) *Deadwood.* Timber produced from dead standing trees.

(5) *Resin pocket.* Cavities in timber containing liquid resin (pitch pocket or gum pocket).

B. Defects due mainly to seasoning

(1) *Check.* A separation of fibres along the grain forming a crack or fissure in the timber, not extending through the piece from one surface to another.

(2) *Ribbing.* A more or less regular corrugation of the surface of the timber caused by differential shrinkage of spring wood and summer wood (crimping).

(3) *Split.* A separation of fibres along the grain forming a crack or fissure that extends through the piece from one surface to another.

(4) *Warp.* A distortion in converted timber causing departure from its original plane. *Cupping* is a curvature occurring in the cross-section of a piece and a *bow* is a curvature of a piece of timber in the direction of its length.

C. Defects due to manufacture (including conversion)

(1) *Chipped grain.* The breaking away of the wood below the finished surface by the action of a cutter or other tool.

(2) *Imperfect manufacture.* Any defect, blemish or imperfection incidental to the conversion or machining of timber, such as a variation in sawing, torn grain, chipped grain or cutter marks.

(3) *Torn grain.* Tearing of the wood below the finished surface by the action of a cutter or other tool.

(4) *Waney edge.* This is the original rounded surface of a tree remaining on a piece of converted timber.

D. Fungal attack

(1) *Fungal decay.* This is decomposition of timber caused by fungi and other micro-organisms, resulting in softening, progressive loss of strength and weight and often a change of texture or colour. Fungi are living plants and require food supply, moisture, oxygen and a suitable temperature. A fungus is made up of cells called *hyphae* and a mass of hyphae is termed *mycelium*. It also contains fruit bodies within which very fine spores are formed. When timber is infected by spores being blown on to it, hyphae are then formed which penetrate the timber and break down the wood as food by means of *enzymes*. The Building Research Establishment[39] distinguishes two main forms of decay in wood according to the colour of the decayed timber (brown rots and white rots).

(2) *Dote.* The early stages of decay characterised by bleached or discoloured streaks or patches in wood, the general texture remaining more or less unchanged. This defect is also known as doaty, dosy, dozy and foxy.

(3) *Dry rot.* This is a serious form of timber decay caused by a fungus, *Merulius lacrymans.* The fungus develops from rust-coloured spores, which can be carried by wind, animals or insects, and throw out minute hollow white silky threads (hyphae). The fungus also produces grey-coloured strands up to 5 mm diameter which can travel considerable distances and penetrate brick walls through mortar joints. The strands throw off hyphae whenever they meet timber and carry the water supply for digestion of the timber, which becomes friable, powdery and dull brown in colour, accompanied by a distinctive mushroom-like smell.[40]

The fungus needs a source of damp timber with a moisture content above 20 per cent, oxygen and a temperature between freezing point and blood heat. To eradicate an outbreak of dry rot, all affected timber and timber for 300 to 400 mm beyond must be cut away and burnt on the site. Surrounding masonry must be sterilised, preferably with a suitable fungicide, such as sodium orthophenylphenate or sodium pentachlorophenate. Ends of sound timber and all replacement timber must be treated with a suitable preservative. The cause of the outbreak must be established and rectified, usually by preventing damp penetration and improving ventilation. The most vulnerable locations are cellars, inadequately ventilated floors, ends of timbers built into walls, backs of joinery fixed to walls and beneath sanitary appliances. The design of timber floors to prevent dry rot is covered in BRE Digest 18.[11]

(4) *Coniophora cerebella (cellar fungus).* This is a fungus which attacks wet timber and is mainly found in badly ventilated basements and bathrooms. The fruit bodies are in sheets, yellow to brown in colour.

(5) *Wet rot.* Originally this was defined as chemical decomposition generally arising from alternate wet and dry conditions, as occur with timber fencing posts at ground level. The decay may be accelerated by fungus attack. More recently[39] the term has been applied to fungus attacks other than dry rot, for instance Coniophora cerebella, Porio vaillantii (white strands) and Paxillus panuoides (pale yellow strands and reddish-brown wood).

E. Insect attack

The life cycle of an insect is in four stages: the egg, the larva, the pupa and the adult. The egg is laid on the surface of the timber in a crack or crevice and the larva when hatched bores into the wood which provides its food. The larva makes a special chamber near the surface and then changes to a pupa. Finally it develops into the adult insect and bores its way out through the surface. The principal wood-boring insects are as follows

(1) *Common furniture beetle or wood worm.* This is about 2½ to 5 mm long and dark brown in colour. Eggs are laid during June to August and larvae subsequently bore into the timber for a year or two, before the beetle emerges, chiefly in June and July leaving an exit hole about 1.6 mm diameter. They prefer old timber, both hardwood and softwood; plywood made from birch or alder and bonded with animal glues is particularly susceptible.[21]

(2) *Death-watch beetle* is about 6 to 9 mm long, dark brown in colour and mainly attacks hardwood, especially in large structural timbers (3 mm diameter holes).

(3) *Lyctus powder-post beetle* is about 5 mm long and reddish-brown in colour. Sapwood of new hardwood is particularly vulnerable (exit holes about 1.6 mm diameter).

(4) *House longhorn beetle* is about 15 mm long and is generally brown or black in colour. It is a most destructive pest on the Continent but, except in some parts of Surrey, is rare in this country. This beetle favours sapwood of new softwood, especially in roof spaces. Exit holes may vary from 6 to 10 mm in diameter.

There are many proprietary preparations available for treating infested timber, most of which contain chemicals such as chlorinated naphthalenes, metallic naphthenates and pentrachlorophenol. Insecticides should be brushed or sprayed over the surface during spring and early summer and injected into exit holes. The treatment should be repeated at least once each year during the summer months when beetles are active until there is no sign of continued activity.

Preservation of Timber

Few timbers are resistant to decay or insect attack for long periods of time, and in many cases the length of life can be much increased by preservative treatment.[43] The need for preservative treatment is largely dependent on the severity of the service environment.[44] The principal protective liquids are toxic oils, such as coal tar creosote; water-borne inorganic salts such as copper/chrome and copper/chrome/arsenic,[6] which are suitable for exterior and interior use; and organic solvent solutions, such as copper and zinc naphthenate,[5] which are also suitable for exterior and interior use.[42]

Preservatives can be applied by non-pressure methods such as brush application, spraying, cold submersion, hot submersion or hot and cold steeping.[41] In the latter process the timber is placed in a tank of preservative, heated to 85 to 95°C expelling air from the cells or pores of the timber; on cooling, preservative is drawn into the pores. For lasting preservation, a pressure method should be used. In the 'full-cell' process, the timber is placed in a closed cylinder and a partial vacuum applied to draw out air from the cells, hot preservative admitted, air pressure applied for one to six hours and a partial vacuum reapplied to remove excess liquid. The 'empty-cell' process is cleaner and more economical and a deeper penetration can be obtained with only limited excess preservative. The timber is subjected to air pressure, the preservative admitted and a higher pressure applied causing the liquid to penetrate the timber and compress air in the cells. When the timber is extracted, the air trapped in the cells forces out excess liquid leaving the cells empty but impregnating the cell walls.[42]

Jointing Timbers

Timber joints can be classified in a number of ways. One method would be to distinguish between carpentry, joinery and general-purpose joints, although some overlapping is bound to occur. BS 1186[22] distinguishes between fixed joints and joints permitting movement in joinery work. Another approach would be to classify the joints according to function: *width* joints for joining up boards such as tongued and grooved; *angle* joints for forming angles; *framing* joints such as mortice and tenon; *lengthening* joints such as scarfed and lapped joints; and *shutting* and *hanging* joints used in hanging doors and windows.

The simplest form of joint is the *butt* joint when adjoining members merely butt one another as with floorboards on occasions. Other methods of connecting floorboards are tongued and grooved (figure 6.1.6), splay rebated tongued and grooved (figure 6.1.6), and ploughed and tongued often incorporating a strip of three-ply as a continuous tongue.

There is a wide range of framing joints from the very complicated tusk tenon joint (figure 6.3.7) sometimes used in trimming work, to the much more widely used mortise and tenon (figure 6.4.12), in which one member is reduced to about one-third of its thickness (the tenon) which is wedged into a recess in the other member (the mortise). Where the penetrating member is not reduced in thickness it is termed a housed joint (figure 6.3.8). A stronger joint is obtained with a dovetail joint (figure 6.3.9). Where one member crosses over another the intersecting joint may be notched (figure 6.4.10) or cogged (figure 6.4.11). The jointing of wall plates in their length or at angles is normally undertaken with halved joints (figure 6.4.7), where each member is halved in depth for the length of the overlap. A stronger joint can be obtained by dovetail-halved (figure 6.4.8) or bevel-halved (figure 6.4.9) joints. Lengthening and birdsmouth joints are described in chapter 7 and other joinery joints in chapters 8, 9 and 10.

FLOOR FINISHINGS

Many factors deserve consideration when selecting a floor finish but not all the factors are of equal importance. Furthermore, requirements vary in different parts of the building. For instance, resistance to oil, grease and moisture is relevant in a kitchen but not a bedroom, and appearance could be important in a lounge but is of little consequence in a store. Some of the principal matters to be considered are

(1) *Durability*. The material must have a reasonable life to avoid premature replacement with resultant extra cost and inconvenience.

(2) *Resistance to wear*. This includes resistance to indentation where the floor has to withstand heavy furniture, fittings or equipment, and resistance to abrasion in buildings subject to heavy pedestrian traffic and movable equipment.

(3) *Economical*. Reasonable initial and maintenance costs, having regard to the class of building and the particular location within the building.

(4) *Resistance to oil, grease and chemicals*. This is particularly important in domestic kitchens, laboratories and some factories.

(5) *Resistance to moisture*. This is important in domestic kitchens, bathrooms, entrance passages and halls, and in some industrial buildings.

(6) *Ease of cleaning*. This is of increasing importance in many classes of building as the labour-intensive cleaning costs continue to rise at a disproportionate rate.

(7) *Warmth*. Some finishes are much warmer than others and this may be an important consideration.

(8) *Nonslip qualities*. These are particularly important in bathrooms and kitchens where floors may become damp.

(9) *Sound absorption*. Some buildings such as hospitals and libraries need floor finishes with a high degree of sound absorption.

(10) *Appearance*. This is an important consideration in many rooms of domestic buildings, although the current tendency to fully carpet rooms may not justify the provision of the more expensive but attractive floor finishes such as wood blocks and strip flooring.

(11) *Resilience*. Some flexibility or 'give' is often desirable.

Some of the more common floor finishes are now described.

Asphalt. The basic aggregates of mastic asphalt flooring are natural rock asphalt[23] or limestone.[24] Asphalt is available in dark colours, principally red, brown and black. It provides a jointless floor which is dustless and impervious to moisture. It is extremely durable but concentrated loads may cause indentations. A thickness of 15 to 20 mm is suitable for light use, increasing to 25 or even 38 mm for heavy duty. Floors over 20 mm thick may be laid directly on new concrete, but an isolating layer of black sheathing felt to BS 747 should be provided for thinner floors.

Pitch mastic. This should comply with either BS 1450[25] for black pitch mastic or BS 3672[26] for coloured pitch mastic flooring. This material is similar to mastic asphalt and is suitable for a wide range of situations. The thickness normally varies between 19 and 25 mm. It is reasonably quiet, warm, resilient, nonslip, dustless, hardwearing, resistant to moisture and fairly low priced.

Rubber latex cement. This is a mixture of Portland cement or high alumina cement, aggregate, fillers and pigment, gauged on the site with a stabilised aqueous emulsion of rubber latex. It is usually laid to a thickness of 6 mm, resists damp, is nonslip, reasonably hardwearing, quiet and warm and can be laid in a variety of colours.

Granolithic concrete. This is primarily used for factory flooring because of its hardwearing qualities. A typical mix is 1:1:2 of Portland cement, sand and granite or whinstone.[27] The best finish is obtained by laying a 10 to 25 mm granolithic topping within three hours of laying the base, to obtain monolithic construction. Otherwise the topping should be at least 40 mm thick and adequately bonded to the base. After compacting the granolithic should be trowelled to a level surface; two hours later it should be re-trowelled to close surface pores and finally trowelled when the surface is hard. The tendency to produce dust can be reduced by applying sodium silicate or magnesium or zinc silicofluoride or by using a sealer such as polyurethane.[27] A nonslip finish can be obtained by sprinkling carborundum over the surface (1.35 kg/m²) and trowelling it in before the granolithic has set. A minimum of seven days is required for curing. It provides a cold, unattractive finish, but at a moderate price.

Terrazzo. This is a decorative form of concrete usually made from white Portland cement and crushed marble aggregate to a mix of 1:2, and may be *in situ* or precast. *In situ* terrazzo can either be laid with all the aggregate incorporated in the mix, or a mix wherein the fine aggregate is spread in position and the larger aggregate is then sprinkled on the surface and beaten and trowelled in. Both monolithic and separate construction from the base methods are used. The following precautions should be taken to prevent cracking

(1) floor divided into panels not exceeding 1 m² in area, with sides in ratio of 3:1, and separated by dividing strips of metal, ebonite or plastics;

(2) water/cement ratio kept as low as possible;

(3) no smaller aggregate than 3 mm used;

(4) terrazzo allowed to dry out slowly, and if possible the building should not be heated for six to eight weeks after the floor finish has been laid.

The material must be thoroughly compacted by tamping, trowelling and rolling, with further trowelling at intervals to produce a dense, smooth surface. After about four days, the surface is ground by machine and finally polished using a fine abrasive stone.[29] It is attractive in appearance, hardwearing, easily cleaned, but is noisy, cold and expensive.

Magnesium oxychloride or magnesite. This is a mixture of calcined magnesite, fillers such as sawdust, wood flour, ground silica, talc or powdered asbestos, to which a solution of magnesium chloride is added. It needs to be laid by specialists and should comply with BS 776.[28] Magnesite flooring is available in various colours to mottled or grained finishes. It is suitable for industrial, commercial or domestic use, but should be laid on a damp-proof membrane as the material tends to soften and disintegrate when wet. The thickness may vary from 10 mm for single-coat work to 50 mm for two or three-coat work. It is reasonably warm, hardwearing, easily cleaned and of good appearance, moderately priced but noisy.

Composition blocks. These are generally about 150 x 60 x 16 mm and are normally made of a mix of cement, wood, gypsum, calcium carbonate, pigment and linseed oil, pressed and cured for three to five months. The blocks are bedded on a 13 mm bed of cement and sand and jointed in a composition material. They are obtainable in a variety of colours and can be laid in similar patterns to wood-block flooring, are moderately priced, warm and quiet, hardwearing and resistant to oils and greases, and soon wear to a nonslip surface, but need wax polishing.

Cork tiles. These are made from granulated cork, compressed and bonded with synthetic resins. Tiles are made to various sizes but 300 x 300 mm are probably the most popular, in thickness varying from 6 to 14 mm depending on the amount of wear they will receive. They provide a warm, quiet and resilient floor finish at a reasonable cost.

Linoleum. This is manufactured from linseed oil, cork, resin gums, fibres, wood flour and pigments applied to a canvas backing. It is best bonded to the base with a waterproof adhesive and for heavy use should be at least 4.5 mm thick. Linoleum forms a resilient, quiet and comfortable floor, to a wide range of colours and patterns, is easily cleaned and wears well.[30]

Rubber tiles and sheets. These are made of natural rubber and various filling compounds. Tiles are normally 3 mm thick for housing work and 4 mm in public buildings. Rubber flooring must be fixed with a suitable adhesive and is available in a variety of colours and patterns. This flooring is quiet, warm, resilient, nonslip, resistant to moisture and hardwearing. Oils, fats and greases have a harmful effect on rubber flooring and it is expensive.[31] A variety of adhesives for use with flooring materials are listed in BS 5442.[49]

Thermoplastic tiles. These are made from asbestos fibres, mineral fillers, pigments and a thermoplastic binder and can be obtained in a variety of colours and patterns. The most common size is 250 X 250 x 2.5 or 3 mm thick. The tiles become flexible when heated and are fixed with a special adhesive to a cement and sand screed. They are moderately warm, quiet and resistant to water, oil and grease.[32]

Vinyl asbestos tiles. These were a later development of thermoplastic tiles and are made of PVC resin, asbestos fibre, powdered mineral fillers and pigments.[33] They are particularly resistant to grease and hence well suited for use in kitchens and will withstand underfloor heating at moderate temperatures.

Flexible PVC tiles and sheet. These contain a substantial proportion of plasticised PVC resin, which gives added flexibility and resilience. A common size of tile is 300 x 300 x 2 mm thick and some possess realistic imitations of expensive traditional finishes, such as terrazzo and marble. They should comply with BS 3261.[45] BS 5085[50] specifies the requirements for flexible PVC flooring with a wear layer, an underlayer and a backing of needle-loom felt or cellular PVC.

Clay tiles. Two types of tile are specified in BS 1286[34]: floor quarries (12.5 to 32 mm thick) and ceramic floor tiles (9.5 to 19 mm thick). When bedding tiles a separate layer of building paper or polythene sheet should be laid over the base, to permit variations in movement between the tiles and the base.[35] The tiles are soaked in clean water and laid on a bed of cement, lime and sand (1:½:4 or 5) 13 mm thick. Alternatively a thicker bed (40 mm) can be used without a separate layer. Adhesives should be those specified in CP 202.[36] Tiles should be laid on a bitumen bed where subjected to high temperatures and an expansion joint should be provided around the perimeter of tiled floors. The tiles are obtainable in a variety of colours and sizes (75 x 75 mm to 150 x 150 mm and 12.7 mm thick). They are useful for kitchens and toilets, being hardwearing, easily cleaned and resistant to water, oils and greases; but they are also cold, noisy and relatively expensive.

Concrete tiles. These are made from coloured cement and hardwearing aggregate set on a normal concrete backing, varying in size from 150 x 150 mm to 500 x 500 mm in thicknesses ranging from 15 to 40 mm.[46] They are usually laid on a bed of cement mortar on a separating layer, as for clay tiles. They have attractive finishes in a wide range of colours, are hard, durable, dustless, easily cleaned and resistant to damp, but are also cold, noisy and attacked by oils.

Wood blocks. These are made in lengths of 150 to 375 mm, widths of up to 88 mm and in thicknesses of 19 to 38 mm, with 25 mm as the most common thickness.[37] Various interlocking systems are used, the commonest being the tongue and groove and the dowel. The blocks are usually laid on a screeded concrete floor and dipped in a mastic or a cold latex bitumen emulsion adhesive.[38] They are often laid to herringbone or basket-weave patterns with the surface of the blocks sealed and waxed. Wood blocks provide a warm, quiet, resilient, attractive, hardwearing floor, but are expensive.

REFERENCES

1. *BRE Digest 222*: Fill and hardcore. HMSO (1972)
2. *BRE Digest 54*: Damp-proofing solid floors. HMSO (1971)
3. *BRE Digest 174*: Concrete in sulphate-bearing soils and groundwaters. HMSO (1975)
4. The Building Regulations 1976 SI 1676. HMSO (1976)
5. BS 5056: 1974 Copper naphthenate wood preservatives
6. BS 3452: 1962 Copper/chrome waterborne wood preservatives and their application
7. BS 4072: 1974 Wood preservation by means of waterborne copper/chrome/arsenic compositions
8. DOE Advisory leaflet 34: Thermal insulation. HMSO (1975)
9. *BRE Digest 104*: Floor screeds. HMSO (1979)
10. BS 882: Coarse and fine aggregates from natural sources; Part 2: 1973 Metric units
11. *BRE Digest 18*: Design of timber floors to prevent dry rot. HMSO (1973)
12. BS 493: Airbricks and gratings for wall ventilation; Part 2: 1970 Metric units
13. *BRE Digest 103*: Sound insulation of traditional dwellings Part 2. HMSO (1979)
14. *BRE Digest 143*: Sound insulation: basic principles. HMSO (1976)
15. BS 4: Structural steel sections; Part 1: 1972. Hot-rolled sections
16. BS 881, 589: 1974 Nomenclature of commercial timbers, including sources of supply
17. Princes Risborough Laboratory. Handbook of hardwoods. HMSO (1972)
18. Princes Risborough Laboratory. Handbook of softwoods. HMSO (1977)
19. DOE Advisory leaflet 5: Laying floor screeds. HMSO (1975)
20. BS 565: 1972 Glossary of terms relating to timber and woodwork
21. DOE Advisory leaflet 42: Woodworm. HMSO (1975)
22. BS 1186: Quality of timber and workmanship in joinery; Part 1: 1971 Quality of timber, Part 2: 1971 Quality of workmanship
23. BS 1162, 1418 and 1410: 1973 Mastic asphalt for building (natural rock asphalt aggregate)
24. BS 988, 1076, 1097 and 1451: 1973 Mastic asphalt for building (limestone aggregate)
25. BS 1450: 1963 Black pitch mastic flooring
26. BS 3672: 1963 Coloured pitch mastic flooring
27. *BRE Digest 47*: Granolithic concrete, concrete tiles and terrazzo flooring. HMSO (1970)
28. BS 776: Materials for magnesium oxychloride (magnesite) flooring, Part 2: 1972 Metric units
29. CP 204: *In situ* floor finishes, Part 2: 1970 Metric units
30. BS 810: 1966 Sheet linoleum (calendered types), cork carpet and linoleum tiles
31. CP 203: 1969 Sheet and tile flooring (cork, linoleum, plastics and rubber), Part 2: 1972 Metric units
32. BS 2592: 1973 Thermoplastic flooring tiles
33. BS 3260: 1969 PVC (vinyl) asbestos floor tiles
34. BS 1286: 1974 Clay tiles for flooring
35. *BRE Digest 79*: Clay tile flooring. HMSO (1976)
36. CP 202: 1972 Tile flooring and slab flooring
37. BS 1187: 1959 Wood blocks for floors
38. CP 201: Flooring of wood and wood products, Part 2: 1972 Metric units
39. Princes Risborough Laboratory. Decay of timber and its prevention. HMSO (1976)
40. DOE Advisory leaflet 10: Dry rot and wet rot. HMSO (1975)
41. Princes Risborough Laboratory. Methods of applying wood preservatives. HMSO (1974)
42. *BRE Digest 201*: Wood preservatives: application methods. HMSO (1977)
43. BS 5268: Code of practice for the structural use of timber; Part 5: 1976 Preservation treatments for constructional timber
44. *BRE Digest 156*: Specifying timber. HMSO (1973)
45. BS 3261: Unbacked flexible PVC flooring; Part 1: 1973 Homogeneous flooring
46. BS 1197: Concrete flooring tiles and fittings, Part 2: 1973 Metric units
47. *BRE Digest 145*: Heat losses through ground floors. HMSO (1972)

48. *BRE Digest 208*: Increasing the fire resistance of timber floors. HMSO (1977)

49. BS 5442: Classification of adhesives for construction; Part 1: 1977 Adhesives for use with flooring materials

50. BS 5085: Backed flexible PVC flooring; Part 1: 1974 Needle-loom felt backed flooring; Part 2: 1976 Cellular PVC backing

7 ROOFS

This chapter examines the general principles involved in the design of roofs; the various forms of construction and coverings; the techniques used in roof drainage and insulation; and the materials employed.

GENERAL PRINCIPLES OF DESIGN

Roofs have to perform a number of functions and some of the more important are listed and described.

Weather Resistance

A roof which is not weatherproof will not comply with the Building Regulations[1] (C10) and is of little value. The slope of the roof and lap of the roof coverings must be considered, as well as the degree of exposure. For example, plain tiles are not suitable for use on slopes of less than 40°.

Strength

The roof structure or framework must be of adequate strength to carry its own weight together with the superimposed loads of snow, wind and foot traffic. Code of Practice 3[2] recommends the imposed loadings given in table 7.1.

Table 7.1   Imposed roof loadings

Roof type	Loadings (kN/m^2 of plan area)
Flat roofs (up to 10°) with access for maintenance only, with load of 0.9 kN on 300 mm square	0.75
Flat roofs with access for purposes other than maintenance, with load of 1.8 kN on 300 mm square	1.50
Pitched roof (30° or less)	0.75
Pitched roof (75° or more)	nil
(loadings are interpolated for pitches between 30° and 75°)	

Wind pressure requires special consideration where a light roof covering is laid to a low slope. The greatest suction occurs at slopes below 15°, and this diminishes to around zero at 30°, with maximum suction at the edges of roofs. In conditions of high wind pressure the uplift can exceed the dead weight of the covering and so adequate fixing is necessary to prevent stripping of the coverings.

Durability

The coverings should be able to withstand atmospheric pollution, frost and other harmful conditions. With large concrete roofs and sheet-metal coverings provision must be made to accommodate thermal expansion. There should also be effective means for the speedy removal of rainwater from the roof, which might otherwise cause deterioration of the roof covering.

Fire Resistance

Roofs of certain buildings, including a house in a continuous terrace of more than two houses, shall be covered with a material designated in the Building Regulations.[1] Although in some cases, such as thatch or wood shingles, it has to be a prescribed minimum distance from the boundary, to prevent the possibility of spread of fire to an adjacent building.

Insulation

Thermal insulation of roofs is necessary to reduce heat losses to an acceptable level and to prevent excessive solar heat gains in hot weather, thus ensuring a reasonable standard of comfort within the building. The Building Regulations,[1] clause F3, requires that the thermal coefficient of the roof (including a ceiling), is not more than 0.6 W/m^2°C. Sound insulation on the other hand is rarely an important consideration in roof design.

Appearance

Roof design can have an important influence on the appearance of a building, both in regard to the form and shape of the roof, and as to the colour and texture of the covering material. Plain tiles are often better suited in scale to small domestic buildings than the larger single lap tiles, whilst clay tiles generally hold their colours better than concrete tiles. Improved appearance often results from the use of roof coverings which are of darker colour than the wall cladding.

CHOICE OF ROOF TYPES

There is a wide range of roof types available and the choice may be influenced by a number of factors. Roofs vary from flat roofs, not exceeding 10° slope, of timber or concrete, covered with one of a number of materials, to pitched roofs of various structural forms which are usually covered with slates or tiles. Factors influencing the choice of roof type follow

Size and shape of buildings. Buildings of simple shape are readily covered with pitched roofs, whereas flat roofs are better suited for irregularly shaped buildings. The clear spans required will also influence the choice of roof.

Appearance. Aesthetic considerations might well dictate a pitched roof for a small building and a flat roof for a large building.

Economics. Both capital and maintenance costs should be considered in selecting a roof type. Some coverings such as zinc will not last the life of the building and replacement costs must be included in the calculations.

Other considerations. Ease with which services can be accommodated in roof space; weatherproofing; condensation; maintenance and similar matters.

PITCHED ROOF CONSTRUCTION

Some of the more common roofing terms are explained by reference to figure 7.1.1. *Hipped* end is where the roof slope is continued around the end of a building, whereas the wall is carried up to the underside of the roof at a *gable* end. *Hip rafters* frame the external angles at the intersection of roof slopes, while *valley rafters* are used at internal angles. The shortened rafters running from hip rafters to plate and from ridge to valley rafters are termed *jack rafters*, while full-length rafters are often called *common rafters*. The bottom portion of the roof overhanging the wall is known as the *eaves*. Where the roof covering overhangs the gable end, it is termed the *verge*. *Purlins* are horizontal roof members which give inter-mediate support to rafters. Rafters are splay cut or bevelled and nailed to the ridge board at the upper end and birdsmouthed and nailed to the wall plate at the lower end (figure 7.1.9). Where purlins or rafters need to be lengthened, a scarfed joint should be used (figure 7.1.10).

The slope of a roof is usually given in degrees, whereas the pitch is the ratio of rise to span. The rise is the vertical distance between the ridge and the wall plate, while the span is the clear distance between walls. In a half pitch or 'square pitched' roof, the span is twice the rise, for example 3.5 m rise with a 7 m span. The minimum pitch or slope is determined by the roof-covering material, as described later in the chapter.

Pitched roofs can be broadly classified into three main categories as follows.

Single Roofs

This is where rafters are supported at the ends only. The simplest form is the *lean-to* or *pent* roof (figure 7.1.2) where one wall is carried up to a higher level than the other and the rafters bridge the space between. It is especially suitable for outbuildings and domestic garages with a span not exceeding 2.5 m. The upper ends of rafters in lean-to roofs are normally supported by a ridge board plugged to the wall or a plate resting on corbel brackets (figure 7.1.3).

Couple roofs (figure 7.1.4). These consist of rafters supported by a ridge board and wall plates. In the absence of a tie, the rafters exert an outward thrust on the walls, and this type of roof should not be used for spans exceeding 3 m. An improvement of this design is to nail ceiling joists to each pair of rafters when the roof is termed a *couple close* roof (figure 7.1.5), and the span may be increased to 4 m.

Collar roofs (figure 7.1.6). These may be used for spans up to 5 m. Each pair of rafters is framed up with a horizontal collar or collar tie, which should not be placed higher than one-third of the vertical height from wall plate to ridge. Collars should preferably be jointed to rafters with a suitable timber connector to give increased strength. This form of roof is valuable when additional headroom is needed.

The sizes of rafters, ceiling joists and purlins are dependent on the dead load to be carried, and the span and spacing of the members. Suitable sizes can be obtained from schedule 6 of the Building Regulations.[1]

Double Roofs

When the span exceeds 5 m, single roofs are no longer suitable as their use would entail excessively large and uneconomical timbers. A typical double or purlin roof is shown in figure 7.1.7, in which purlins are used at about 3 m

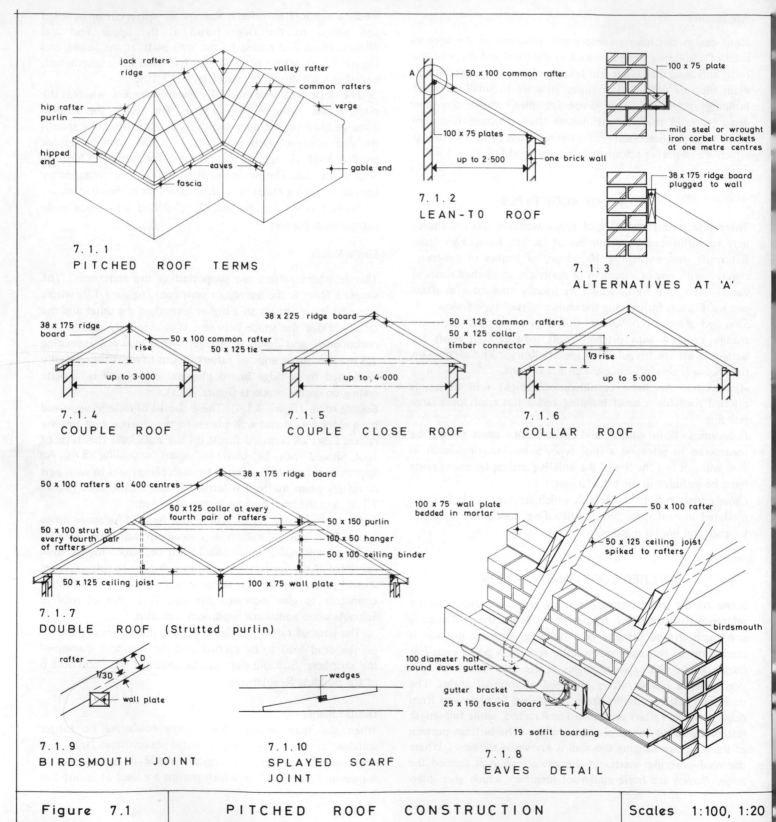

7.1.1
PITCHED ROOF TERMS

jack rafters
ridge
valley rafter
common rafters
hip rafter
purlin
verge
hipped end
eaves
gable end
fascia

7.1.2
LEAN-TO ROOF

50 x 100 common rafter
A
100 x 75 plates
up to 2·500
one brick wall

7.1.3
ALTERNATIVES AT 'A'

100 x 75 plate
mild steel or wrought iron corbel brackets at one metre centres
38 x 175 ridge board plugged to wall

7.1.4
COUPLE ROOF

38 x 175 ridge board
50 x 100 common rafter
rise
up to 3·000

7.1.5
COUPLE CLOSE ROOF

38 x 225 ridge board
50 x 125 tie
up to 4·000

7.1.6
COLLAR ROOF

50 x 125 common rafters
50 x 125 collar
timber connector
1/3 rise
up to 5·000

7.1.7
DOUBLE ROOF (Strutted purlin)

38 x 175 ridge board
50 x 100 rafters at 400 centres
50 x 125 collar at every fourth pair of rafters
50 x 150 purlin
50 x 100 strut at every fourth pair of rafters
100 x 50 hanger
50 x 100 ceiling binder
50 x 125 ceiling joist
100 x 75 wall plate

7.1.8
EAVES DETAIL

100 x 75 wall plate bedded in mortar
50 x 100 rafter
50 x 125 ceiling joist spiked to rafters
birdsmouth
100 diameter half round eaves gutter
gutter bracket
25 x 150 fascia board
19 soffit boarding

7.1.9
BIRDSMOUTH JOINT

rafter
D
1/3D
wall plate

7.1.10
SPLAYED SCARF JOINT

wedges

| Figure 7.1 | PITCHED ROOF CONSTRUCTION | Scales 1:100, 1:20 |

centres to give intermediate support to rafters. The purlins are in turn supported by struts which bear on loadbearing partitions or ceiling beams, or by stepped internal partitions. To further strengthen and stiffen the roof, collars may be provided at each fourth pair of rafters. For larger spans, the introduction of hangers at every fourth rafter with connections to the purlin and ceiling binder, which is in turn nailed to the ceiling joists, provides additional strengthening of the roof.

Figure 7.1.8 illustrates a typical form of construction at the eaves. The feet of the rafters are splay cut to receive the fascia and soffit boarding. The soffit boarding may be tongued to the fascia and both soffit and fascia are nailed to the rafters.

Trussed or Framed Roofs

These are used over large spans (in excess of 7.5 m), where further support to purlins is needed in the form of roof trusses.

King and queen post timber trusses

The original type of timber roof truss consisted of large timber members bolted and strapped together at joints. Although these have been superseded by TRADA (Timber Research and Development Association) timber trusses and steel strusses, the student should be aware of the earlier trusses and the principles on which they were based. The king post truss (figure 7.3.1) was used for spans of 6 to 9 m, whilst the queen post truss (figure 7.3.2) was used for spans of from 9 to 12.75 m. The mansard truss (figure 7.3.3) incorporated both types of truss and enabled accommodation to be provided in the roof space. The members of roof trusses were arranged so that their centre lines intersected at joints.

The king post truss (figure 7.3.1) consisted of a triangular frame which supported the ridge and purlin; these spanned from truss to truss and carried the common rafters and roof covering. The roof load was primarily transmitted through principal rafters to the walls below, and the feet of the principal rafters were strapped and bolted to a tie beam to prevent them spreading under load. A strut was introduced under the midpoint of the principal rafter to prevent it sagging, and the central post or king post prevented the tie beam from sagging and effectively stiffened and strengthened the truss.

Figure 7.3.4 shows the detailed arrangements for a queen post truss which is similar to a king post but has two vertical posts (queen posts), which are strutted apart at their heads by a straining beam. Principal rafters are supported by two purlins and the feet of queen posts are held in position by tenons, straps and a straining sill.

TRADA roof trusses

The Timber Research and Development Association has designed standard timber roof trusses which are strong and effective and dispense with the need for support from internal loadbearing partitions, allowing greater flexibility in internal planning. Figure 7.2.1 shows a TRADA *timber truss* with a slope of $30°$ and a span of 11 m, capable of supporting slates or tiles weighing not more than 73 kg/m^2 on slope, together with a superimposed load including snow of 73.2 kg/m^2 measured on plan, and a ceiling load of 48.8 kg/m^2. Trusses will be spaced at 1.8 m centres with common rafters at 450 mm centres. Timber is to be S2–50 grade to CP 112.[3] The trusses can be made up in sections for transporting and then be assembled on site.

Joints between truss members are mainly formed with timber connectors complying with BS 1579,[4] as they increase considerably the strength of a bolted joint. The connectors must be adequately sheradised or galvanised to give protection against corrosion. The most common form of connector is the *toothed plate* (figure 7.2.2) consisting of round or possibly square metal plates with projecting teeth around the edge. Another type of connector is the *split ring* (figure 7.2.3) comprising a steel ring, 63 to 100 mm diameter, cut at one point, and the split so formed permits the ring to adapt itself to any movement of the timber and thus maintain a tight connection. Split rings possess a higher bearing resistance than toothed plates. In both cases the connectors are fitted into precut grooves formed with a special tool. Joints can also be formed with *truss plates* (figures 7.2.5 and 7.2.6). One variety has projecting teeth, while another contains predrilled holes to take 32 mm sheradised nails. The plates are generally made from galvanised mild steel sheet about 1 to 2 mm thick and are fitted to each side of the joint (see figure 7.2.6).[12]

TRADA trussed rafters have been designed for the shallower slopes ($10°$, $15°$, $20°$ and $25°$), serving spans of 5.1, 6.0, 6.9 and 8.1 m. Figure 7.2.4 shows a truss of this type suitable for a 6 m span and a lightweight roof covering, such as woodwool slabs and felt. This form of roof covering is supported directly on the trusses fabricated prior to erection spaced at 600 mm centres, without the need for purlins, common rafters and ceiling joists. The trusses at slopes of $20°$ and $25°$ can carry tiles or slates, but if the load exceeds 50 kg/m^2 on slope the trusses must be spaced closer together. These have largely superseded the TRADA roof trusses. Permissible spans for trussed rafters are detailed in BRE Digest 147,[30] and they do require adequate lateral bracing.

A TRADA trussed rafter suitable for a *monopitch* roof with a slope of $10°$ and a span of 6 m is illustrated in figure 7.2.7. It is designed to carry a roof covering of the type previously described (woodwool slabs and felt). All the designs are based on the use of S2–50 grade timber complying with CP 112.[3]

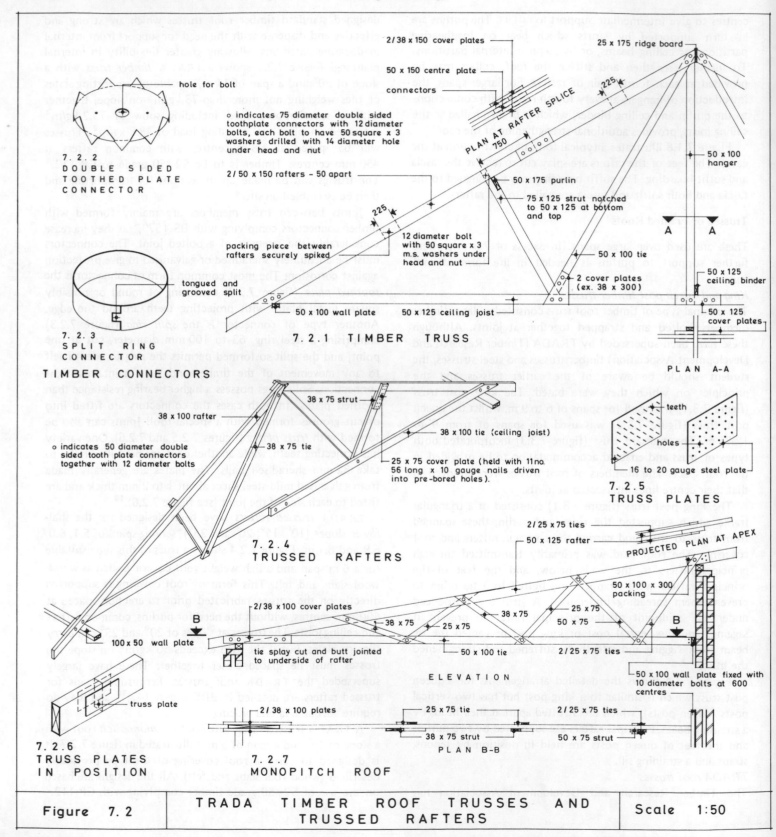

hole for bolt

7.2.2
D O U B L E   S I D E D
T O O T H E D   P L A T E
C O N N E C T O R

tongued and
grooved split

7.2.3
S P L I T   R I N G
C O N N E C T O R

TIMBER CONNECTORS

o indicates 50 diameter double
sided tooth plate connectors
together with 12 diameter bolts

7.2.4
T R U S S E D   R A F T E R S

100 x 50 wall plate

truss plate

7.2.6
TRUSS PLATES
IN POSITION

2/38 x 150 cover plates

50 x 150 centre plate

connectors

o indicates 75 diameter double sided
toothplate connectors with 12 diameter
bolts, each bolt to have 50 square x 3
washers drilled with 14 diameter hole
under head and nut

2/50 x 150 rafters – 50 apart

225

packing piece between
rafters securely spiked

12 diameter bolt
with 50 square x 3
m.s. washers under
head and nut

PLAN AT RAFTER SPLICE

750

50 x 175 purlin

75 x 125 strut notched
to 50 x 125 at bottom
and top

50 x 100 tie

2 cover plates
(ex. 38 x 300)

50 x 100 wall plate   50 x 125 ceiling joist

7.2.1 TIMBER TRUSS

38 x 75 strut

38 x 100 rafter

38 x 100 tie (ceiling joist)

38 x 75 tie

25 x 75 cover plate (held with 11no.
56 long x 10 gauge nails driven
into pre-bored holes).

2/38 x 100 cover plates

38 x 75   25 x 75

tie splay cut and butt jointed
to underside of rafter

2/38 x 100 plates

7.2.7
MONOPITCH ROOF

25 x 175 ridge board

225

50 x 100
hanger

50 x 125
ceiling binder

50 x 125
cover plates

PLAN A-A

teeth

holes

16 to 20 gauge steel plate

7.2.5
TRUSS PLATES

PROJECTED PLAN AT APEX

50 x 125 rafter

50 x 100 x 300
packing

25 x 75

50 x 75

50 x 100 tie   2/25 x 75 ties

2/25 x 75 ties

38 x 75 strut   50 x 75 strut

ELEVATION

50 x 100 wall plate fixed with
10 diameter bolts at 600
centres

PLAN B-B

| Figure 7.2 | TRADA   TIMBER   ROOF   TRUSSES   AND   TRUSSED   RAFTERS | Scale   1:50 |

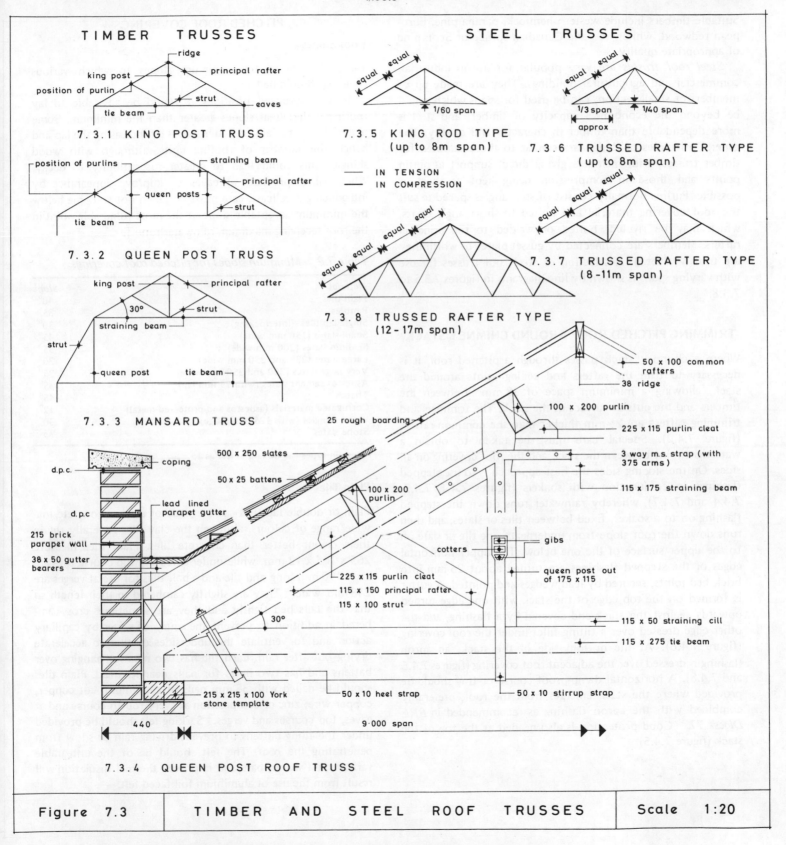

TIMBER TRUSSES

7.3.1 KING POST TRUSS

ridge
king post
principal rafter
position of purlin
strut
eaves
tie beam

7.3.2 QUEEN POST TRUSS

position of purlins
straining beam
principal rafter
queen posts
strut
tie beam

7.3.3 MANSARD TRUSS

king post
principal rafter
30°
strut
straining beam
strut
queen post
tie beam

STEEL TRUSSES

7.3.5 KING ROD TYPE
(up to 8m span)

equal   equal
1/60 span

IN TENSION
IN COMPRESSION

7.3.6 TRUSSED RAFTER TYPE
(up to 8m span)

equal   equal
1/3 span   1/40 span
approx

7.3.7 TRUSSED RAFTER TYPE
(8-11m span)

equal   equal   equal

7.3.8 TRUSSED RAFTER TYPE
(12-17m span)

equal   equal   equal

7.3.4 QUEEN POST ROOF TRUSS

coping
d.p.c.
d.p.c
215 brick parapet wall
38 x 50 gutter bearers
500 x 250 slates
50 x 25 battens
lead lined parapet gutter
100 x 200 purlin
225 x 115 purlin cleat
115 x 150 principal rafter
115 x 100 strut
30°
25 rough boarding
50 x 10 heel strap
215 x 215 x 100 York stone template
440

50 x 100 common rafters
38 ridge
100 x 200 purlin
225 x 115 purlin cleat
3 way m.s. strap (with 375 arms)
115 x 175 straining beam
gibs
cotters
queen post out of 175 x 115
115 x 50 straining cill
115 x 275 tie beam
50 x 10 stirrup strap
9·000 span

| Figure 7.3 | TIMBER AND STEEL ROOF TRUSSES | Scale 1:20 |

Suitable timbers include western hemlock, parana pine, European redwood, whitewood and Canadian spruce or Scots pine of appropriate quality.

Steel roof trusses are very popular for use in industrial, commercial and agricultural buildings. They are made up of members of small section, can be used for spans which would be beyond the economical capacity of timber, and steel is more dependable than timber in character and quality. The general principles of design are similar to those adopted for timber trusses, with members giving direct support at purlin points and those in compression being kept as short as possible. Purlins normally consist of steel angles spaced to suit the roof covering material and bolted to short angle cleats, which may be riveted, bolted or welded to the principal rafters. Members are connected by gusset plates to which they are usually riveted. A selection of steel roof trusses for use with varying spans is shown by line diagrams in figures 7.3.5 to 7.3.8.

TRIMMING PITCHED ROOF AROUND CHIMNEY STACK

Where a chimney stack passes through a pitched roof, it is necessary to trim the rafters and ceiling joints around the stack, allowing a minimum space of 38 mm[1] between the timbers and the outside surface of the stack. The trimmers and trimming rafters are 25 mm thicker than the common rafters (figure 7.4.2). Special care must be taken to obtain a watertight joint between the stack and the roof covering on all sides. On the sloping sides the best approach is to use stepped flashings in conjunction with soakers (figures 7.4.1, 7.4.3, 7.4.4 and 7.4.7), whereby rainwater runs down the stepped flashing on to a soaker, fixed between tiles or slates, and then runs down the roof slope from underneath one tile or slate on to the upper surface of the one below. The upper horizontal edges of the stepped flashings are turned about 25 mm into brick bed joints, secured by lead wedges and pointed. A gutter is formed on the top edge of the stack with one edge turned upwards against the stack and covered by a flashing, and the other edge dressed over a tilting fillet under the roof covering (figure 7.4.6). At the bottom side of the stack, an apron flashing is dressed over the adjacent roof covering (figures 7.4.6 and 7.4.8). A horizontal damp-proof course or tray should be provided where the stack emerges from the roof, preferably combined with the apron flashing as recommended in *BRE Digest 77*.[5] Good protection is also needed at the top of the stack (figure 7.4.5).

PITCHED ROOF COVERINGS

Roof Slopes

Table 7.2 shows the minimum slopes to which various coverings should be laid.

In very exposed positions it would be advisable to lay roofing materials at slopes greater than the minimum. Some materials can be laid to flatter slopes by increasing the lap and hence the number of roofing units, although with wood shingles this could lead to greater susceptibility to decay. When using sprockets at eaves to improve appearance by introducing a bellcast, care must be taken not to come below the minimum acceptable slope at the most vulnerable point in the roof receiving maximum rainwater runoff.

Table 7.2   Minimum slopes for pitched roof coverings

Material	Slope
Plain tiles	40°
Pantiles	35°
Single lap tiles (interlocking)*	30°
Small slates (150 mm wide)	45°
Medium slates (200 mm wide)	35°
Large slates (225 and 250 mm wide)	30°
Very large slates (300 and 350 mm wide)	25°
Asbestos cement slates (with 75 mm lap)	35°
Thatch	45°
Corrugated materials (asbestos and protected metal)	12°
Wood shingles (with 125 mm gauge)	30°
Stone slates	33°

*Certain types can be laid as flat as 15°

Plain Tiles

Plain or double lap tiles can be made of clay or concrete in a wide range of colours, although the clay tiles generally retain their colour better than concrete tiles. Plain tiles measure 265 x 165 x 12 mm, while under-eaves and top-course tiles are each 190 mm long and tile-and-a-half tiles for use at verges are 248 mm wide.[6] They are slightly cambered in their length so that the tails bed tightly and they are sometimes cross-cambered in addition to prevent the entry of water by capillary action and to ventilate the undersides of tiles to accelerate drying out after rain. Each tile has two nibs for hanging over battens and has two holes for nails near its head. Plain tiles should be nailed with 38 mm nails of aluminium, cut copper, copper wire, zinc or composition at every fourth course and at eaves, top courses and verges.[7] Sarking felt should be provided under the tiling battens to prevent driving rain or snow from penetrating the roof. The felt should be of the untearable variety on a woven base and improved thermal insulation will result from the use of aluminium foil faced felt.

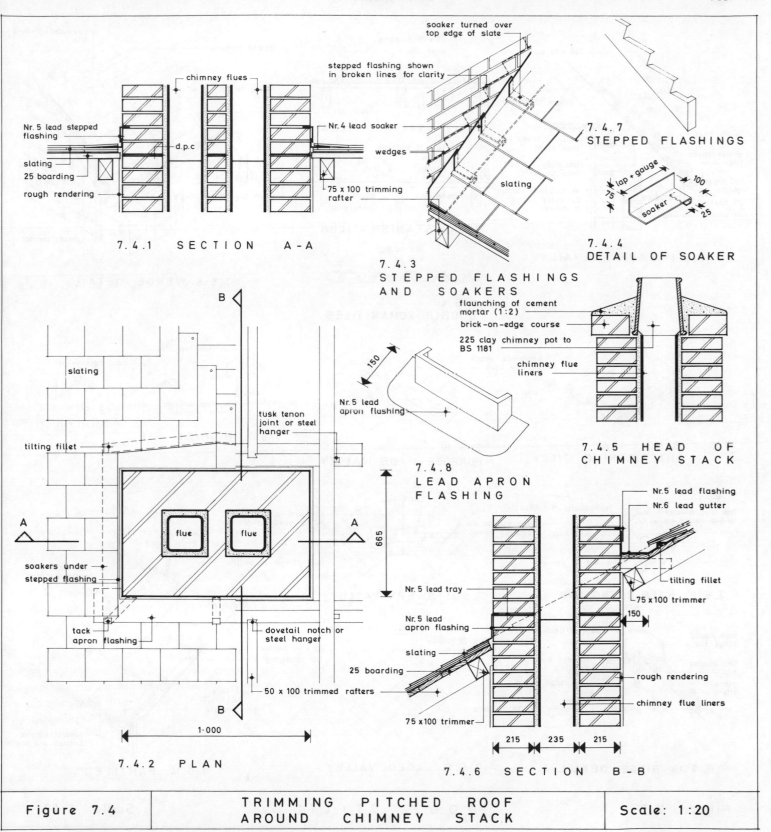

soaker turned over
top edge of slate

stepped flashing shown
in broken lines for clarity

Nr. 5 lead stepped
flashing

Nr. 4 lead soaker

slating

25 boarding

rough rendering

chimney flues

d.p.c

wedges

75 x 100 trimming
rafter

slating

7.4.1 SECTION A-A

**7.4.3
STEPPED FLASHINGS
AND SOAKERS**

**7.4.7
STEPPED FLASHINGS**

lap + gauge

100

75

soaker

25

**7.4.4
DETAIL OF SOAKER**

flaunching of cement
mortar (1:2)

brick-on-edge course

225 clay chimney pot to
BS 1181

chimney flue
liners

**7.4.5 HEAD OF
CHIMNEY STACK**

B

slating

tilting fillet

tusk tenon
joint or steel
hanger

A

flue

flue

A

665

soakers under
stepped flashing

tack
apron flashing

dovetail notch or
steel hanger

50 x 100 trimmed rafters

B

1·000

150

Nr. 5 lead
apron flashing

**7.4.8
LEAD APRON
FLASHING**

Nr.5 lead flashing

Nr.6 lead gutter

tilting fillet

75 x 100 trimmer

150

Nr. 5 lead tray

Nr. 5 lead
apron flashing

slating

25 boarding

75 x 100 trimmer

rough rendering

chimney flue liners

215

235

215

7.4.2 PLAN

7.4.6 SECTION B-B

| Figure 7.4 | TRIMMING   PITCHED   ROOF
AROUND   CHIMNEY   STACK | Scale: 1:20 |

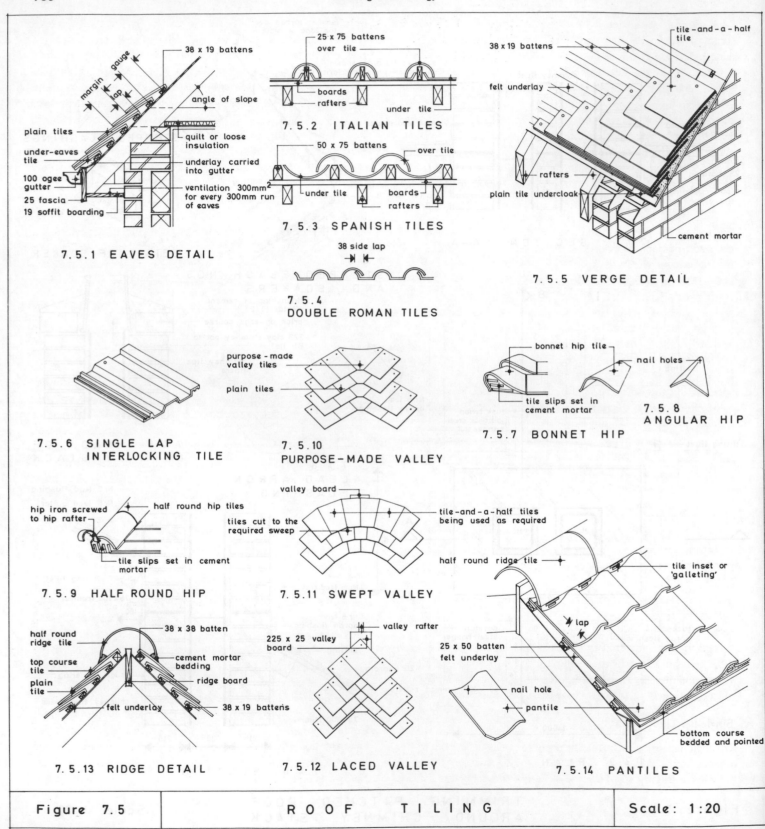

7.5.1 EAVES DETAIL
- 38 x 19 battens
- gauge
- margin
- lap
- angle of slope
- plain tiles
- under-eaves tile
- quilt or loose insulation
- underlay carried into gutter
- 100 ogee gutter
- 25 fascia
- 19 soffit boarding
- ventilation 300mm² for every 300mm run of eaves

7.5.2 ITALIAN TILES
- 25 x 75 battens over tile
- boards
- rafters
- under tile

7.5.3 SPANISH TILES
- 50 x 75 battens
- over tile
- under tile
- boards
- rafters

7.5.4 DOUBLE ROMAN TILES
- 38 side lap

7.5.5 VERGE DETAIL
- tile–and–a–half tile
- 38 x 19 battens
- felt underlay
- rafters
- plain tile undercloak
- cement mortar

7.5.6 SINGLE LAP INTERLOCKING TILE

7.5.10 PURPOSE–MADE VALLEY
- purpose–made valley tiles
- plain tiles

7.5.7 BONNET HIP
- bonnet hip tile
- nail holes
- tile slips set in cement mortar

7.5.8 ANGULAR HIP

7.5.9 HALF ROUND HIP
- hip iron screwed to hip rafter
- half round hip tiles
- tile slips set in cement mortar

7.5.11 SWEPT VALLEY
- valley board
- tiles cut to the required sweep
- tile–and–a–half tiles being used as required

7.5.13 RIDGE DETAIL
- half round ridge tile
- top course tile
- plain tile
- felt underlay
- 38 x 38 batten
- cement mortar bedding
- ridge board
- 38 x 19 battens

7.5.12 LACED VALLEY
- valley rafter
- 225 x 25 valley board

7.5.14 PANTILES
- half round ridge tile
- tile inset or 'galleting'
- lap
- 25 x 50 batten
- felt underlay
- nail hole
- pantile
- bottom course bedded and pointed

| Figure 7.5 | ROOF TILING | Scale: 1:20 |

A typical eaves detail is shown in figure 7.5.1 to which some of the more commonly used roofing terms have been added. The *lap* is the amount by which the tails of tiles in one course overlap the heads of tiles in the next course but one below, and for plain tiles should not be less than 65 mm or 75 mm in exposed positions. Shorter under-eaves tiles are provided at the eaves to maintain the lap. *Gauge* is the distance between centres of battens and is calculated by the formula: gauge = length of tile minus lap/2, hence the gauge of plain tiles to a 65 mm lap = $(265 - 65)/2 = 100$ mm. The *margin* is the exposed area of each tile on the roof and the length of the margin is the same as the gauge.

A typical *verge* detail is shown in figure 7.5.5 using tile-and-a-half tiles to maintain the bond bedded on and pointed in cement mortar on an undercloak of plain tiles. The tiles overhang the wall by 50 to 75 mm to give protection against the weather. Half-round tiles are commonly used to cover *ridges* (figure 7.5.13) and these are bedded in cement mortar. Exposed ends are filled solid with mortar and often incorporate horizontal tile slips. Alternatively, segmental or angular ridge tiles may be used.

Hips of plain tiled roofs may be covered in a variety of ways. Half-round tiles, similar to those used at ridges, are popular and they require a galvanised or wrought-iron hip iron or bracket fixed to the bottom of the hip rafter to give support (figure 7.5.9). Other alternatives include bonnet hip tiles (figure 7.5.7) which are bedded at the tails, and angular hip tiles (figure 7.5.8) which are bedded at the heads of tiles.

Valleys are a particularly vulnerable part of a roof as the slope is several degrees less than that of the general roof surface and they have to provide a channel for water converging on it from two slopes. One approach is to form an open gutter with a timber sole covered with metal which is dressed over a tilting fillet and under the tiles on each side of the gutter. Purpose-made valley tiles (figure 7.5.10) provide a sound and attractive finish to a valley and do not require nailing or bedding. Swept valleys (figure 7.5.11) consist of tiles cut to the required sweep on a board and strip of felt not less than 600 mm wide, laid the full length of the valley and turned into the gutter at the bottom end. Laced valleys (figure 7.5.12) have a board not less than 225 mm wide laid in the valley and the plain tiling radiates from a tile-and-a-half tile placed obliquely in the middle of the valley in each course.

Single-lap Tiles

In single-lap tiling each tile overlaps the head of the tile in the course below and there is also a side lap, the dimensions of which are usually fixed by the design of the tile. Thus there is only one thickness of tile on the greater part of the roof with two thicknesses as the ends and sides of each tile. Single-lap tiles can be laid at a flatter slope than double lap or plain tiles, as they are obtainable in larger sizes and the actual inclination of the tiles for any given roof slope is greater. Some of the more common types of single-lap tile are illustrated: Italian tiles (figure 7.5.2); Spanish tiles (figure 7.5.3); double Roman tiles (figure 7.5.4); and pantiles (figure 7.5.14). Pantiles are the oldest form of single-lap tile and are used extensively in East Anglia. There are many forms of interlocking tiles manufactured in concrete complying with BS 473, 550,[8] a common size being 416 x 330 mm — a typical tile is shown in figure 7.5.6. The principal advantages of single-lap tiles over plain tiles is that they give a lighter roof covering and permit a flatter slope of roof. They are, however, more difficult to replace, loss of a tile results in water penetration and they are not so adaptable to complicated roof designs.

Vertical Tile Hanging

Vertical tiling may be fixed either direct to a wall, to battens or to battens and counter-battens fixed to the wall face, to give added interest to a façade of a building. If vertical tiling is used as a facing for timber-framed construction, a suitable underlay must be provided in the form of a breather membrane which permits limited transfer of moisture vapour, or a vapour barrier, whichever is appropriate. The lap of tiles must be not less than 32 mm, tiles must be nailed with two nails to each tile, and the bottom edge of vertical tiling must finish with an undercourse as described for eaves of pitched roofs.[7]

Slates

Although slates have been superseded in many parts of the country by clay or concrete tiles and other forms of roofing material, they are still used predominantly in slate-producing districts. They vary from 255 x 150 mm to 610 x 355 mm in 27 different sizes.[9] Each slate is secured by two nails, at the head or centre of the slate, and the nails may be yellow metal, copper, aluminium alloy or zinc and they vary in length from 32 to 63 mm according to the weight of slate. Zinc nails are unsuitable for industrial or coastal districts. A typical eaves detail is shown in figure 7.6.1, from which it will be apparent that the lap is the amount by which the tails of slates in one course overlap the heads of slates in the next course but one below, as for plain tiles. It is customary to centre nail all but

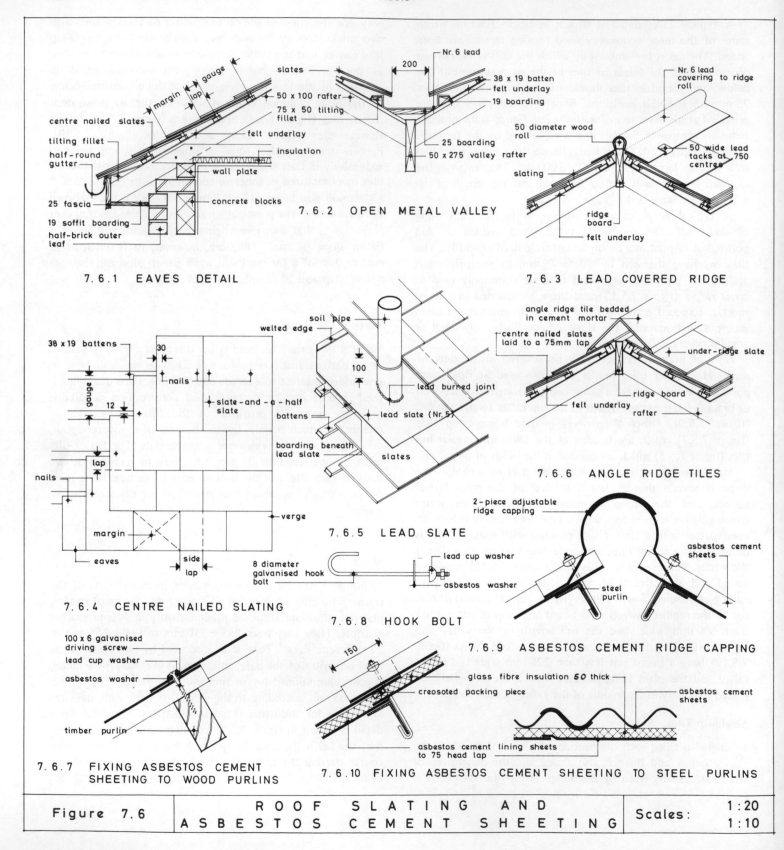

7.6.1 EAVES DETAIL

7.6.2 OPEN METAL VALLEY

7.6.3 LEAD COVERED RIDGE

7.6.4 CENTRE NAILED SLATING

7.6.5 LEAD SLATE

7.6.6 ANGLE RIDGE TILES

7.6.7 FIXING ASBESTOS CEMENT SHEETING TO WOOD PURLINS

7.6.8 HOOK BOLT

7.6.9 ASBESTOS CEMENT RIDGE CAPPING

7.6.10 FIXING ASBESTOS CEMENT SHEETING TO STEEL PURLINS

| Figure 7.6 | ROOF SLATING AND ASBESTOS CEMENT SHEETING | Scales: | 1:20 1:10 |

the smallest slates as there is a tendency for the larger head-nailed slates to lift in high winds. The main advantage claimed for head-nailed slates is that there are two thicknesses of slate covering the nails, but this involves the use of a larger number of slates and they are not so easily repaired. Nails should be not less than 30 mm from the edges and 25 mm from the heads of slates. The gauge is the distance between the nail holes in one slate from those in the adjoining slate, and for centre-nailed slates = (length − lap)/2, whereas for head-nailed slates it is [length − (lap + 25 mm)]/2. Thus, taking 460 mm long slates head nailed, the gauge becomes [460 − (75 + 25)]/2 = 180 mm.

Figure 7.6.4 illustrates centre-nailed slates abutting a verge incorporating slate-and-a-half slates. The slates are arranged to bond so that side joints in one course are over the centre of slates in the course below. The slates may be nailed to battens fixed direct to rafters, to boarding fixed to rafters, to battens on boarding, or to battens on counter-battens. The main advantage of counter-battens is that any rain driven between the slates can run down the sarking felt between the sloping battens. The short slates at eaves are head nailed and they should overhang the gutter by 50 mm.

Ridges and hips may be covered by half-round, segmental or angular tiles (figure 7.6.6), preferably Staffordshire blue or tinted terracotta, or be formed with a 50 mm diameter wood roll covered with Nr. 5 or Nr. 6 lead sheet, 450 to 500 mm wide, lapped 75 mm at joints and secured by screws and lead tacks (figure 7.6.3). The most common form of valley for slated roofs is the open metal valley illustrated in figure 7.6.2. The metal (lead, copper or zinc) should extend for at least 38 mm beyond the tilting fillets and should be nailed at both edges. Alternative valley treatments include swept or laced valleys which require highly skilled and experienced slaters, and mitred and secret valleys, neither of which are entirely satisfactory.[7] The best form of treatment at abutments is the use of soakers and cover flashings (figure 7.4.3). Perforations for pipes and similar fittings should be made weathertight by dressing over and under the slating or tiling a lead slate to which a lead sleeve is burned or soldered (figure 7.6.5). The sleeve should be bossed up around the pipe and suitably sealed at the top. Alternatively copper, 'nuralite' or 'zincon' can be used for this purpose.

Asbestos Cement Slates

Asbestos cement slates are supplied in sizes the width of which is half their length, and they are suitable for use at a slope of not less than 35° if laid with a 75 mm lap. They are sometimes laid to a diagonal pattern. Their life depends on the degree of acid pollution of the atmosphere to which they are exposed. Asbestos cement slates are centre nailed with two copper wire nails to each slate, and the tails are prevented from lifting by a copper rivet passing through the tail and between the edges of the two slates of the course below. These slates are so light that rafters can be spaced up to 750 mm apart.[7]

Wood Shingles

Wood shingles are normally obtained in red cedar as it is very durable and resistant to fungus attack, although the BRE Princes Risborough Laboratory has recommended that they should be pressure impregnated with a copper—chrome—arsenic preservative. Cedar shingles weather rapidly to silver grey, or they may be dipped in oil to retain the natural brown colour. They are imported in lengths varying from 390 to 410 mm and widths of from 75 to 300 mm, and they are normally laid in random widths, to a minimum slope of 30° with a 125 mm gauge. Each shingle is nailed twice in its length with a minimum side lap of 40 mm and a gap of 3 mm between shingles in the same course. As shingles are light in weight rafters are often placed at 600 mm centres or more. One or two undercourses are used at eaves, hips may be cut and mitred with sheet metal soakers, ridges may be covered with narrow width shingles or a lead roll may be used, and valleys are best formed with open sheet-metal gutters and cut shingles. Shingles form an attractive finish but are more expensive than clay tiles and some fire risk is involved.

Thatch

Thatching with reed from the Norfolk Broads is traditional in East Anglia. It provides a most attractive finish to steep roofs of at least 45° and preferably 55°. Reed is the most durable thatching material with a life of 50 to 75 years, but ridges are made of softer sedge with a life of about 20 years. Reed is laid about 300 mm thick with high thermal insulation properties, but it is more prone to fire damage than other coverings, is liable to attack by birds and vermin, is costly and is best confined to simple roofs. Rafters spaced at about 400 mm centres support battens (25 x 19 mm at 225 mm centres) which carry the reed. Eaves and verges have tilting or tilter boards to tighten the reed. The reed is used in bundles as cut, laid butt downwards, and tapering upwards in courses in a similar manner to tiling. The first course is tied in position but each subsequent course is held in place with a hazel rod or sway (about 19 mm diameter and 2.00 to 2.25 m long) laid across and fixed with hooks driven through the reed into the rafters. The ridge is formed of sedge bundles bent over a roll and down both roof slopes, laid alternate ways to correct the taper of the bundles, to a thickness of about 75 mm.

Sheet Coverings

Sheet coverings are available in a number of different materials and are particularly well suited for garages, stores, agricultural and industrial buildings. Probably the most commonly used material is asbestos cement which is made of fibreised asbestos and Portland cement; the natural colour is grey but it is also obtainable in other colours. Sheets are normally corrugated about 50 mm deep in various lengths up to 4.60 m. They are generally laid with an end lap of 150 mm and the side lap varies with the design. Asbestos cement sheets are fixed to wood purlins with galvanised drive screws (figure 7.6.7) or to steel-angle purlins with hook bolts (figures 7.6.8 and 7.6.10). The bolts or screws should be placed at the top of corrugations and lead cup washers used since they grip the sheet tightly, to form a watertight joint. Special fittings are made for use at ridges (figure 7.6.9), hips, corners and eaves. Asbestos cement sheeting is unattractive in appearance, and although it is incombustible and light in weight, the surface softens with weathering and the material becomes increasingly brittle with age and is unlikely to have a life exceeding 30 years. A reasonable standard of thermal insulation can be obtained by providing a 50 mm layer of glass fibre insulation supported on flat asbestos cement lining sheets (figure 7.6.10).

To overcome the brittleness of asbestos cement sheets other materials have been produced, such as sheets incorporating a core of corrugated steel covered with layers of asbestos and bitumen. This combines the strength of steel with the corrosion-resisting properties of asbestos. Corrugated galvanised steel is inclined to be fairly shortlived, noisy, subject to condensation and rusting at bolt-holes, and is not well suited for most purposes. Aluminium sheets, of alloy NS3-H complying with BS 1470[10] are useful for roofing purposes. They are corrosion resistant, of light weight and reasonably good appearance, and their reflective value has some thermal insulation properties. The minimum recommended slope is 15° and the sheets are fixed in a similar manner to other corrugated sheet materials.[11]

FLAT ROOF CONSTRUCTION

A flat roof may be the only practicable form of roof for many large buildings or those of complicated shape and can be a more economical proposition than a pitched roof. It does

unfortunately constitute a common area for premature failure in modern building. Most flat roof failures could have been avoided if the design principles now outlined had been adhered to. The technical options are described in BRE Digest 221.[35]

Movement. Continuous coverings on flat roofs are much more susceptible to the effect of movement than the small units on pitched roofs. All forms of roof construction are liable to thermal movement and deflection under load, and in the case of a timber roof can subject the roof finish to considerable strain. Timber construction is subject to moisture movement, concrete to drying shrinkage and new brickwork to expansion. Hence only those roof finishes which are able to withstand some movement should be used on the more flexible types of roof, and asphalt, which is ill suited to accept movement, should not be used on timber roofs. Upstands should not be less than 150 mm high and if movement between the vertical and horizontal sections is possible, a separate metal or semi-rigid asbestos bitumen sheet flashing should be provided to cap the top of the turned-up edge of the covering. Adequate expansion joints should also be provided, consisting of an upstand not less than 150 mm high with a metal capping.

Falls. The retention of water on most forms of roof covering is undesirable and constitutes a common cause of failure. This often results from ponding caused by inadequate falls. The Codes of Practice relating to flat roof coverings (CP 143 and 144) recommend a finished fall of 1 in 80, but after making allowance for building inaccuracies and structural deflection, a fall of 1 in 40 would be more appropriate.[12]

Insulation. The standards of insulation prescribed in The Building Regulations[1] should be regarded as minima. Care must be taken to keep all insulating materials dry as they cease to be effective and may deteriorate when wet.

Solar protection. With higher standards of insulation, solar reflective treatment is a necessity for asphalt and bitumen felt roofs. White asbestos tiles or white spar chippings are useful, whilst concrete tiles may be justified if the roof is accessible and likely to receive regular use. Upstands which cannot be treated with chippings should receive an applied reflective coating, such as metal foil on a felt backing. Moreover, solar reflective finishes prolong the life of the roof covering.

Condensation. The humidity and vapour pressure are normally higher inside an occupied building than outside. Water vapour will usually penetrate most of the internal surfaces of a building and when the outside temperature is lower, it may condense at some point in the roof structure, often on the surface of insulation.[33] Where the insulation is immediately under the roof covering it must be placed on a vapour barrier, often consisting of bitumen felt laid and lapped in hot bitumen and turned past the edges of the insulation to meet the roof covering. With timber roofs it is necessary to cross-ventilate the spaces between joists,[36] and where the roof extends over a cavity wall the cavity should be sealed at the top to prevent moist air entering the roof.[18] Where cavity barriers restrict the crossflow of air in a flat roof structure, cowl type ventilators penetrating through the roof surface are needed.[34]

Timber Roofs

Timber is reasonably adaptable, easily worked and hence quite popular for small buildings, such as domestic garages and stores. Timber flat roofs generally consist of roof joists, 50 mm thick and spaced at 400 to 450 mm centres, carrying tapering firring pieces to give the necessary fall and support to the boarding which carries the roof covering (figures 7.7.1 and 7.7.3). It is good practice to provide galvanised steel straps at intervals, screwed to the roof joists and built into the brickwork to prevent the possibility of the roof being lifted in a high wind. The fall is normally within the range of 1 in 80 to 1 in 40, with the surface water usually drained to an eaves gutter (figures 7.7.1 and 7.7.2) or a wall gutter (figure 7.7.5). The roof joists generally bridge the shortest span and the boarding is nailed at right angles to them, although the boarding or its grain should preferably follow the fall to avoid warping boards retarding the flow of water. Square-edged boarding is sometimes used but tongued and grooved boarding or blockboard give a much sounder job. Each board should be nailed with two nails to each joist with the nailheads well punched down below the surface of the boarding. The moisture content of the boarding should not exceed 15 per cent and the roof space should be ventilated (at least 300 mm² free opening for every 300 mm run of eaves) as shown in figure 7.7.1, to prevent fungal growth and condensation. The timber may be treated with preservative but creosote must not be used with built-up felt roofs as it is injurious to bitumen.

Built-up felt roofing. This is now a very popular covering to this type of roof owing to the high cost of metal coverings. It normally consists of three layers of felt laid breaking joint and bonded together with a hot bitumen adhesive. On boarded roofs the bottom layer of felt is random nailed with 19 mm galvanised clout nails at about 150 mm centres with each length of felt overlapping the adjoining sheet by at least 50 mm. To secure a good standard of thermal insulation, 50 mm thick flaxboard or compressed strawboard may be used, supported on roof joists at about 1.80 m centres (figure 7.7.1). On flaxboard or compressed strawboard, the joints between boards should be taped, and the bottom layer of felt, probably asbestos based, either random nailed with 45 mm aluminium serrated nails or bedded in mastic. The second and top or cap sheets are fixed with adhesive, staggering the joints in successive sheets. The roof surface can be finished with limestone, granite or other suitable chippings on an adhesive coating, to protect the cap sheet, provide additional fire resistance and increase solar reflection. Typical verge details are shown in figures 7.7.4 and 7.7.6, finishing about 50 mm above the roof surface to prevent rainwater discharging over the edge. The welted apron is nailed with 19 mm nails at about 50 mm centres.[13] A vapour check should be placed immediately above the ceiling finish, in the form of aluminium foil or polythene.

Concrete Roofs

Reinforced concrete roofs are constructed in a similar manner to reinforced concrete floors and may be solid, hollow pot or self-centering. Concrete roof slabs are often reinforced with steel bars in both directions, with the larger bars following the span, and should have bearings on walls of at least 100 mm. The slab is generally finished level and the fall obtained with a screed (figure 7.7.7), possibly one with a lightweight aggregate to improve thermal insulation. Breather vents (figure 7.7.9), may be provided to remove trapped air and moisture from under the roof covering. To prevent ceiling staining, it is good practice to fix the ceiling to treated battens, with a vapour check between them (figure 7.7.10). A typical concrete roof slab screeded to falls and insulated is illustrated in figure 7.7.7, while suitable verge and eaves details are shown in figures 7.7.8 and 7.7.11.

FLAT ROOF COVERINGS

Asphalt

Mastic asphalt to BS 1162[15] or BS 988[16] is highly suitable for covering concrete roof slabs, but because of its high coefficient of thermal expansion, it is generally necessary to separate the asphalt from any substrate by an isolating membrane of sheathing felt.[17] Two layers of asphalt are always necessary and the total finished thickness should be not less than 20 mm with joints staggered at least 150 mm at laps. Where the roof is likely to be subject to more than maintenance traffic, it is best finished with solar reflective asbestos cement or concrete tiles. An insulating membrane such as glass fibre or cork board should be placed above the concrete deck and vapour barrier and below the sheathing felt and asphalt. Asphalt skirtings at upstands should be 13 mm thick in two coats to a minimum height of 150 mm above the finished level of the asphalt flat. The top edge of the skirting should be tucked into a chase 25 x 25 mm and pointed in cement mortar, and be masked with a lead, copper or aluminium flashing, with its tail finishing at least 75 mm above the roof surface to avoid capillary attraction.

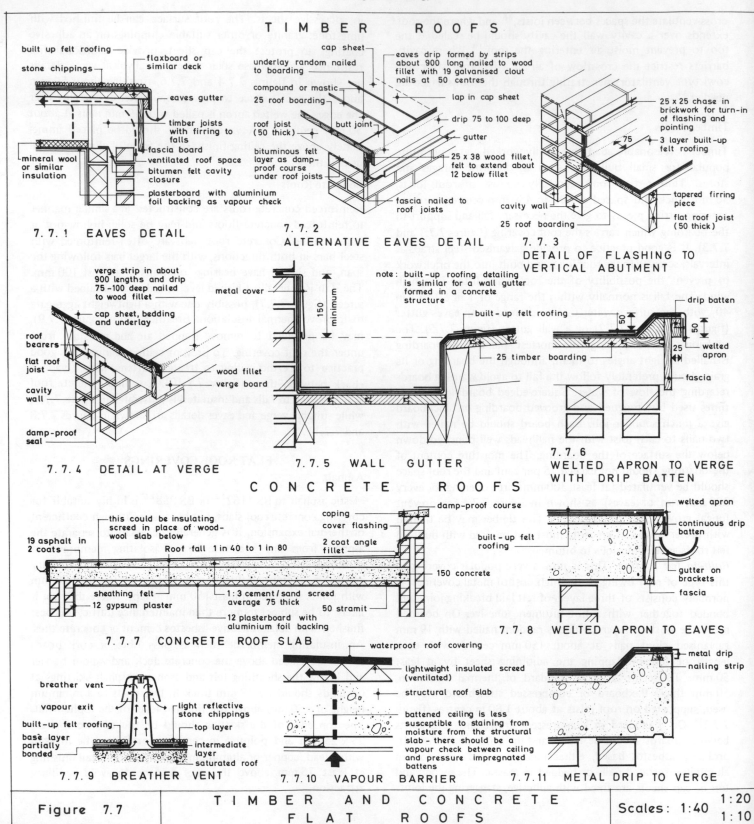

TIMBER ROOFS

7.7.1 EAVES DETAIL

built up felt roofing
stone chippings
flaxboard or similar deck
eaves gutter
timber joists with firring to falls
fascia board
ventilated roof space
bitumen felt cavity closure
plasterboard with aluminium foil backing as vapour check
mineral wool or similar insulation

7.7.2 ALTERNATIVE EAVES DETAIL

cap sheet
underlay random nailed to boarding
compound or mastic
25 roof boarding
roof joist (50 thick)
butt joint
bituminous felt layer as damp-proof course under roof joists
eaves drip formed by strips about 900 long nailed to wood fillet with 19 galvanised clout nails at 50 centres
lap in cap sheet
drip 75 to 100 deep
gutter
25 x 38 wood fillet, felt to extend about 12 below fillet
fascia nailed to roof joists

7.7.3 DETAIL OF FLASHING TO VERTICAL ABUTMENT

25 x 25 chase in brickwork for turn-in of flashing and pointing
75
3 layer built-up felt roofing
tapered firring piece
cavity wall
25 roof boarding
flat roof joist (50 thick)

7.7.4 DETAIL AT VERGE

verge strip in about 900 lengths and drip 75-100 deep nailed to wood fillet
cap sheet, bedding and underlay
roof bearers
flat roof joist
cavity wall
wood fillet
verge board
damp-proof seal

7.7.5 WALL GUTTER

metal cover
150 minimum
wood fillet

note: built-up roofing detailing is similar for a wall gutter formed in a concrete structure

7.7.6 WELTED APRON TO VERGE WITH DRIP BATTEN

drip batten
50
50
built-up felt roofing
25 timber boarding
welted apron
25
fascia

CONCRETE ROOFS

7.7.7 CONCRETE ROOF SLAB

this could be insulating screed in place of wood-wool slab below
Roof fall 1 in 40 to 1 in 80
19 asphalt in 2 coats
sheathing felt
12 gypsum plaster
1:3 cement/sand screed average 75 thick
12 plasterboard with aluminium foil backing

coping
cover flashing
2 coat angle fillet
damp-proof course
150 concrete
50 stramit

7.7.8 WELTED APRON TO EAVES

built-up felt roofing
welted apron
continuous drip batten
gutter on brackets
fascia

7.7.9 BREATHER VENT

breather vent
vapour exit
built-up felt roofing
base layer partially bonded
light reflective stone chippings
top layer
intermediate layer
saturated roof

7.7.10 VAPOUR BARRIER

waterproof roof covering
lightweight insulated screed (ventilated)
structural roof slab
battened ceiling is less susceptible to staining from moisture from the structural slab – there should be a vapour check between ceiling and pressure impregnated battens

7.7.11 METAL DRIP TO VERGE

metal drip
nailing strip

Figure 7.7	TIMBER AND CONCRETE FLAT ROOFS	Scales: 1:40	1:20
			1:10

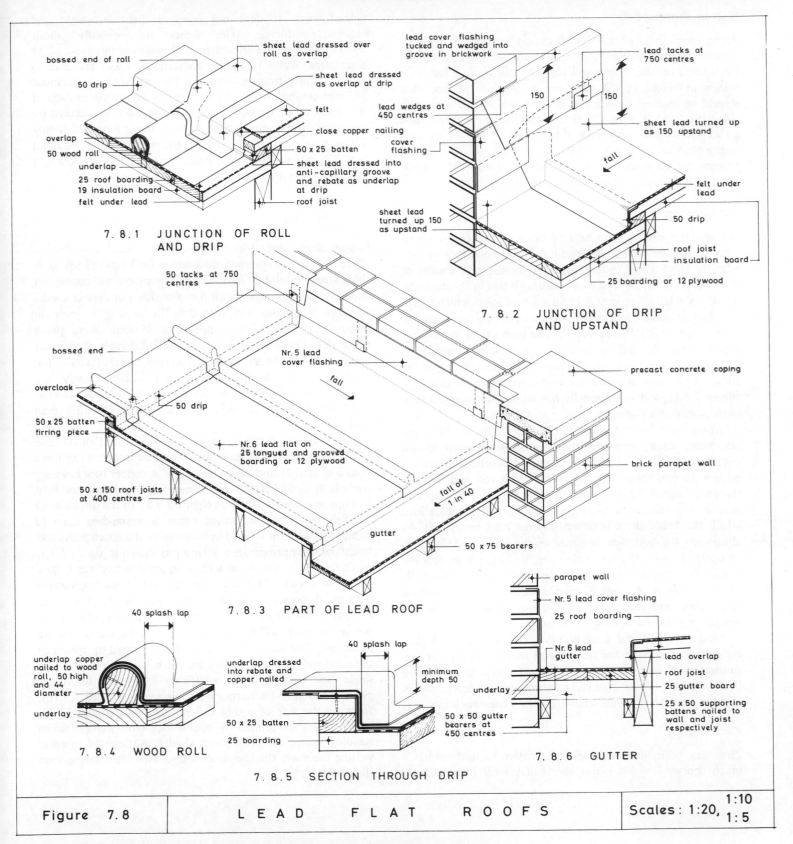

bossed end of roll

50 drip

sheet lead dressed over roll as overlap

sheet lead dressed as overlap at drip

felt

overlap

50 wood roll

underlap

25 roof boarding

19 insulation board

felt under lead

close copper nailing

50 x 25 batten

sheet lead dressed into anti-capillary groove and rebate as overlap at drip

roof joist

7.8.1 JUNCTION OF ROLL AND DRIP

lead cover flashing tucked and wedged into groove in brickwork

lead tacks at 750 centres

lead wedges at 450 centres

cover flashing

150   150

sheet lead turned up as 150 upstand

fall

sheet lead turned up 150 as upstand

felt under lead

50 drip

roof joist

insulation board

25 boarding or 12 plywood

7.8.2 JUNCTION OF DRIP AND UPSTAND

50 tacks at 750 centres

bossed end

overcloak

50 x 25 batten firring piece

50 x 150 roof joists at 400 centres

Nr. 5 lead cover flashing

fall

50 drip

Nr.6 lead flat on 25 tongued and grooved boarding or 12 plywood

fall of 1 in 40

gutter

50 x 75 bearers

precast concrete coping

brick parapet wall

7.8.3 PART OF LEAD ROOF

40 splash lap

underlap copper nailed to wood roll, 50 high and 44 diameter

underlay

7.8.4 WOOD ROLL

40 splash lap

underlap dressed into rebate and copper nailed

minimum depth 50

50 x 25 batten

25 boarding

7.8.5 SECTION THROUGH DRIP

parapet wall

Nr.5 lead cover flashing

25 roof boarding

Nr.6 lead gutter

lead overlap

roof joist

underlay

25 gutter board

50 x 50 gutter bearers at 450 centres

25 x 50 supporting battens nailed to wall and joist respectively

7.8.6 GUTTER

| Figure 7.8 | L E A D   F L A T   R O O F S | Scales: 1:20, 1:10, 1:5 |

Built-up Bitumen Felt

Three layers are essential for all except temporary buildings, bonded with hot bitumen.[14] Upstands and skirtings are formed by turning up the second and top layers of roofing felt for a minimum height of 150 mm over an angle fillet, and they should be masked by a metal or semi-rigid asbestos/bitumen sheet (SRABS) flashing. On timber roofs the base may consist of 25 mm tongued and grooved boarding, or 12 mm plywood where the spacing of roof joists does not exceed 400 mm and 15 mm for spacings between 400 and 600 mm. An insulating membrane and a vapour barrier are also required.

Lead Sheeting

A lead flat is divided into bays by drips, at a spacing of about 2.50 m, and the bays subdivided into further divisions by rolls with the spacing of 675 mm being determined by the width of sheets (fige 7.8.3).[19] The roof boarding is laid in the direction of the roof fall of at least 1 in 80 on roof joists which bridge the shortest span. The fall is obtained by firring pieces on top of the roof joists (figure 7.8.3). The lead sheets are copper nailed at the top and one side only to allow movement with changes of temperature. Rainwater discharges into a parallel gutter positioned between the end joist and a parapet wall (figure 7.8.6), and this normally has an outlet through a hole in the wall into a rainwater head.

Wood rolls of the form illustrated in figure 7.8.4 are commonly used, with one sheet being nailed with 25 mm copper nails to the roll and the adjoining sheet passing over it with a 38 mm splash lap. A junction of a roll and drip are shown in figure 7.8.1. Drips are normally 50 mm deep and may incorporate an anti-capillary groove and a rebate into which the lower sheet is nailed (figures 7.8.1 and 7.8.5). At abutments the lead sheet is turned up the wall face 150 mm as an upstand and a 150 mm wide cover flashing passes over the top of the upstand to form a watertight joint (figure 7.8.2). The top edge of the cover flashing is tucked and wedged into a brick joint and lead tacks at 750 mm centres prevent the bottom edge of the flashing from curling.

Lead is ductile and flexible, easily cut and shaped, highly resistant to corrosion and has a long life. It is however of low strength, high cost, very heavy in weight, creeps on all but the flattest slopes, can be attacked by damp cement mortar and is supplied in smallish sheets entailing many expensive joints.

Zinc Sheeting

Zinc may be used as a cheaper alternative to lead but has a much shorter life and is not very suitable for heavy indust-

rialised areas. It does not creep, is light in weight and reasonably ductile. The sheets are normally about 2400 x 900 mm and should have a minimum thickness of 14 zinc gauge. Wood or batten rolls (figure 7.9.1) are used to provide joints running with the fall of the roof, at about 885 mm centres, where a zinc capping covers the turn-up of the sheets on either side of the roll, and these are secured by zinc clips at 750 mm centres.[20] Drips may be beaded (figure 7.9.3) or welted (figure 7.9.2) and are provided at about 2.25 m centres. Cover flashings (figure 7.9.4) are used at abutments as with lead roofs.

Copper Sheeting

Copper sheet and strip for flat roofs should conform to BS 2870[21] and a common thickness is 0.61 mm (23 SWG). A thin, stable insoluble film (patina) forms on the copper on exposure to air, consisting of a combination of copper oxide, sulphate, carbonate and chloride. The coating is green in colour and improves the appearance in addition to giving protection. Copper sheet is very tough and durable, readily cut and bent, is light in weight does not creep, and resists corrosion reasonably well. On the other hand it is fairly costly, the smallish sheets involve many joints and electrolytic corrosion may take place if it comes into contact with metals other than lead.

The minimum fall for a copper roof is 1 in 60. Drips should be used on roofs of 5° slope or less, should be spaced at not more than 3 m centres and be 63 mm deep. A copper roof covering consists of a number of sheets joined along the edges, and held by clips inserted in the folds (figure 7.9.6). In the direction of the fall the joints are raised either as a standing seam (a double-welted joint formed between sides of adjacent bays and left standing approximately 20 mm) as shown in figure 7.9.7, or dressed to a wood roll with a separate welted cap (figure 7.9.5), at about 525 mm centres, the actual spacing varying with the thickness of the sheet. The choice of standing seams or rolls is influenced by the architectural treatment required, but where roofs will be subject to foot traffic wood rolls are preferable. The joints across the fall are formed by means of flattened welted seams. Where the pitch is greater than 45°, a single lock welt seam is used (figure 7.9.9), and for flatter pitches, a double-lock cross welt is essential (figure 7.9.10). Welts on the side of a drip can be single lock (figure 7.9.8). Where sheets are jointed by double-lock cross welts between standing seams, the cross joints should be staggered to avoid welting too many thicknesses of copper into the standing seam (figure 7.9.6).[22]

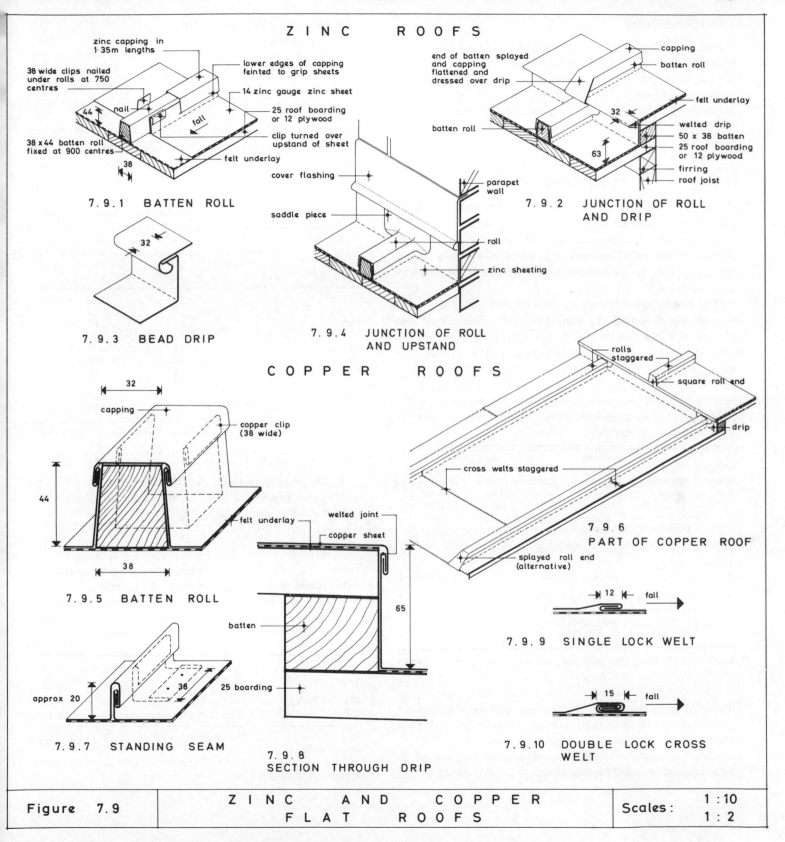

Z I N C   R O O F S

7.9.1 BATTEN ROLL

zinc capping in 1·35m lengths

lower edges of capping feinted to grip sheets

38 wide clips nailed under rolls at 750 centres

14 zinc gauge zinc sheet

nail

44

fall

25 roof boarding or 12 plywood

clip turned over upstand of sheet

38 x 44 batten roll fixed at 900 centres

felt underlay

38

7.9.3 BEAD DRIP

32

7.9.2 JUNCTION OF ROLL AND DRIP

end of batten splayed and capping flattened and dressed over drip

capping

batten roll

felt underlay

batten roll

32

welted drip

50 x 38 batten

63

25 roof boarding or 12 plywood

firring

roof joist

7.9.4 JUNCTION OF ROLL AND UPSTAND

cover flashing

saddle piece

parapet wall

roll

zinc sheeting

C O P P E R   R O O F S

7.9.5 BATTEN ROLL

32

capping

copper clip (38 wide)

44

felt underlay

38

7.9.6 PART OF COPPER ROOF

rolls staggered

square roll end

drip

cross welts staggered

splayed roll end (alternative)

7.9.7 STANDING SEAM

approx 20

38

7.9.8 SECTION THROUGH DRIP

welted joint

copper sheet

batten

25 boarding

65

7.9.9 SINGLE LOCK WELT

12

fall

7.9.10 DOUBLE LOCK CROSS WELT

15

fall

| Figure   7.9 | Z I N C   A N D   C O P P E R   F L A T   R O O F S | Scales: | 1 : 10
1 : 2 |

Aluminium Sheeting

A common size of sheet is 1800 x 610 mm and the thickness is often 0.90 mm (20 SWG). The method of fixing and jointing is similar to copper, using batten rolls. An alternative is to use extruded aluminium alloy rolls.[23] The strength of aluminium is increased by the addition of alloys, although they reduce its ductility.

ROOF DRAINAGE

For roof drainage calculations it is usual to assume a rate of rainfall of 75 mm per hour. Storms of higher intensity up to 150 mm/h or more do occur, but they are of short duration. A simple method of determining the sizes of eaves gutters and downpipes has been formulated by the Building Research Establishment.[24]

The rate of run-off from a roof is the product of the design rate of rainfall (usually 75 mm/h) and the effective roof area (1 mm of rainfall on an area of 1 m^2 is 1 litre of water). To allow for wind pressure the effective roof area should be taken as the plan area plus half the elevation area.

Table 7.3   Flow capacities (litres/second) for level gutters with outlet at one end

Nominal gutter size (mm)	True half-round gutter (1 and 2)	Nominal half-round (segmental) gutter (3 and 4)	Ogee gutter Asbestos cement (1)	Pressed steel (2)	Aluminium or cast iron (3 or 4)
75	0.4	0.3	—	—	—
100	0.8	0.7	1.1	0.9	0.5
115	1.1	0.8	1.5	1.4	0.7
125	1.5	1.1	1.8	1.7	0.8
150	2.3	1.8	2.7	2.6	not standard

Notes: (1) Asbestos cement to BS 569 (1974)
          (2) Pressed steel to BS 1091 (1963)
          (3) Aluminium to BS 2997 (1958)
          (4) Cast iron to BS 460 (1964)

Source: *BRE Digest 188: Roof drainage, Part 1*[24] *and CP 308: 1974*[32]

Table 7.3 gives flow capacities for level gutters of half-round, segmental and ogee section, with an outlet at one end. Flow capacities can be adjusted for other situations

Hence the rate of run-off can be multiplied by one of the following factors where appropriate

angle within 2 m of the outlet
    sharp-cornered x 1.2
    round-cornered x 1.1
angle within 2 to 4 m of the outlet
    sharp-cornered x 1.1
    round-cornered x 1.05

In practice it may be more convenient to calculate the flow load per metre run of eaves. Table 7.4 shows suitable downpipe sizes.

Table 7.4   Minimum downpipe sizes for various eaves gutters

Half-round gutter size (mm)	Sharp-(SC) or round-cornered (RC) outlet	Outlet diam. at one end of gutter	Outlet diam. not at one end of gutter
75	SC	50	50
	RC	50	50
100	SC	63	63
	RC	50	50
115	SC	63	75
	RC	50	63
125	SC	75	89
	RC	63	75
150	SC	89	100
	RC	75	100

Source: *BRE Digest 188: Roof drainage, Part 1*[24] *and CP 308: 1974*[32]

If the designed rate of rainfall selected is 75 mm/h, the gutter capacity given in table 7.3 results in the effective areas to be drained as given in table 7.5.

Table 7.5   Effective areas drained by level eaves gutter with outlet at one end at a rate of rainfall of 75 mm/h

Size mm	True half-round m^2	Nominal half-round m^2	Aluminium or cast iron ogee m^2	Pressed steel ogee m^2
75	20.1	14.7	—	—
100	39.5	31.4	23.2	43.0
115	53.7	39.5	31.4	64.5
125	71.5	50.2	39.5	80.5
150	111.1	85.8	Not standard	125.3

Note: For other rates of rainfall intensity the areas will be inversely proportional

Source: *CP 308: 1974*[32]

BRE Digest 189[24] covers methods of determining the sizes of gutters of other profiles and sizes.

Eaves Gutters

Eaves gutters are normally either half-round (figure 7.1.8), ogee (figure 7.5.1) or moulded in section and are made in a variety of materials — cast iron, asbestos cement, plastics, pressed steel, aluminium, and wrought copper and wrought zinc, together with the necessary fittings (stop ends, angles and outlets).

Cast iron gutters. These are supplied in 1800 mm lengths to BS 460[25] and in sizes varying from 75 to 150 mm, with a shallow socket at one end to receive the adjoining length. The joint is made with red lead or a cold caulking compound and is secured by a small bolt. Half-round gutters are supported by brackets at 900 mm centres and these are screwed to the fascia or to the tops or sides of rafter feet, while ogee gutters are screwed through the back at 600 mm centres to the fascia.

Asbestos cement gutters. These are supplied in 1800 mm lengths to BS 569[26], and in sizes from 75 to 200 mm, screwed together and jointed with a special jointing compound, and often fixed by galvanised steel brackets to feet of rafters or to fascia boards.

Plastic gutters. These are made in unplasticised PVC (polyvinyl chloride) to BS 4576,[29] are black, white or grey in colour and are subject to some distortion and slight colour change in use. Their expected life is 20 years or more. They are made in a variety of sections in lengths of 1.8 m and 3.6 m which are clipped together and usually supported by vinyl brackets.

Pressed steel galvanised gutters. These are made in light gauge to BS 1091[27] in sizes ranging from 75 to 150 mm in both half-round and ogee sections, in lengths of 900, 1200 and 1800 mm. Gutters are bolted and jointed with either red lead and putty or a cementitious cold caulking compound. Heavy pressed valley, box and half-round gutters are made for use in industrial buildings.

Aluminium gutters. These are made by casting, extrusion or pressing from sheet, in a variety of sections and sizes to BS 2997.[28] They give good performance with most roof coverings, except copper, with very little maintenance. Aluminium gutters are jointed with bituminous mastic and supported every 1800 mm by aluminium or galvanised mild steel brackets.

Wrought copper and wrought zinc gutters. These require little maintenance and are light in weight. They are made in various sizes up to 125 mm and are supported by stays at 375 mm centres and brackets at 750 mm centres.

Precast concrete eaves gutters. These have a minimum internal diameter of 125 mm. The units are jointed with mastic or a suitable mortar and they should be lined with a nonferrous metal or other suitable protective coating. Other types of gutter include valley gutters and parapet gutters, which may be tapering (figure 7.3.4) or parallel (figures 7.7.5 and 7.8.6).

Downpipes

Downpipes convey rainwater from roof gutters to underground drains, often through a rainwater gully at ground level. When used with projecting eaves they generally require a swan-neck consisting of a fitting with two bends to negotiate the soffit. Flat roof parapet gutters often discharge into rainwater heads at the top of downpipes.

Cast iron spigot and socket downpipes. These are made to BS 460[25] in 1800 mm lengths and diameters from 50 to 150 mm. Pipes are generally unjointed, but joints may be filled with red and white lead putty. Fixing of pipes is normally performed by nails through ears and distance pieces into hardwood plugs built into walls. Cast iron pipes are usually painted or coated with composition.

Asbestos cement spigot and socket downpipes. These are made to BS 569[26] in diameters of 50 to 150 mm and lengths from 1.80 to 3.00 m. Joints are usually left unfilled but are otherwise jointed with cement and sand (1:2). Pipes are fixed with galvanised mild steel ring clips to keep them 38 mm clear of the wall.

Plastics downpipes. Made in unplasticised PVC to BS 4576[29] in lengths of 2 m and 4 m and diameters of 63, 68 and 75 mm. They are jointed with rubber sealing rings.

Pressed steel galvanised light gauge downpipes. These are made to BS 1091,[27] are of limited durability and best confined to temporary work. They are made in lengths of 900, 1200 and 1800 mm and in diameters from 50 to 100 mm. Joints are left loose and pipes are fixed by galvanised pipe nails through ears on the pipes, keeping them 38 mm clear of the wall.

Aluminium downpipes. Made to BS 2997[28] in a variety of shapes, thicknesses and sizes. Joints may be loose or caulked and fixing is by pipe nails through ears.

Wrought copper and wrought zinc downpipes. These are made in lengths of 1.8, 2.1 and 2.4 m and in diameters of 50 to 100 mm. Joints are made in telescoped form without any jointing compound. Pipes are fixed through ears, keeping them at least 38 mm from the wall.

Wire balloons of galvanised steel, aluminium or copper should be inserted in gutter outlets to prevent blockages occurring in downpipes.

THERMAL INSULATION

The Building Regulations[1] require the thermal transmittance coefficient of the roof and ceiling of a dwelling to be not more than 0.6 W/m^2 °C and schedule II of the Regulations specifies various forms of construction which will meet this requirement. The insulation can take various forms

(1) Boards or slabs: woodwool, compressed strawboard, corkboard, fibre insulating board, asbestos insulating board, expanded polystyrene, and insulating gypsum plasterboard.

(2) Mats and quilts: mineral wool (glass, rock or slab) eel grass, cellulose acetate fibre or other fibrous materials enclosed between sheets of waterproof paper.

(3) Loose fills: gypsum granules, exfoliated vermiculite, nodulated slag wool and nodulated polystyrene.

(4) Aluminium foil: single or double-sided paper reinforced or combined corrugated and flat aluminium foil.

(5) Insulating screeds: made with lightweight aggregates such as vermiculite, expanded clay, foamed slag or sintered pulverised fuel ash or aerated or cellular concrete.

With pitched roofs, it is more economical to position the insulating membrane across the ceiling rather than in the plane of the rafters. Insulation fixed to rafters is generally in the form of boards or slabs, whilst that at ceiling level is usually a quilt laid over the ceiling joists or loose fill between them. Any tanks or pipes above the ceiling insulation will need insulating treatment.

With timber flat roofs, board or slab insulation may be provided under or over the roof joists or possibly both (figures 7.7.1 and 7.8.1). In the case of concrete flat roofs, insulation can take the form of a lightweight insulating screed or as permanent formwork of woodwool slabs or fibreboard under the concrete slab (figure 7.7.7). In all cases it it necessary to provide a vapour barrier, preferably of bitumen felt, on the warm side of the insulation to prevent water vapour condensing on the underside of the flat roof covering and causing saturation inside the roof structure.

PRINCIPAL MATERIALS USED IN ROOFS

Clay Roofing Tiles

These are principally manufactured in districts producing clay bricks from well-weathered or well-prepared clay or marl. The process of manufacture is similar to that for bricks and they may be machine-made or hand-made, rough or smooth in texture, and even or mottled in colour. Tiles should be free from particles of lime or fire cracks, true in shape, dense, tough, show a clean fracture when broken and be well burnt throughout.[6]

Concrete Roofing Tiles

These are made from fine concrete with the addition of a colouring pigment and may have a textured surface. They have a dense structure which is highly resistant to lamination and frost damage. Concrete tiles are required by BS 473[8] to be true to shape and show a uniform structure on fracture.

Slates

Slates are a type of rock which can be split into very thin layers. They form an excellent roof-covering material of good durability, impermeability and lightweight. BS 680[9] prescribes that they shall be of reasonably straight cleavage, ring true when struck, and the grain shall run longitudinally. Uniform length slates shall be to one of the sizes listed in the standard. The principal sources and colours of roofing slates in this country are as follows

Location	Colours
North Wales	Mainly blue, blue-purple, blue-grey and grey
South Wales	Green, silver grey and rustic
Cornwall	Grey and grey-green
Westmorland	Various shades of green; rough texture
North Lancashire	Soft blue-grey
Scotland (Argyllshire)	Blue

Asphalt

This is a mixture of bitumen and inert mineral matter, and may be lake asphalt from the West Indies or rock asphalt from central Europe. Most asphalt used in roofing is *mastic asphalt*, which is a synthetic substitute for natural asphalt, and is made from bitumen and fillers.

Bitumen Felts

These are covered by BS 747[31] which recognises three main categories of felt used in built-up roofing: class 1 — fibre base; class 2 — asbestos base; class 3 — glass fibre base. There are various types of felt in each category; for instance, fibre base felt is subdivided into

(1B) fine granule surfaced — coated on both sides and suitable for use as lower layers of built-up roofing;

(1E) mineral surfaced — finished with mineral granules on upper side and fine sand on other and suitable as external layer of built-up roofing;

(1F) reinforced — contains layer of jute hessian embedded in coating on one side of felt to strengthen it and may have an aluminium foil face; suitable for use under tiles and slates.

Metals used in Roofing

Lead. A grey, soft, heavy, malleable nonferrous metal, smelted from lead ores, impurities removed, heated and cast into sheets or pigs. Most lead used in building work is 'milled' by rolling into sheets of the required thickness. It is particularly useful for flashings, but its wider use is restricted by its high cost.

Zinc. A light grey nonferrous metal, made by smelting zinc ores. It is relatively strong and ductile, but is not very resistant to polluted atmospheres. Its use in roofing is restricted by its limited durability but it is used widely as a protective coating to other metals.

Copper. A reddish-brown, nonferrous metal which is made by smelting copper ores. It soon obtains a greenish protective coating on exposure to the atmosphere, and is durable, tough and ductile, and does not creep on slopes.

Aluminium. A silvery white nonferrous metal obtained from bauxite by electrical processes, used principally as an alloy. It is light in weight, resistant to corrosion, fairly soft and reasonably ductile. It is used for structural members, windows and doors, roof coverings and wall claddings, rainwater goods and for thermal insulation.

REFERENCES

1. The Building Regulations 1976. SI 1676. HMSO (1976)
2. CP 3: chapter V: Loading; Part 1: 1967 Dead and imposed loads and Part 2: 1972: Wind loads
3. CP 112: The structural use of timber, Part 2: 1971 (Metric units); Part 3: 1973 Trussed rafters for roofs of dwellings
4. BS 1579: 1960 Connectors for timber
5. *BRE Digest 77*: Damp-proof courses. HMSO (1971)
6. BS 402: Clay plain roofing tiles and fittings, Part 2: 1970 (Metric units)
7. BS 5534: 1978 Code of practice for slating and tiling
8. BS 473, 550: Concrete roofing tiles and fittings Part 2: 1971 (Metric units)
9. BS 680: Roofing slates, Part 2: 1971 (Metric units)
10. BS 1470: 1972 Wrought aluminium and aluminium alloys for general engineering purposes; plate, sheet and strip
11. CP 143: Sheet roof and wall coverings, Part 1: 1958 Aluminium corrugated and troughed sheet (Metric units)
12. DOE Advisory leaflet 68: Timber connectors. HMSO (1974)
13. CP 144, Part 3: 1970 Built-up bitumen felt
14. *BRE Digest 8*: Built-up felt roofs. HMSO (1970)
15. BS 1162, 1418, 1410: 1973 Mastic asphalt for building (natural rock asphalt aggregate)
16. BS 988, 1076, 1097, 1451: 1973 Mastic asphalt for building (limestone aggregate)
17. *BRE Digest 144*: Asphalt and built-up felt roofings: durability. HMSO (1972)
18. DOE Advisory leaflet 79: Vapour barriers. HMSO (1976)
19. CP 143: Sheet roof and wall coverings, Part 11: 1970 Lead (Metric units)
20. CP 143: Sheet roof and wall coverings, Part 5: 1964 Zinc
21. BS 2870: 1968 Rolled copper and copper alloys, sheet, strip and foil
22. CP 143: Sheet roof and wall coverings, Part 12: 1970 Copper (Metric units)
23. CP 143: Sheet roof and wall coverings, Part 15: 1973 Aluminium (Metric units)
24. *BRE Digest 188/189*: Roof drainage. HMSO (1976)
25. BS 460: 1964 Cast iron rainwater goods
26. BS 569: 1973 Asbestos cement rainwater goods
27. BS 1091: 1963 Pressed steel gutters, rainwater pipes, fittings and accessories
28. BS 2997: 1958 Aluminium rainwater goods
29. BS 4576: Unplasticised PVC rainwater goods, Part 1: 1970 Half-round gutters and circular pipe
30. *BRE Digest 147*: Permissible spans for trussed rafters. HMSO (1972)
31. BS 747: 1977 Roofing felts.
32. CP 308: 1974 Drainage of roofs and paved areas
33. *BRE Digest 180*: Condensation in roofs. HMSO (1975)
34. *BRE Digest 218*: Cavity barriers and ventilation in flat and low pitched roofs. HMSO (1978)
35. *BRE Digest 221*: Flat roof design: the technical options. HMSO (1979)
36. *BRE Information Paper 35/79*: Moisture in a timber-based flat roof of cold deck construction. HMSO (1979)

8 WINDOWS

This chapter examines the general principles of design of windows; the construction, detailing, fixing and uses of different types of window; glazing techniques; and the forms and uses of double glazing and double windows.

GENERAL PRINCIPLES OF WINDOW DESIGN

Functions of Windows

The primary functions of a window are to admit light and air into a room in a building. They frequently also provide occupants of the building with an outside view. A number of other factors deserve consideration when designing windows, such as thermal and sound insulation, and avoidance of excessive sunglare and solar heat.

Open Space and Ventilation Requirements

The Building Regulations[1] require a zone or shaft of space open to the sky alongside the window to a habitable room, in order to make adequate provision for the circulation of air in locations where it will afford adequate ventilation to habitable rooms. Under regulation K1(3), where the room has one window only, there shall be a minimum zone of open space outside the window such as to leave adjacent to the window an upright shaft of space wholly open to the sky, the base of the shaft being formed by a plane inclined upwards at an angle of 30° to the horizontal from the wall at the lower window level and having its sides coinciding with the following four vertical planes as illustrated in figure 8.1.1.

(1) An outer plane parallel to the wall which
   (a) is at a distance from the wall of 3.6 m or (subject to a limit of 15 m) one-half the distance between the upper window level and the top of the wall containing the window, whichever is the greatest;
   (b) has a width equal to its required distance from the wall;
   (c) is so located that some part of it is directly opposite some part of the window.

(2) An inner plane which coincides with the external surface of the wall and which
   (a) has a width such that the product of that width

and the window height equals one-tenth of the floor area of the room containing the window;
   (b) is located wholly between the sides of the window or, where it is required to be wider than the window, is so located that it extends across the whole width of the window, and overlaps it on either or both sides.

(3) Two lateral planes joining the corresponding extremities of the inner and outer planes.

Furthermore, under class K4(3) of the Building Regulations[1] any habitable room shall, unless adequately ventilated by mechanical means, have one or more ventilation openings with a total area equal to not less than one-twentieth of the floor area of the room and with some part of it not less than 1.75 m above the floor. This ventilation requirement generally takes the form of opening lights in windows. The total glass area should not be less than one-tenth of the floor area. Ventilation heat loss is examined in BRE Digest 190.[18] BRE Digest 206[2] discusses fresh air requirements and tolerable levels of contamination from various sources and explains the calculation of dilution rates, while BRE Digest 210[3] shows how natural ventilation rates are determined.

Solar Heat and Daylight Admittance

Windows which admit sunlight also admit solar heat, and although heat from the sun is welcome in cool buildings, in excess it can make buildings uncomfortably hot in summer. There is thus an upper limit to the size of windows that can be used without thermal discomfort in sunny spells. Window design is therefore a compromise; if the window size is increased, the daylight illumination and the view through the window are improved but, beyond a certain size, overheating problems arise. The only visual effects limiting window size are glare from sun and sky and loss of privacy, but these can be overcome by fitting internal blinds usually of the venetian type. Although internal blinds can protect occupants from the direct heating effect of sunshine, they do not lower internal temperatures appreciably. External blinds are more effective but high cost and maintenance difficulties usually prevent their provision. In addition, large windows will result in excessive heat loss in cold weather.

BRE Digests 41 and 42[4] explain how the amount of daylight received in buildings is most conveniently expressed in terms of the percentage ratio of indoor-to-outdoor illumination, the ratio being called the *daylight factor*; the method of computation is detailed in the digests. For any given situation, the value of the factor varies with the sky conditions, the size, shape and position of the windows, the effect of any obstructions outside the windows and the reflectivity of the external and internal surfaces. Its value can be determined at the design stage by measurement from a model of the building or, more frequently, by calculation from drawings and other data. The quality and intensity of daylight varies with latitude, season, time of day and local weather conditions. The value is generally based on a heavily overcast sky, although improved conditions are likely to exist for about 85 per cent of normal working time throughout the year.

Positioning and Subdivision of Windows

The glass line or sill level is often about 675 or 750 mm above floor level in living rooms to give maximum vision, possibly increasing to about 900 mm in bedrooms, where rather more privacy is usually required, and 1050 mm in bathrooms and kitchens, where fittings are often located under the windows. The tops of windows should be fairly close to the ceiling to obtain good ventilation and lighting. Horizontal framing members (transoms) and glazing bars should not be positioned at heights where they will restrict the vision of occupants (eye level of persons standing; 1500 to 1600 mm and persons sitting in dining chairs: 1100 to 1150 mm above floor level).

It is now less common to subdivide windows into small panes with glazing bars, as they interfere with the vision of the occupants, make cleaning more difficult and increase painting costs. Some however favour their inclusion on aesthetic grounds, arguing that they give 'scale' to a small dwelling. Where window panes are used, the ratio of width to height should desirably be 2 : 3 to ensure good proportions. Leaded lights give character to a small dwelling but they are costly, break up vision, restrict light and make cleaning more difficult. Diamond-shaped leaded lights increase the risk of leakage.

Window Types

Windows may be classified in two different ways or a combination of them.

(1) The method of opening
   (a) casements which are side hung or top hung on hinges;
   (b) pivot hung either horizontally or vertically;
   (c) double-hung sashes which slide vertically;
   (d) sliding sashes which slide horizontally.

(2) The materials from which they are made: steel; aluminium; timber.

Each of the different types of window are now examined.

WOOD CASEMENT WINDOWS

With casement windows, a solid wood frame is fixed to the edges of the opening and this receives the glazed casements which may be side hung or top hung. The window frame is sometimes subdivided; the vertical divisions are known as mullions and the horizontal members as transoms (figure 8.1.2). Casements may be subdivided into smaller areas by glazing bars. BS 644, Part 1[5] prescribes the quality, design and construction of outward-opening wood casement windows, and CP 153[17] details timber species and protective treatments.

The British Standard prescribes minimum effective standards, and manufacturers often produce casement windows of heavier sections. British Standard wood casement windows are made in widths of 438, 641, 1226, 1810 and 2394 mm and heights of 768, 1073, 1226, 1378 and 1530 mm.

Figure 8.1.2 illustrates a typical two-light wood casement window (the two refers to the number of lights in the width of the window). The window opening is spanned externally by a brick-on-edge arch backed by a reinforced concrete lintel. As the flat arch has little strength it is supported by a mild steel angle, often 75 x 75 x 6 mm, with ends built into the brickwork and the exposed edge painted for protection. Other variations would be to use a precast reinforced concrete boot lintel as shown in figure 8.2.2 or a galvanised steel lintel (figure 8.2.1). The window frame may be fixed to the sides of brick jambs as in figure 8.1.2 or be set behind recessed jambs to give additional protection from the weather. With cavity walls, the frame is often set about 38 mm back from the outer wall face and so receives little protection from the weather. The joint between the frame and the brick jambs should be sealed with mastic. The wood frame or jamb may be fixed to the brickwork by screwing or nailing to hardwood plugs let into mortar joints or by right-angled galvanised steel cramps, with one leg screwed to the back of the frame and the other built into a mortar joint. The external sill in figure 8.2.1 consists of a wood sill, preferably of hardwood to improve its weathering qualities, which overhangs the face of the brickwork. Where the window is set well back from the outer wall face, it is necessary to incorporate a subsill below the main sill. The subsill could be made of timber, precast concrete, stone, bricks or a double course of roofing tiles.[14]

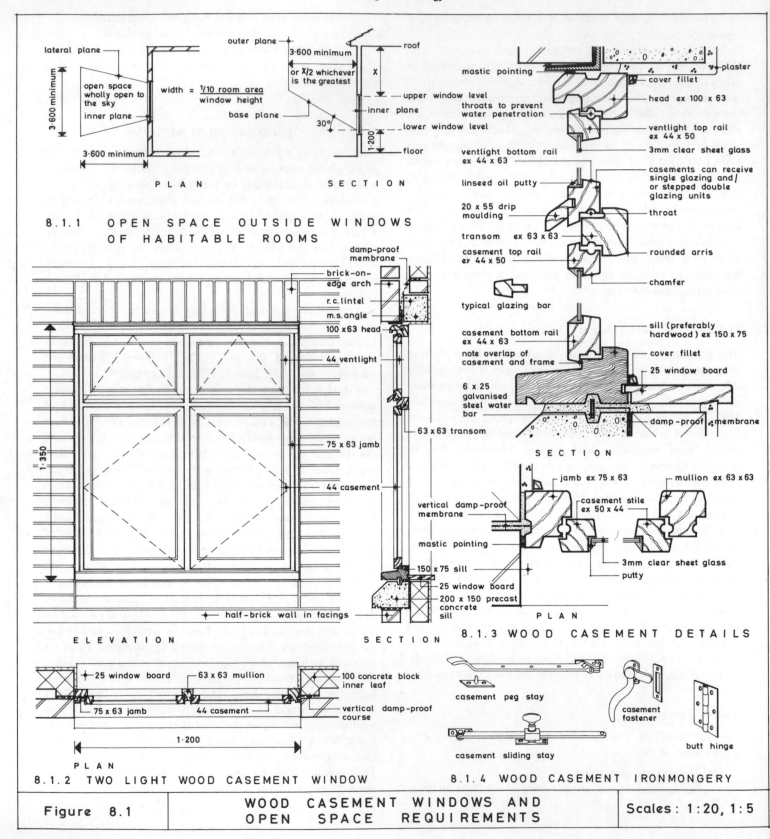

lateral plane
open space wholly open to the sky
inner plane
3·600 minimum
3·600 minimum
width = 1/10 room area / window height
PLAN

outer plane
3·600 minimum / or X/2 whichever is the greatest
base plane
30°
X
1·200
roof
upper window level
inner plane
lower window level
floor
SECTION

8.1.1   OPEN SPACE OUTSIDE WINDOWS OF HABITABLE ROOMS

damp-proof membrane
brick-on-edge arch
r.c. lintel
m.s. angle
100 x 63 head
44 ventlight
63 x 63 transom
75 x 63 jamb
44 casement
1·350
half-brick wall in facings
ELEVATION

vertical damp-proof membrane
mastic pointing
150 x 75 sill
25 window board
200 x 150 precast concrete sill
SECTION

25 window board
63 x 63 mullion
75 x 63 jamb
44 casement
100 concrete block inner leaf
vertical damp-proof course
1·200
PLAN

8.1.2 TWO LIGHT WOOD CASEMENT WINDOW

mastic pointing
plaster
cover fillet
head ex 100 x 63
throats to prevent water penetration
ventlight top rail ex 44 x 50
3mm clear sheet glass
ventlight bottom rail ex 44 x 63
casements can receive single glazing and/ or stepped double glazing units
linseed oil putty
20 x 55 drip moulding
throat
transom ex 63 x 63
casement top rail ex 44 x 50
rounded arris
chamfer
typical glazing bar

casement bottom rail ex 44 x 63
note overlap of casement and frame
sill (preferably hardwood) ex 150 x 75
cover fillet
25 window board
6 x 25 galvanised steel water bar
damp-proof membrane
SECTION

jamb ex 75 x 63
mullion ex 63 x 63
casement stile ex 50 x 44
vertical damp-proof membrane
mastic pointing
3mm clear sheet glass
putty
PLAN

8.1.3 WOOD CASEMENT DETAILS

casement peg stay
casement fastener
casement sliding stay
butt hinge

8.1.4 WOOD CASEMENT IRONMONGERY

| Figure 8.1 | WOOD CASEMENT WINDOWS AND OPEN SPACE REQUIREMENTS | Scales : 1:20, 1:5 |

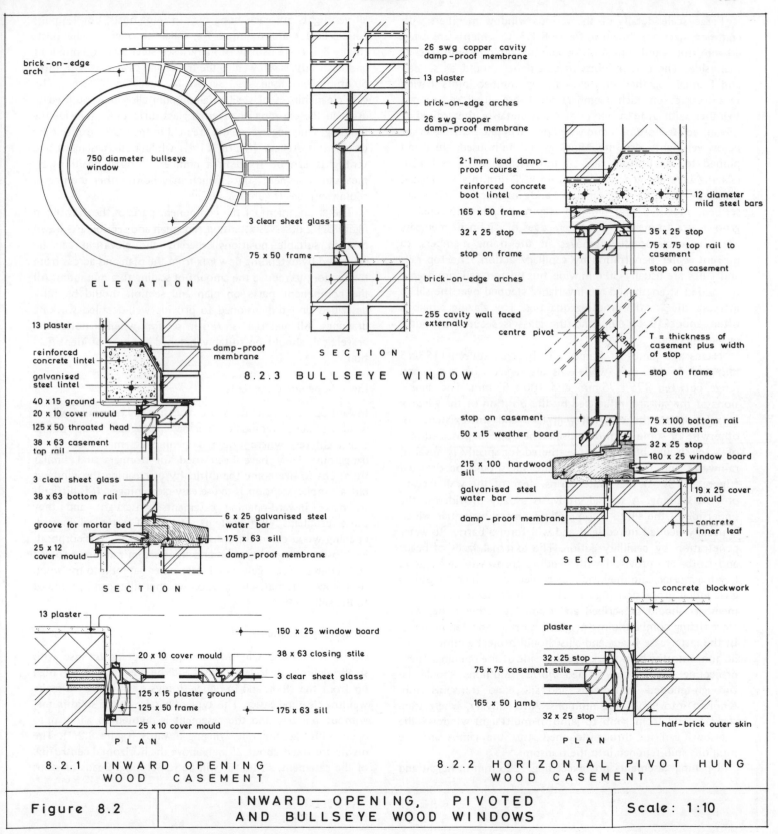

brick-on-edge arch

750 diameter bullseye window

ELEVATION

26 swg copper cavity damp-proof membrane

13 plaster

brick-on-edge arches

26 swg copper damp-proof membrane

3 clear sheet glass

75 x 50 frame

brick-on-edge arches

255 cavity wall faced externally

SECTION

8.2.3 BULLSEYE WINDOW

2·1 mm lead damp-proof course

reinforced concrete boot lintel

165 x 50 frame

32 x 25 stop

stop on frame

12 diameter mild steel bars

35 x 25 stop

75 x 75 top rail to casement

stop on casement

centre pivot

T = thickness of casement plus width of stop

stop on frame

stop on casement

50 x 15 weather board

215 x 100 hardwood sill

galvanised steel water bar

damp-proof membrane

75 x 100 bottom rail to casement

32 x 25 stop

180 x 25 window board

19 x 25 cover mould

concrete inner leaf

SECTION

13 plaster

reinforced concrete lintel

galvanised steel lintel

40 x 15 ground

20 x 10 cover mould

125 x 50 throated head

38 x 63 casement top rail

damp-proof membrane

3 clear sheet glass

38 x 63 bottom rail

groove for mortar bed

25 x 12 cover mould

6 x 25 galvanised steel water bar

175 x 63 sill

damp-proof membrane

SECTION

13 plaster

20 x 10 cover mould

125 x 15 plaster ground

125 x 50 frame

25 x 10 cover mould

150 x 25 window board

38 x 63 closing stile

3 clear sheet glass

175 x 63 sill

PLAN

8.2.1 INWARD OPENING WOOD CASEMENT

concrete blockwork

plaster

32 x 25 stop

75 x 75 casement stile

165 x 50 jamb

32 x 25 stop

half-brick outer skin

PLAN

8.2.2 HORIZONTAL PIVOT HUNG WOOD CASEMENT

| Figure 8.2 | INWARD—OPENING, PIVOTED AND BULLSEYE WOOD WINDOWS | Scale: 1:10 |

Large scale details of the various window members with common sizes are shown in figure 8.1.3. Casements are often 44 mm thick, and consist of top and bottom rails with stiles at each side. The corner joints of casements should be scribed and framed together with close-fitting combed joints, having two tongues on each member, which are glued and pinned together with at least one nonferrous metal or sherardised or galvanised steel pin or 6 mm wood peg. Where casements occur below ventlights, drip mouldings should be housed, glued and pinned to the bottom rails of the ventlights (figure 8.1.3). Glazing bars are generally about 44 x 22 mm in size, rebated on both sides to receive glass, and intersecting joints are scribed, mortised and tenoned (figure 8.1.3). Anti-capillary grooves are formed in the outer edges of casement members opposite corresponding grooves in the frame members, to prevent water penetration by capillary action. The top rails and stiles are usually 50 mm wide, but bottom rails are deeper for added strength and are invariably stepped over the sill to increase the weatherproofing qualities. The glazing rebate is often about 19 x 9 mm, and the glass is secured by putty or glazing beads.

Heads and jambs of frames vary in size between 115 x 75 mm and 75 x 63 mm, while sills are larger with a common range between 175 x 75 mm and 100 x 63 mm. The dimensions of the sill are influenced by the position of the window in relation to the external wall face and the existence or otherwise of some form of subsill. The upper exposed surface of a sill should be suitably weathered for speedy removal of rainwater and the underside of the sill should be throated (figure 8.1.3) so that water drips clear of the wall face to prevent staining. A second groove is usually provided to the underside of the sill to accommodate a metal water bar, which is usually bedded in red lead and will form a barrier to water penetration by capillary action. The external faces of heads and jambs are provided with bedding grooves to give a good key for mortar, and the head may project over the casement to give added protection (figure 8.1.3). All corner joints of these members should be scribed and framed together either with close-fitting combed joints or glued mortise and tenon joints. In the latter case heads and sills should project a minimum of 38 mm into the brickwork at either side of the opening. These projecting pieces are termed *horns*. Mullions should be through-tenoned into heads and sills, and transoms stub-tenoned into jambs and mullions. Alternatively, where transoms extend from jamb to jamb of multi-light windows, the transoms can be through-tenoned into the jambs and the mullions stub-tenoned into the transoms.[5]

Opening casements not exceeding 1226 mm in height and all ventlights are each hung on a pair of butts, but 1½ pairs (three butts) are required for taller casements. The butts (figure 8.1.4) are usually of sherardised steel 50 or 63 mm long and each butt is screwed to the frame and casement with three 50 mm sherardised steel screws to each flap. Alternatively, easy-clean hinges of steel or aluminium alloy may be used to assist in the cleaning of external glass surfaces from inside the building. Opening casements are held in the closed position by casement fasteners (figure 8.1.4), while both casement and ventlights can be fixed in a number of open positions by means of casement stays, which may be of either the peg or sliding varieties (figure 8.1.4).

The arrangement of the component parts of the drawing in figure 8.1.2 deserves attention. The plan and section are drawn first in suitable positions whereby the elevation can be produced by extending upwards from the plan and across from the section, to reduce the amount of scaling to a minimum. All the component parts on plan and section should be fully described and dimensioned to provide well-detailed working drawings. All materials shown in section, whether horizontal or vertical, should be suitably hatched for ease of identification.

Inward-opening Casements

Inward-opening casements are occasionally provided where windows abut verandahs, narrow passageways or public thoroughfares, where outward-opening casements could be dangerous. They have been used for dormers and similar situations to overcome the difficulty of cleaning the windows but a simpler solution is to use easy-clean hinges. It is difficult to make inward-opening casements watertight and they cause problems with curtains. Details of a typical inward-opening wood casement are shown in figure 8.2.1, incorporating closing or meeting stiles and the bottom rails are throated and rebated over a galvanised steel water bar let into the wood sill below. Alternatively, a wood weather fillet can be screwed to the sill.

Pivot-hung Casements

Centre pivot-hung windows are quite popular for use with small windows to toilets, larders and the like, in positions high up from the floor and in upper storeys of buildings to make window-cleaning easier. This type of window has a solid frame, without rebates, and the casement is pivoted to allow it to open with the top rail swinging inwards (figure 8.2.2). The pivots are fixed about 25 mm above the horizontal centre line of the casement, so that it will be self-closing. Beads or stops are fixed to certain sections of frame to replace the rebate and to form a stop for the casement. As shown in figure 8.2.2 the external stop to the upper part is fixed to the frame and the lower part to the casement; this is reversed for internal stops where the internal stop to the upper part is fixed to the casement and the lower part to the frame. The stops must be cut in the correct positions to enable the casement to open and close freely. A convenient way to obtain the cut lines on the stops is to draw a circle around the pivot point having a radius equal to the thickness of the casement plus the width of one stop plus 3 mm. A weather board about 50 x 15 mm in size may be fitted to the bottom rail of the casement (figure 8.2.2) to direct rainwater away from the foot of the window and to help stiffen the stops fixed to the outside of the casement. Pivot-hung casements are generally hung with adjustable friction pivot hinges, with the initial opening controlled by a cranked roller armstay, and there is usually a four-point locking system. When fully reversed the casement can be locked by a catch and held for cleaning purposes. It is also possible for casements to be pivoted vertically.

Horizontal Sliding Casements

Casements can be fitted to slide horizontally in a similar manner to sliding doors. They can be fitted with rollers to the bottom rails running on a brass track fixed to the sill and with the top rail sliding in a guide channel incorporated in the frame. Cheaper methods incorporate fibre or plastic gliders and tracks.

French Casements

When casements extend to the floor so that they can be used as doors to give access to balconies, gardens and the like, they are termed *French casements*. French casements may be provided singly or in pairs and frequently open inwards. They may have sidelights added to give additional light to rooms and entrance halls. The bottom or kicking rail is normally not less than 200 mm wide and the glass should be fixed with beads to reduce the risk of breakage.

Bullseye Windows

Circular or bullseye windows are sometimes used as features to give added interest to an elevation. A typical bullseye window is illustrated in figure 8.2.3 in the form of a fixed light with the glass bedded into the frame. The opening is formed with brick-on-edge arches with a copper vertical damp-proof course sealing the cavity. An additional damp-proof membrane is inserted above the window to disperse water in the cavity at this point. The internal reveals to bullseye windows are usually plastered and decorated.

DOUBLE-HUNG SASH WINDOWS

Sash windows are those in which the sashes slide up and down and they normally consist of two sashes, placed one above the other. Where both sashes open they are termed *double-hung sash windows* and if the top sash only opens it is known as a *single-hung sash window*. BS 644[5] describes how the sashes may either be hung from cased frames with counterbalancing weights, usually about 50 mm diameter of iron (figure 8.3.1), or from solid frames with spring devices (figure 8.3.2). With cased frames, two-light windows have boxed centre mullions and three-light windows have fixed side lights and solid mullions, with the cords passing over the sidelights. With solid frames, three-light windows normally have fixed sidelights, only the centre sashes being made to open.

The cased frame to accommodate the weights consists of an inner and outer lining, pulley stile and back lining. The parting bead between the two sashes is housed into the pulley stile. The construction of the head is often similar to that of the sides, with the omission of the back lining. Figure 8.3.1 shows how the sashes can slide past one another, the lower sash being placed on the inside. The outer lining, parting bead and inner or staff bead form the recesses in which the sashes slide. The sashes are hung to cords or chains which pass over a pulley, often 44 mm diameter in cast iron with steel axles and brass bushes, in the pulley stiles and to which the weights are attached.

To allow the removal of the sashes for adjustment or the renewal of broken cords, the inner lining is stopped and a removable bead nailed to it. When the inside bead is removed the sash can be swung inwards. Similarly, by removing the parting bead the top sash can be taken out. A pocket is formed in the pulley stile to provide access to the weights. To prevent the weights colliding or the cords becoming entangled, a parting slip is inserted in the cased frame, and is suspended from the soffit lining by a wedge passing through it. An inner and removable bead is fixed all round the frame, and a deeper bead (draught stop) is often fixed to the sill (figure 8.3.2). This permits some ventilation to be obtained between the meeting rails without having an opening at the bottom of the window. The meeting rails comprise the top rail of the lower sash and the bottom rail of the upper sash and they are kept as shallow as possible to cause the least possible obstruction. They are thicker than the other sash members to accommodate the space occupied by the parting bead, and are

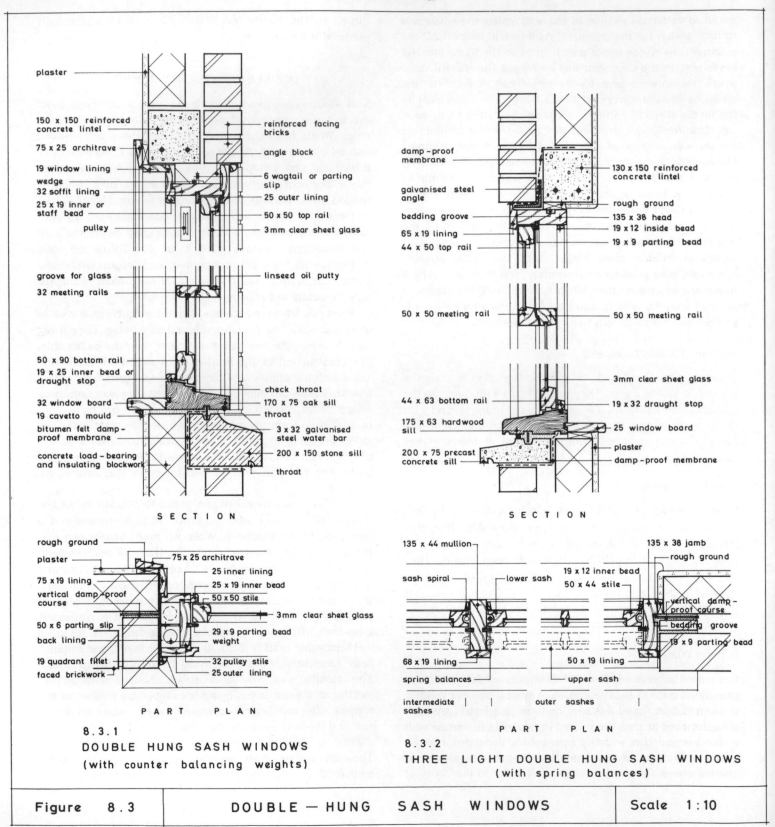

plaster

150 x 150 reinforced concrete lintel

75 x 25 architrave

19 window lining wedge

32 soffit lining

25 x 19 inner or staff bead

pulley

reinforced facing bricks

angle block

6 wagtail or parting slip

25 outer lining

50 x 50 top rail

3mm clear sheet glass

groove for glass

32 meeting rails

linseed oil putty

50 x 90 bottom rail

19 x 25 inner bead or draught stop

32 window board

19 cavetto mould

bitumen felt damp-proof membrane

concrete load-bearing and insulating blockwork

check throat

170 x 75 oak sill

throat

3 x 32 galvanised steel water bar

200 x 150 stone sill

throat

S E C T I O N

rough ground

plaster

75 x 19 lining

vertical damp-proof course

50 x 6 parting slip

back lining

19 quadrant fillet

faced brickwork

75 x 25 architrave

25 inner lining

25 x 19 inner bead

50 x 50 stile

3mm clear sheet glass

29 x 9 parting bead

weight

32 pulley stile

25 outer lining

P A R T   P L A N

8.3.1

DOUBLE HUNG SASH WINDOWS
(with counter balancing weights)

damp-proof membrane

galvanised steel angle

bedding groove

65 x 19 lining

44 x 50 top rail

130 x 150 reinforced concrete lintel

rough ground

135 x 38 head

19 x 12 inside bead

19 x 9 parting bead

50 x 50 meeting rail

50 x 50 meeting rail

3mm clear sheet glass

44 x 63 bottom rail

175 x 63 hardwood sill

200 x 75 precast concrete sill

19 x 32 draught stop

25 window board

plaster

damp-proof membrane

S E C T I O N

135 x 44 mullion

sash spiral

68 x 19 lining

spring balances

intermediate sashes

lower sash

135 x 38 jamb

rough ground

19 x 12 inner bead

50 x 44 stile

vertical damp-proof course

bedding groove

19 x 9 parting bead

50 x 19 lining

upper sash

outer sashes

P A R T   P L A N

8.3.2

THREE LIGHT DOUBLE HUNG SASH WINDOWS
(with spring balances)

| Figure   8.3 | DOUBLE — HUNG   SASH   WINDOWS | Scale   1 : 10 |

splayed and rebated on their adjoining faces to fit tightly together when the window is closed. A rebated splay drawn in the wrong direction would keep the sashes permanently closed. The lower sash is normally provided with sash lifts on the bottom rail for raising, and a sash fastener on the meeting rails provides a means of fastening the sashes in the closed position.

A number of balancing devices are now available to dispense with cords and expensive cased or boxed frames. Often spring balances with a metal case receive a rustless steel tape which winds onto a revolving drum leaving its free end to be fixed to a sash. The required degree of balance is obtained by adjusting the spring drum. Another form of spring balance consists of a torsion spring and helical rod enclosed in a metal tube, with the rod passing through a nylon bush which causes the spring to wind or unwind as the sash moves. The barrel of the balance is housed in either the sash or frame.

Sash windows vary between 750 and 1200 mm in width and 1350 and 2100 mm in height and, for stability, the width of each sash should not exceed twice its height. Panes in sash windows are rarely less than 225 mm wide x 300 mm high. The dimensions and rather formal character of sash windows generally make them unsuitable for small dwellings, although they are often considered to be superior to casements both on aesthetic and functional grounds, through better control of ventilation. One major disadvantage of sash windows in the past has been high maintenance costs stemming from the replacement of broken sash cords, but the use of chains, albeit noisy, and the more recently introduced spring balances have overcome this problem.

DEFECTS IN WOOD WINDOWS

Building Research Establishment Digest 73[6] describes how window joinery in newly built houses has sometimes given cause for complaint. Decay is particularly marked in ground-floor windows, especially in kitchens and bathrooms, and the lower parts of the windows (sills, bottoms of jambs and mullions, and lower rails of opening lights) are most vulnerable. Decay is generally of the 'wet rot' variety resulting from the use of timber of low natural resistance. Hence it is advisable to either use well-seasoned heartwood from timber, which is naturally resistant to decay, or timber treated with preservative, followed in both cases by two coats of aluminium primer. Surfaces should be designed to shed rainwater satisfactorily on the outside and condensation internally. The use of flimsy sections in window joinery, particularly where weather conditions are severe, should be avoided.

METAL WINDOWS

Steel Windows

Steel windows are fabricated from hot-rolled steel sections, mitred and welded at the corners, while subdividing bars are hot-tenon riveted to the frames and each other. Steel windows are hot-dip galvanised to BS 729[8] to resist corrosion, and some manufacturers supply windows with decorative coatings such as nylon or silicon polyester applied over the galvanising. Windows may be coupled together by the use of mullions (vertical coupling bars) and transoms (horizontal coupling bars) as shown in figures 8.4.4 and 8.4.5. Steel windows for domestic and similar buildings normally meet the requirements of BS 990[7] in the module 100 range where windows can be fixed, side, top and bottom hung or be reversible horizontally pivoted casements, and can vary in width from 500 to 1800 mm and in height from 200 to 1500 mm. All steel windows are supplied complete with fittings (hinges, handles and stays), but the design and quantity of fittings may vary between manufacturers. Handles and stays are usually of brass, zinc-based or aluminium alloy to various finishes. Optional accessories include safety devices, ventilators, remote controls, insect screens and pressed metal sills. Glazing should be carried out in accordance with CP 152[9] using metal casement putty, although galvanised steel or aluminium glazing bars may be used for large windows.

Steel windows are fixed to brick walls with metal or wire lugs (figure 8.4.3), to concrete with fibre plugs or screwed to timber surrounds (figure 8.4.2). The use of timber surrounds (figure 8.4.1) as described in BS 1285[10] improves the appearance of steel windows considerably, although they do increase the cost. It is necessary to point with mastic the gap between the metal window and the adjoining masonry or wood surround, and between wood surround and masonry to ensure a watertight joint in these vulnerable positions. Figure 8.4.1 also illustrates the use of an external tile subsill and an internal quarry tile sill. Steel windows with their lighter and thinner sections cause less obstruction than wood windows and they are not subject to warping, rot or insect attack. They do however need careful handling on the site, ample protection from corrosion by a good paint film and must not be subject to loads.[11] Steel sections are accurately rolled to provide a close fit between members, but additional draught-proofing or weatherstripping can be obtained by inserting a neoprene weatherstrip in a groove inside the opening frame.

Aluminium Windows

Aluminium windows are supplied in a variety of forms: fixed; bottom, side and top-hung casements; horizontally and

METAL CASEMENTS

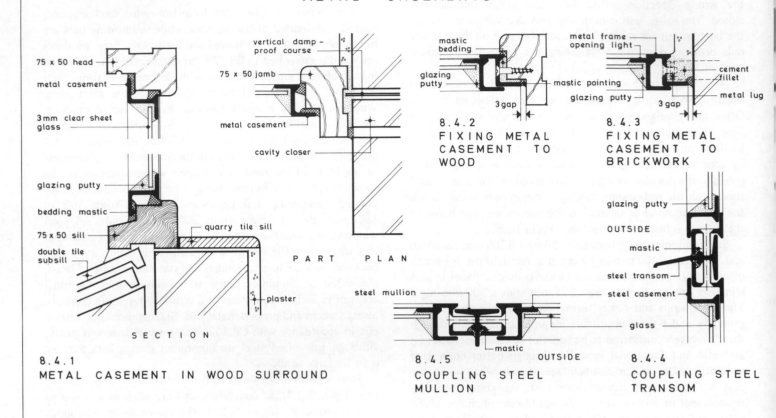

75 x 50 head

metal casement

3mm clear sheet glass

glazing putty

bedding mastic

75 x 50 sill

double tile subsill

plaster

S E C T I O N

8.4.1
METAL CASEMENT IN WOOD SURROUND

vertical damp-proof course

75 x 50 jamb

metal casement

cavity closer

P A R T   P L A N

mastic bedding

glazing putty

3 gap

8.4.2
FIXING METAL CASEMENT TO WOOD

metal frame opening light

mastic pointing

glazing putty

3 gap

cement fillet

metal lug

8.4.3
FIXING METAL CASEMENT TO BRICKWORK

quarry tile sill

steel mullion

mastic

OUTSIDE

8.4.5
COUPLING STEEL MULLION

glazing putty

OUTSIDE

mastic

steel transom

steel casement

glass

8.4.4
COUPLING STEEL TRANSOM

DOUBLE GLAZING AND DOUBLE WINDOWS

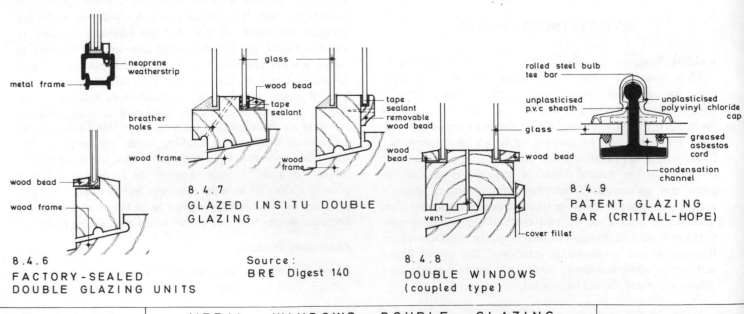

metal frame

neoprene weatherstrip

breather holes

wood bead

wood frame

8.4.6
FACTORY-SEALED DOUBLE GLAZING UNITS

glass

wood bead

tape sealant

wood frame

8.4.7
GLAZED INSITU DOUBLE GLAZING

Source:
BRE Digest 140

tape sealant

removable wood bead

wood frame

wood bead

vent

cover fillet

8.4.8
DOUBLE WINDOWS (coupled type)

rolled steel bulb tee bar

unplasticised p.v.c sheath

glass

wood bead

unplasticised polyvinyl chloride cap

greased asbestos cord

condensation channel

8.4.9
PATENT GLAZING BAR (CRITTALL-HOPE)

| Figure 8.4 | METAL WINDOWS, DOUBLE GLAZING AND DOUBLE WINDOWS | Scales 1:4, 1:2 |

vertically pivoted, including reversible pivoted windows; horizontal and vertical sliding windows; and horizontally louvred ventilating windows – all should comply with BS 4873.[12] The windows are fabricated from aluminium alloy extrusions,[12] and are provided with a standard range of hardware. The structural members can be single web or tubular sections which are mechanically jointed, cleated or welded. Glass may be inserted in glazing gaskets or non-setting compound with clip-on or screw-fixed glazing beads. The glass louvres in ventilating windows fit into blade holders of polypropylene or aluminium. Finishes to aluminium windows include anodic oxidised, anodised, and stoved organic. Top and bottom-hung casements vary from 600 to 1800 mm in width and 300 to 1800 mm in height, whereas horizontal sliding windows have widths varying from 900 to 2400 mm and heights from 500 to 1500 mm. Casement, sliding and pivoted windows can be coupled with mullion and transom bars to form co-ordinated assemblies. Aluminium windows need careful handling to avoid scratching surface coatings or bending members. Aluminium windows have slender sections, are attractive, durable and require only limited maintenance, but they may be double the initial cost of similar type wood and steel windows. Maintenance treatments are described in CP 153.[17]

GLASS AND GLAZING

Glass

Glass is one of our oldest materials but through modern research work it has been possible to alter the properties of glass to make it a more versatile material. The constituent materials are normally sand, soda ash, limestone, dolomite, felspar, sodium sulphate and cullet (broken glass), which are mixed, melted and refined; the glass is then either drawn, cast or rolled, annealed, possibly polished, and cut to the required sizes. BS 952[13] recognises and describes a number of different types of glass.

Transparent glasses. These transmit light and permit clear vision through them; they include sheet glass and clear plate glass. *Sheet glass* has natural fire-finished surfaces and as the two surfaces are never perfectly flat and parallel, there is always some distortion of vision and reflection. It varies in thickness from 2 to 6.75 mm, can be coloured and is supplied in four qualities:

(1) ordinary glazing quality (OQ) for general glazing purposes;
(2) selected glazing quality (SQ);
(3) special selected quality (SSQ) for high-grade work such as pictures and cabinets;
(4) horticultural: an inferior quality.

For panes exceeding 1 m² clear sheet glass should be at least 4 mm thick.

Float or *polished plate glass* has flat and parallel surfaces providing clear undistorted vision and reflection, produced either by grinding and polishing or by the float process. Generally, clear float glass has superseded polished plate glass in thicknesses up to 25 mm. It is normally 6 mm thick and is supplied in three qualities:

(1) glazing quality (GG);
(2) selected glazing quality (SG) and is also suitable for mirrors and bevelling;
(3) silvering quality (SQ) used for high class mirrors and wherever a superfine glass is required.

Translucent glasses. These transmit light with varying degrees of diffusion so that vision is not clear. They include rough-cast glass (textured on one surface), rolled glass, (narrow parallel ribs on one surface), fluted and ribbed glass (wider flutes or ribs), reeded glass (various patterns of ribs or flutes), and cathedral and figured rolled glass (one surface textured and other patterned). Shallow patterns give a partial degree of diffusion while the deeper patterns almost completely obscure.

Opal glasses. These may be white or coloured and have light-scattering properties due to the inclusion of small particles in the glass.

Glasses for special purposes. Probably the most important is *wired glass* with a wire mesh embedded in it, which holds the glass together on fracture, and is well suited for rooflights and similar situations. Georgian wired cast glass is a translucent glass with rough-cast finish and contains electrically welded 12 mm square wire mesh; Georgian polished wired glass is a transparent glass similarly wired but with two ground and polished surfaces; and hexagonal wired cast glass is a translucent glass with rough-cast finish containing 22 mm hexagonal wire mesh.

Prismatic glass is a translucent rolled glass, one surface of which consists of parallel prisms, while *heat-absorbing glass* is almost opaque to infra-red radiation and usually has a bluish-green tint. *Heat-resisting sheet glass* has a low coefficient of expansion and greater resistance to changes of temperature; *toughened glass* has increased resistance to external forces; *laminated safety glass* is less likely to cause severe cuts on fracture; and *mirror glass* is clear plate glass silvered on one face. *Leaded lights* are panels consisting of small pieces of glass held together with lead cames; while copper sections are used in *copper lights*, and these are sometimes used in the windows of small houses to give 'scale'.

Hollow glass blocks can be used in non-loadbearing partitions to permit light transmittance. The blocks or bricks are available in two standard sizes, 240 x 240 x 80 mm and 190 x 190 x 80 mm. They are hollow, translucent glass units with various patterns moulded on their interior or exterior faces, or on both; they are normally jointed in a weak-gauged mortar such as 1:1:8.

Glazing

Prior to glazing, timber rebates should be cleaned, primed and painted with one coat of oil paint, and metal rebates cleaned and primed in accordance with the recommendations of CP 152.[9] Glass should be cut to allow a small clearance at all edges and then be back-puttied, by laying putty along the entire rebates and bedding the glass solidly, sprigged for timber rebates (using small square nails without heads) and pegged for metal rebates, and neatly front puttied, taking care to ensure that the putty does not appear above the sight lines. Putty used for timber rebates should be linseed oil putty conforming to BS 544,[15] while that for metal rebates should be an approved metal casement putty. Glass to doors, screens and borrowed lights is best bedded in wash leather or plastics and held in place by wood beads fixed with brass screws in cups.

DOUBLE GLAZING AND DOUBLE WINDOWS

Various double glazing and double window systems are available stemming from the increasing demand for improved heat insulation in buildings. They range from simple 'do it yourself' *in situ* systems (figure 8.4.7) to the more sophisticated factory-produced hermetically sealed double-glazing units (figure 8.4.6) and from double-rebated frames (figure 8.4.7) to openable coupled casements and sashes (figure 8.4.8) or separate secondary windows. The optimum width of air space for vertical double glazing is usually taken as 20 mm, although widths down to 12 mm are almost equally effective.[16] An air space 12 to 20 mm wide halves the thermal transmittance, thus with normal exposure the thermal transmittance (U) of single glazing is 5.6 and that of double glazing 2.8 W/m^2 °C. For sound insulation, however, a minimum air space of 150 mm is required, although 200 to 300 mm is desirable.

Double glazing can reduce the risk of condensation on the glass because the surface exposed to the room is warmer than single glazing and is more likely to be above the prevailing dewpoint temperature, but it is less effective with high humidities, as in kitchens, and low standards of heating. Double glazing results in some reduction in light transmission as it could be about 70 per cent compared with 85 to 90 per cent for single glazing with clear glass up to 6 mm thick.

Factory-sealed Units

Sealed units with fused all-glass edges are being increasingly used as well as those with a welded glass-to-metal seal (figure 8.4.6), but where use is made of flat sheets of glass bonded to spacing strips and sealed, the edges must be kept dry. Some manufacturers produce double-glazing units to fit into the rebates of standard wood or metal sections, without the use of beads. Stepped units are used where the frames are too small or unsuitable for enlargement. Airspace widths normally vary from 3 to 20 mm.[13]

Single-frame Double-glazing Systems Sealed *in situ*

Various arrangements are available including glazing to double rebated frames or fixing a second line of glazing with wood or plastic face beads, generally to existing frames (figure 8.4.7). No matter how well the glazing seals are made, the cavities cannot be expected to remain airtight indefinitely. The seals may disintegrate under movement and shrinkage, allowing water vapour to enter the air space and to condense on the inside of the outer glazing. The wood exposed to the air space should be painted or varnished to reduce the evaporation of moisture from the timber into the air space and breather holes should be provided at the rate of one 6 mm diameter hole per 0.5 m^2 of window (figure 8.4.7). The holes should be plugged with glass fibre or nylon to exclude dust and insects.[16]

Coupled or Sliding Double Sashes

These may be of wood or metal pivoted with openable coupled sashes (figure 8.4.8), or sliding with pairs of metal sliders in the same frame. The inner sashes should be well sealed when closed together, and in the coupled pivoted type, a ventilating slot is often left around the periphery of the outer sash, to ventilate the air space externally when the sashes are closed (figure 8.4.8), and so reduce the risk of condensation.

PATENT GLAZING

Patent glazing is used extensively for roof and vertical glazing, particularly in industrial and commercial buildings, on account of its high durability and light transmittance properties. The recommended minimum slope when used in roofs is 15°. A typical patent glazing bar is illustrated in figure 8.4.9, incorporating a rolled steel bulb tee bar sheathed in unplasticised PVC to form internal condensation channels and an unplasticised PVC cap to form a watertight joint with the glass. The glass is seated on greased asbestos cord. Another form of patent glazing bar is of aluminium alloy with aluminium wings or cap to provide a watertight joint. Requirements for patent glazing are specified in BS 5516.[19]

REFERENCES

1. The Building Regulations 1976. SI 1676. HMSO (1976)
2. *BRE Digest 206*: Ventilation requirements. HMSO (1977)
3. *BRE Digest 210*: Principles of natural ventilation. HMSO (1978)
4. *BRE Digests 41 and 42*: Estimating daylight in buildings. HMSO (1970/1969)
5. BS 644 Part 1: 1951 Wood casement windows, Part 2: 1958 Wood double-hung sash windows
6. *BRE Digest 73*: Prevention of decay in external joinery. HMSO (1978)
7. BS 990: Steel windows, Part 2: 1972 Metric units
8. BS 729: 1971 Hot-dip galvanised coatings on iron and steel articles
9. CP 152: 1972 Glazing and fixing of glass for buildings
10. BS 1285: 1963 Wood surrounds for steel windows and doors
11. DOE Advisory leaflet 12: Standard metal windows — fixing, pointing and glazing. HMSO (1976)
12. BS 1470: 1972 Wrought aluminium and aluminium alloys — plate, sheet and strip; BS 1474: 1972 Wrought aluminium alloys — bars, extruded round tube and sections; BS 4873: 1972 Aluminium alloy windows
13. BS 952: Glass for glazing, Part 1: 1978 Classification
14. BS 5642: 1978 Sills and copings
15. BS 544: 1969 Linseed oil putty for use in wooden frames
16. *BRE Digest 140*: Double glazing and double windows. HMSO (1972)
17. CP 153: Windows and rooflights, Part 2: 1970 Durability and maintenance
18. *BRE Digest 190*: Heat losses from dwellings. HMSO (1976)
19. BS 5516: 1977 Code of practice for patent glazing

9 DOORS

Consideration of general design principles is followed by an examination of the various types of door and their uses and the constructional techniques employed, together with the main characteristics of frames and linings. Finally, metal doors and ironmongery are investigated.

GENERAL PRINCIPLES OF DESIGN OF DOORS

Doors form an important part of joinery work, which can be defined as 'the art of preparing and fixing the wood finishings of buildings'. Carpentry work primarily makes use of sawn or unwrought timber, whilst joinery work embraces almost entirely planed or wrought timber, including hardwoods as well as softwoods. Most joinery work is exposed to view and is usually painted or polished.

The main principles to be observed in the construction of doors and framing of joiner's work generally, may be summarised as follows

(1) Timber should be dry and well seasoned with a moisture content within the following limits prescribed in BS 1186 Part 1.[1]

Type of joinery	per cent
External joinery	17
Internal joinery (with intermittent heating)	15
Internal joinery (with continuous heating: 12 to 18°C)	12
Internal joinery (with continuous heating: 20 to 24°C)	10
Internal joinery (close to heat source)	8

(2) Timber should be free from serious defects as listed in BS 1186,[1] such as excessive deviation from straightness of grain; large checks, splits and shakes; knots except sound, tight, small knots; pitch pockets; decay and insect attack.

(3) The work should consist of timber which is suitable for the particular situation; for example, hemlock and whitewood are suitable for internal doors but unsuitable for external doors.[1]

(4) The joints between timbers should permit movement due to variations in temperature or humidity without exposing open joints.

(5) The faces of members joined shall be flush with one another unless the design requires otherwise.

(6) The haunch in a tenon joint or the tongue in a dowel joint shall be a push fit in its groove.

Finished Sizes

It is common to specify the sizes out of which a joinery member is to be worked, and 3 mm should then be allowed for each wrought face. Thus a door frame specified as 100 x 75 mm (nominal) will have a finished size of 94 x 69 mm and a door with a nominal thickness of 38 mm will actually be 32 mm. Specification clauses for joinery work should distinguish between nominal and actual sizes. Full-size and 1:5 joinery details should be drawn to finished sizes, whereas nominal sizes can be used for smaller scale drawings.

DOOR TYPES

A variety of matters need consideration when deciding on the type of door and the associated constructional work around the door opening. This is now illustrated by reference to the front entrance door of a good quality dwelling house.

(1) Size of door to be adequate for all needs including passage of perambulators and furniture, probably 900 x 2100 mm opening.

(2) Adequate strength and durability, panelled or flush, 44 to 50 mm thick, well constructed and hung on adequate butt hinges.

(3) Attractive appearance: careful design of door including mouldings and door furniture.

(4) Weatherproofing qualities: consider provision of water bar in threshold, weatherboard to bottom rail of door, throats to frame and mastic pointing of joint between frame and reveal.

(5) Type of timber and finish: painted softwood or polished hardwood and whether glazing is required to give natural light in house.

(6) Adequate frame to support door and nature and extent of mouldings to frame.

(7) Treatment of reveals: for example, internal, whether plain or panelled linings or plastered and decorated; external, brick, stone or rendered.

(8) Head of opening: consider various alternatives; provision of arch, boot lintel, steel lintel, among other possibilities, and whether a fanlight is desirable.

(9) Threshold and steps: consider alternatives, such as brick, concrete, stone, terrazzo and clay tiles.

There is a wide range of door types available each with their own particular uses, and they can be broadly classified as panelled, flush and matchboarded. The majority of doors used in domestic work are standard doors complying with BS 459,[2] and made of timber and plywood conforming to BS 1186.[1] The usual range of sizes of doors and doorsets (doors and frames), as detailed in BS 4787,[15] follows

	Overall size of door opening (mm)	Door size (mm)
(1)	900 x 2100	826 x 2040 x 40 or 44 thick
(2)	800 x 2100	726 x 2040 x 40 thick
(3)	700 x 2100	626 x 2040 x 40 thick
(4)	600 x 2100	526 x 2040 x 40 thick

Type (1) doors are especially suitable for external doors, type (2) doors for most internal doors, type (3) doors may be useful for cloakrooms and cupboards and type (4) doors for small cupboards.

Panelled Doors

Panelled doors are usually described by the number of panels which they contain and which may vary from one to six as shown in figure 9.1, and the thickness and finish to the edges of the framing (stiles and rails). Panelled doors are framed by joining the members where they intersect by dowels (figure 9.2.4) or mortises and tenons (figure 9.2.3). Common finished sizes for framing are 94 x 34 or 44 mm stiles, muntins (intermediate vertical members) and top and intermediate rails, and 194 x 34 or 44 mm bottom and lock rails, with a minimum plywood panel thickness of 6 mm for doors with more than one panel and 9 mm for single panel doors. Panels are framed into grooves in the rails and stiles and panels should be about 2 mm smaller in height and width than the overall distance between grooves.[2] Mouldings to the edges of panel openings may take one of several forms; typical solid mouldings are shown in figures 9.1.10 and 9.2.1, while figure 9.2.2 shows a separate mould which is planted or nailed to the frame, but this constitutes a less satisfactory finish. In high-class work, doors may be provided with solid moulded and raised panels where the centre portion of the panel is thicker than the edges or margins (figure 9.1.10), or bolection moulding (figure 9.1.11) where the moulding projects beyond the face of the framing and covers any shrinkage in the panels. A flush panel with a recessed moulding incorporated in it is described as bead and butt (figure 9.1.9), and this gives a strong door and conceals the joint between the panel and frame. Openings for glazing are often rebated and moulded out of the solid (figure 9.1.2 and 9.1.4), whereas fully glazed doors (figure 9.1.5), are usually provided with separate mitred glazing beads. Some exterior doors (figure 9.1.4) are prepared to receive letter plates to BS 2911[3] with apertures 250 x 38 mm. It is advisable to provide weatherboards and water bars (figure 9.2.1) to external doors to ensure adequate protection against driving rain. Doors either hung singly or in pairs with substantial glazed areas (figures 9.1.5 and 9.1.6) are referred to as *wood casement doors* or *French casements*, and are useful for increasing light transmittance within a dwelling and providing access to balconies, patios and the like. A typical contemporary hardwood domestic front entrance door is illustrated in figure 9.1.12.

Mortise and tenon joints between framed members. These are illustrated in figure 9.2.3. BS 459[2] recommends that top and bottom rails and at least one other rail should be through mortised and tenoned. Other intermediate rails, muntins and glazing bars should be stub-tenoned to the maximum depth possible, often about 25 mm. Haunchings, to prevent rails twisting, should be not less than 10 mm deep and no tenon should be within 38 mm of the top or bottom of the door. Through tenons are wedged, and glue is applied to the tenon and shoulders before being inserted in the mortise. The thickness of the tenon should equal one-third of the stile and its width should not exceed five times this thickness or a maximum of 125 mm, whichever is the least.

Dowelled joints. These (figure 9.2.4) are becoming increasingly common in machine-made doors as they are cheaper than mortice and tenon joints and are usually equally satisfactory. The dowels are usually of hardwood but may be of the same material as the framing members.[2] They should have a minimum size of 16 x 122 mm, be slightly grooved to give a key for the glue and be equally spaced at distances not exceeding 56 mm centre to centre. There should be at least three dowels for lock and bottom rails, two for top rails and one for intermediate rails. The lowest dowel in a bottom rail should not be less than 44 mm from the bottom of the door.

Panelled doors have to some extent been superseded by flush doors, which with their large smooth surfaces are devoid

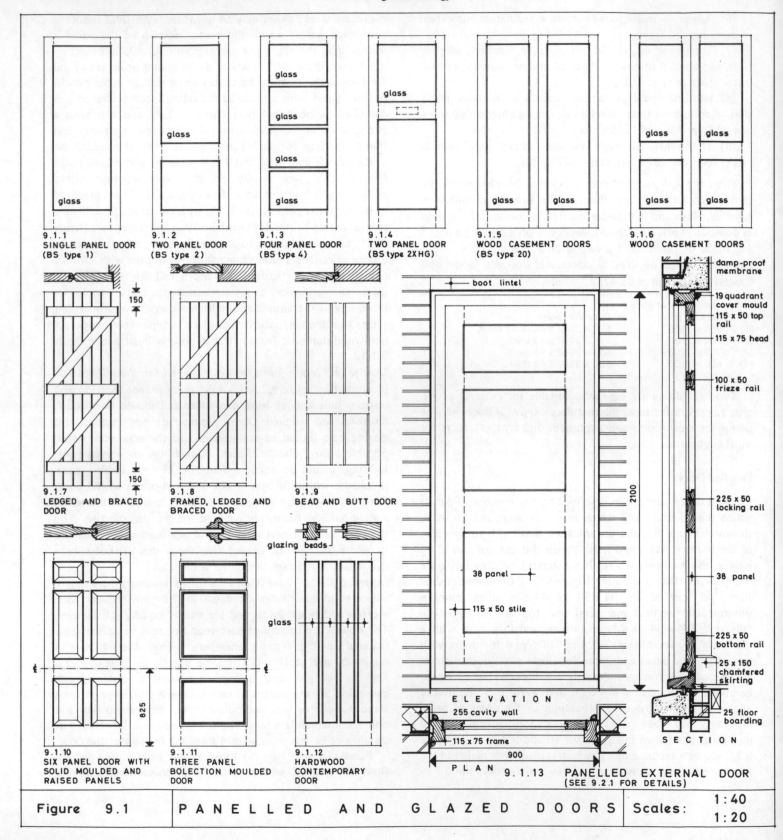

9.1.1
SINGLE PANEL DOOR
(BS type 1)

9.1.2
TWO PANEL DOOR
(BS type 2)

9.1.3
FOUR PANEL DOOR
(BS type 4)

9.1.4
TWO PANEL DOOR
(BS type 2XHG)

9.1.5
WOOD CASEMENT DOORS
(BS type 20)

9.1.6
WOOD CASEMENT DOORS

9.1.7
LEDGED AND BRACED
DOOR

9.1.8
FRAMED, LEDGED AND
BRACED DOOR

9.1.9
BEAD AND BUTT DOOR

9.1.10
SIX PANEL DOOR WITH
SOLID MOULDED AND
RAISED PANELS

9.1.11
THREE PANEL
BOLECTION MOULDED
DOOR

9.1.12
HARDWOOD
CONTEMPORARY
DOOR

9.1.13   PANELLED EXTERNAL DOOR
(SEE 9.2.1 FOR DETAILS)

ELEVATION

PLAN

SECTION

boot lintel

255 cavity wall

115 x 75 frame

900

2100

damp-proof membrane

19 quadrant cover mould

115 x 50 top rail

115 x 75 head

100 x 50 frieze rail

225 x 50 locking rail

38 panel

225 x 50 bottom rail

25 x 150 chamfered skirting

25 floor boarding

38 panel

115 x 50 stile

glazing beads

glass

150

150

825

| Figure   9.1 | P A N E L L E D   A N D   G L A Z E D   D O O R S | Scales: 1:40 / 1:20 |

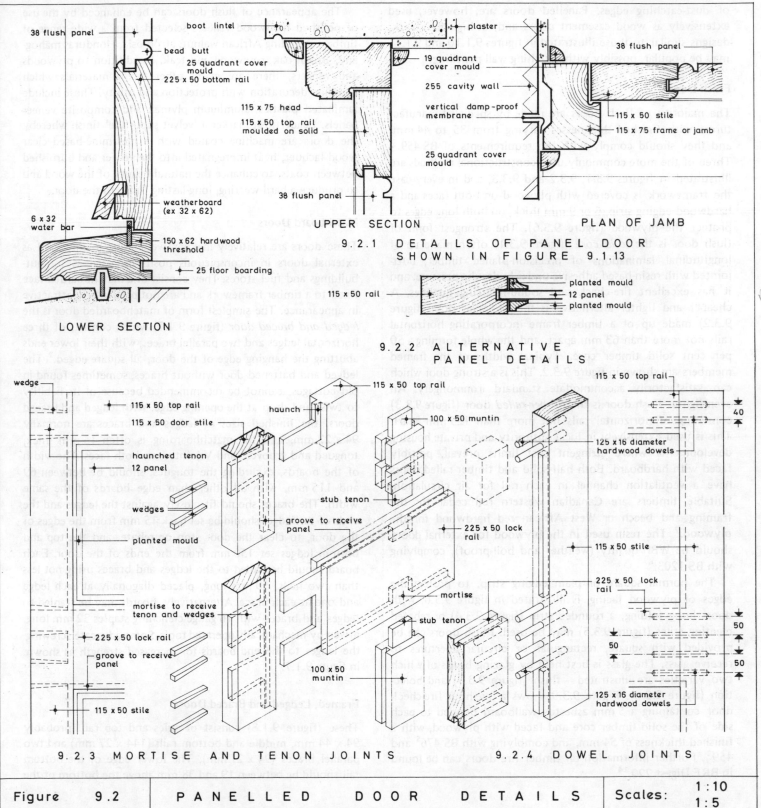

38 flush panel

bead butt

25 quadrant cover mould

225 x 50 bottom rail

boot lintel

6 x 32 water bar

weatherboard (ex 32 x 62)

150 x 62 hardwood threshold

25 floor boarding

LOWER SECTION

plaster

19 quadrant cover mould

255 cavity wall

vertical damp-proof membrane

115 x 75 head

115 x 50 top rail moulded on solid

25 quadrant cover mould

38 flush panel

UPPER SECTION

38 flush panel

115 x 50 stile

115 x 75 frame or jamb

PLAN OF JAMB

9.2.1 DETAILS OF PANEL DOOR SHOWN IN FIGURE 9.1.13

115 x 50 rail

planted mould

12 panel

planted mould

9.2.2 ALTERNATIVE PANEL DETAILS

wedge

115 x 50 top rail

115 x 50 door stile

haunched tenon

12 panel

wedges

planted mould

mortise to receive tenon and wedges

225 x 50 lock rail

groove to receive panel

115 x 50 stile

115 x 50 top rail

haunch

100 x 50 muntin

stub tenon

groove to receive panel

225 x 50 lock rail

mortise

stub tenon

100 x 50 muntin

9.2.3 MORTISE AND TENON JOINTS

115 x 50 top rail

40

125 x 16 diameter hardwood dowels

115 x 50 stile

225 x 50 lock rail

50

50

50

125 x 16 diameter hardwood dowels

9.2.4 DOWEL JOINTS

| Figure 9.2 | PANELLED DOOR DETAILS | Scales: 1:10 1:5 |

of dust-catching edges. Panelled doors are, however, used extensively as wood casement doors and, in high-class work, designs similar to those illustrated in figures 9.1.10 and 9.1.11 may be popular, possibly with matching wall panelling.

Flush Doors

The majority of flush doors are made by specialist manufacturers with finished thicknesses varying from 35 to 44 mm, and they should comply with the requirements of BS 459.[2] Three of the more commonly used constructional methods are illustrated in figures 9.3.1, 9.3.2 and 9.3.3, and in every case the framework is covered with plywood on both faces and a hardwood edging strip, 6 or 9 mm thick, on both long edges to protect the plywood (figure 9.3.6). The strongest form of flush door is the *solid core* (figure 9.3.1) often made up of longitudinal laminations of precision-planed timber, butt-jointed with resin-based adhesive under hydraulic pressure, and it has excellent fire-check and sound-reducing qualities. A cheaper and lighter alternative is the *half-solid* door (figure 9.3.2) made up of a timber frame incorporating horizontal rails not more than 63 mm apart, and the whole forming a 50 per cent solid timber core. Typical widths of the framed members are shown in figure 9.3.2. This is a strong door which can satisfactorily accommodate standard ironmongery. An even lighter flush door is the *timber-railed* door (figure 9.3.3) consisting of horizontal rails not more than 125 mm apart. This is used extensively in local authority and private housing developments where stringent cost limits prevail, possibly faced with hardboard. Both half-solid and timber-railed doors have a ventilation channel in each rail for air circulation. Suitable timbers are Canadian western red cedar for the framing and beech or West African red hardwood for the plywood.[7] The resin used in the plywood for external doors should be WBP quality (weather and boil-proof), complying with BS 1203.[6]

The normal square or plain edging strip, to protect the edges of plywood facing, is illustrated in figure 9.3.6. With doors hung folding, a rounded strip (figure 9.3.4) or rebated meeting stiles (figure 9.3.5) may be used. Flush doors can be provided with square, rectangular or circular apertures to receive glass. The glass is best held by glazing beads of which two varieties are illustrated — flush (figure 9.3.7) and bolection (figure 9.3.8). Figure 9.3.9 shows a one-hour fire-check door containing a 5 mm asbestos wallboard bonded to each side of the solid timber core and faced with plywood, with a finished thickness of 54 mm, and complying with BS 476[5] and 459[2]. Further information on timber fire doors can be found in BRE Digest 220.[14]

The appearance of flush doors can be enhanced by the use of polished hardwood veneers selected from a wide range of timbers including African walnut, afromosia, Honduras mahogany, iroko, oak, rosewood and teak. In addition to plywoods and veneers, there are many other facing materials which combine decoration with protection and utility. These include laminated plastics, aluminium plymax and composite veneer panels. Leaderflush market a 'velvet superfine' finish whereby the doors are machine coated with a melamine-based clear wood lacquer, heat impregnated into the veneer and burnished between coats, to enhance the natural beauty of the wood and to produce a hard-wearing, long-lasting finish to the door.

Matchboard Doors

These doors are relatively inexpensive and are mainly used as external doors in inconspicuous positions, such as for out-buildings and fuel stores. They consist of a matchboarded face fixed to a timber framework and are not particularly attractive in appearance. The simplest form of matchboarded door is the *ledged and braced door* (figure 9.1.7) which consists of three horizontal ledges and two parallel braces, with their lower ends abutting the hanging edge of the door, all square edged.[2] The ledged and battened door without braces, sometimes found in old cottages, cannot be recommended because of its liability to twist and drop at the opening edge. With ledged and braced doors, the finished sizes of ledges and braces are normally 94 x 22 mm, and the matchboarding is often 15 mm thick, tongued and grooved, and V-jointed on both faces. The width of the boards, excluding the tongues, should be between 69 and 115 mm, with all other than edge boards of the same width. The braces should fit closely against the ledges and the ends of the ledges should be set back 15 mm from the edges of the door, to clear the door stop or rebate, and the top and bottom ledges set 150 mm from the ends of the door. Each board should be nailed to the ledges and braces using not less than two nails, 50 mm long, placed diagonally at each ledge and one at each brace. Alternatively, boards may be stapled to ledges and braces with 16 gauge clenching staples 32 mm long, driven by mechanically operated tools. Some advocate screwing the ledges to the end boards for increased strength as shown in figure 9.1.7.

Framed, Ledged and Braced Doors

These (figure 9.1.8) consist of stiles and top rail (probably 94 x 44 mm, middle and bottom rails (144 x 27 mm) and two parallel braces (94 x 27 mm). The lower edge of the bottom rail should be between 19 and 38 mm above the bottom of the

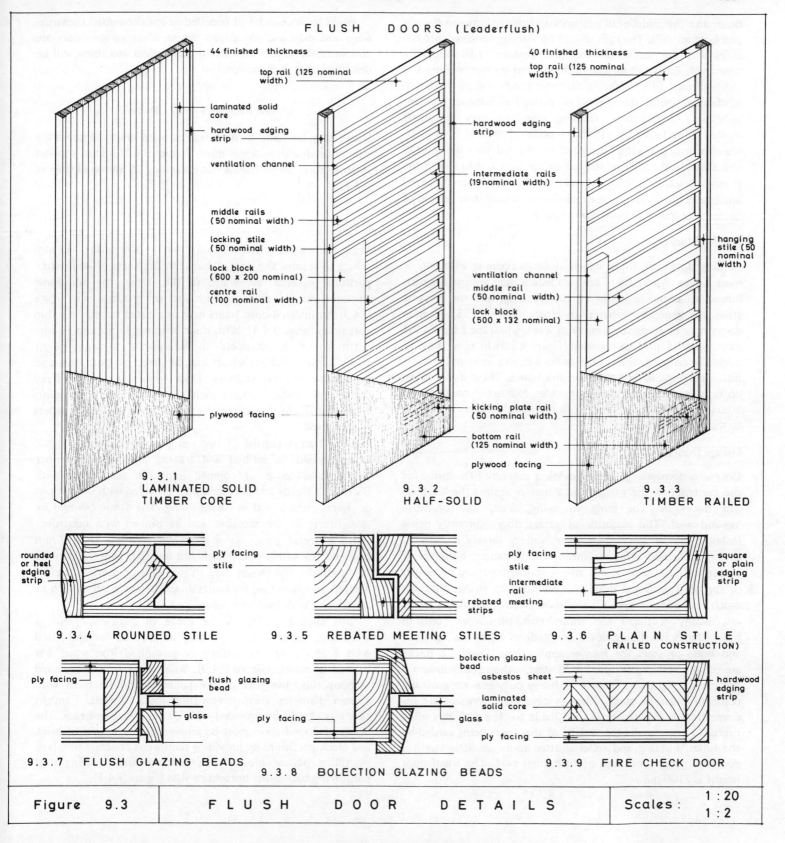

FLUSH DOORS (Leaderflush)

44 finished thickness

top rail (125 nominal width)

laminated solid core

hardwood edging strip

ventilation channel

middle rails (50 nominal width)

locking stile (50 nominal width)

lock block (600 x 200 nominal)

centre rail (100 nominal width)

plywood facing

9.3.1
LAMINATED SOLID
TIMBER CORE

40 finished thickness

top rail (125 nominal width)

hardwood edging strip

intermediate rails (19 nominal width)

ventilation channel

middle rail (50 nominal width)

lock block (500 x 132 nominal)

kicking plate rail (50 nominal width)

bottom rail (125 nominal width)

plywood facing

9.3.2
HALF-SOLID

hanging stile (50 nominal width)

9.3.3
TIMBER RAILED

rounded or heel edging strip

ply facing

stile

9.3.4 ROUNDED STILE

ply facing

stile

rebated strips

intermediate rail meeting

9.3.5 REBATED MEETING STILES

ply facing

stile

intermediate rail

square or plain edging strip

9.3.6 PLAIN STILE
(RAILED CONSTRUCTION)

ply facing

flush glazing bead

glass

9.3.7 FLUSH GLAZING BEADS

bolection glazing bead

ply facing

9.3.8 BOLECTION GLAZING BEADS

bolection glazing bead

asbestos sheet

laminated solid core

glass

ply facing

hardwood edging strip

9.3.9 FIRE CHECK DOOR

| Figure 9.3 | F L U S H   D O O R   D E T A I L S | Scales: | 1:20
 1:2 |

door, and the middle rail positioned centrally between the top and bottom rails. The rails should be through-tenoned into the stiles and the tenons to top and bottom rails should be haunched. Each tenon should be secured by two wedges and either pinned with a 10 mm diameter hardwood pin or 6 mm nonferrous metal star dowel, or bedded in weather-resistant adhesive. Alternatively, framing members may be dowel jointed using two 15 x 115 mm wooden dowels at each joint. Matchboarding is either tongued or rebated into the top rail and stiles, and nailed or stapled to the rails and braces in the manner described for ledged and braced doors. Framed, ledged and braced doors are more expensive but much stronger and of better appearance than unframed doors.

Folding Doors

Folding doors are those which close an opening with two or more leaves. Where there are two leaves they are usually each hinged to opposite jambs of the opening, and the meeting stiles are generally rebated to close (figure 9.3.5). Double doors may be hung to swing both ways, when the edges of the meeting stiles must be rounded (figure 9.3.4) to allow them to pass. The hanging stiles of the doors will also be rounded to fit into a corresponding hollow on the frame. These doors swing on centres, the lower being connected to a spring contained in a box set in the floor (floor spring) and which acts as a check to the swing of the door.

Garage Doors

Doors to domestic garages provide a measure of security for the contents of the garage and a barrier against the weather. The doors may be hung to swing, slide, fold or move 'up-and-over'. The majority of garage doors currently being installed are of the 'up-and-over' variety largely because of their simple and effective means of operation. Methods of suspension vary widely, but all are counterbalanced by weight or spring. Traditional double doors hung to swing, although much less popular are still available. Horizontal sliding leaves are usually sectional and slide 'round-the-corner'. Bottom roller tracks for sliding doors are made of nylon for light use and brass or steel for heavier applications. Overhead tracks are usually rolled galvanised mild steel, although aluminium is suitable for light use. Wheels for heavy-duty gear are generally made of brass or steel although nylon is often suitable for domestic use. Bottom tracks are liable to clog with dirt while overhead gear entails the weight of the doors being carried by the lintel. Folding and roller shutter doors are little used for domestic garages because of their high cost. The usual door height is 2100 mm.

Wood leaves can be of panelled or matchboarded construction, and may include glazed panels. Most garage doors are now manufactured from steel or aluminium and these will be described later in the chapter.

DOOR FRAMES AND LININGS

Doors may be hung to solid frames or to linings. Frames are mainly used with external doors (figure 9.4.7) and to support internal doors in partitions not exceeding 75 mm thick (figure 9.4.12).

Door Frames

Door frames are often 100 x 75 mm in size with a 12 mm rebate to receive the door. Where the thickness of an internal partition exceeds the width of the frame, the additional thickness may be made up with wrought grounds as in figure 9.4.3. An internal door frame may be rebated to receive a thin partition (figure 9.4.4). With door openings in 50 and 62 mm partitions, it is advisable to incorporate a storey-height frame (figure 9.4.2), which can be fixed at both top and bottom to increase stability. Door frames are usually fixed with three metal anchors each side built into brick joints (figure 9.4.7), but on occasions fixing bricks or timber pallets may be used.

Door frames consist of two uprights (jambs) and a head, which should be scribed and framed together with either mortise and tenon or combed joints, in accordance with BS 1567.[8] Heads and sills should be provided with projections or 'horns' not less than 38 mm long. The tenons should be close-fitting in the mortises and be pinned with corrosion-resisting metal pins, star-shaped and not less than 7 mm diameter, or with wood dowels not less than 9 mm diameter. Alternatively, the tenons may fit into tapered mortises and be wedged. In either case the joints should be made either with an adhesive or with lead base paint.

Sills should preferably be made of hardwood with a weathered upper surface and may be grooved for and fitted with a 25 x 6 mm sherardised or galvanised iron water bar bedded in mastic (figure 9.4.9). When door frames are used without sills, the feet of the jambs should be fixed with 12 mm diameter corrosion-resisting metal dowels. Fanlight openings should be provided with glazing fillets or beads. The heads of door frames must be relieved of the weight of brick and block partitions by providing reinforced concrete lintels or by filling hollow blocks with concrete and inserting steel reinforcing bars in the bottom cavities (figure 9.4.1).

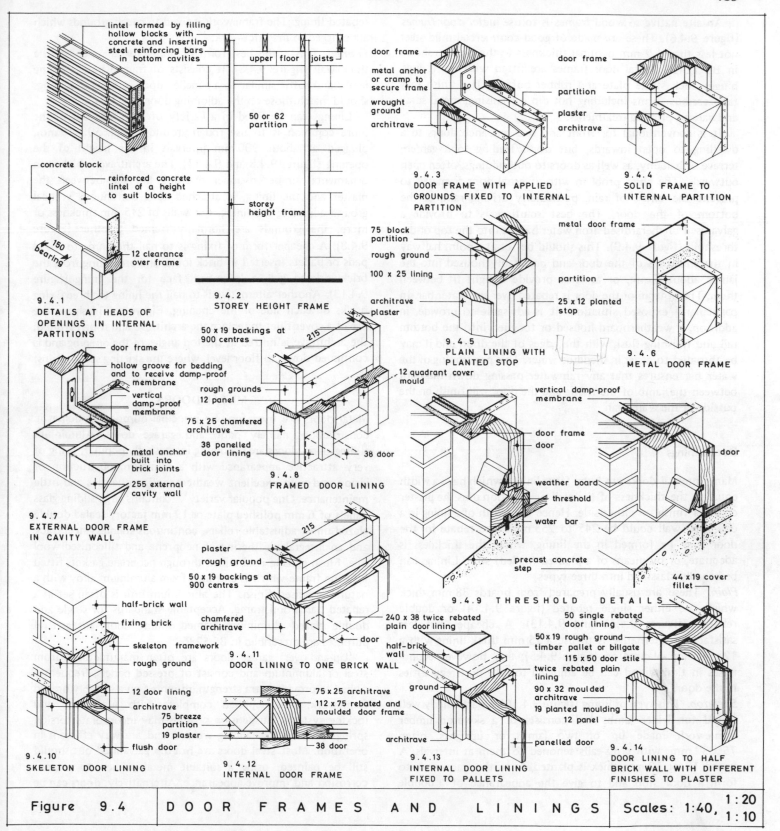

Figure 9.4 DOOR FRAMES AND LININGS Scales: 1:40, 1:20, 1:10

An alternative to wood frames is to use *metal door frames* (figure 9.4.6). These are made of good commercial mild steel not less than 1.2 mm nominal thickness to the profiles shown in BS 1245.[9] Metal door frames are fitted with fixing lugs, hinges, lock strike plate and rubber buffers. The finish may take various forms including hot dip galvanising, metal spray and various priming treatments.

It is conventional for front and rear entrance doors to a dwelling to open inwards, but doors leading to a garden, terrace or balcony, as well as doors to outbuildings, often open outwards. The chief problem with external door frames is to prevent the entry of rain, particularly driving rain, at the bottom of the door. The best solution is to provide a galvanised or sherardised iron water bar set into the top of the threshold (figure 9.4.9). This should be set at a point halfway in the thickness of the door and preferably housed into the jambs about 6 mm, or at least provide a tight fit between them. The bottom of the door is rebated over the water bar. In particularly exposed situations it is advisable to provide, in addition, a weatherboard housed or tongued into the bottom rail and finishing flush with the edges of the door, and it may be throated to assist in shedding water. This positioning of the water bar ensures that any rainwater passing down the joint between the jamb of the frame and the door will finish on the outside of the water bar.

Door Linings

Many internal doors are hung from linings which have a width equal to the thickness of the wall or partition plus the plaster or other finish on either side. Hence the width of a lining in a one-brick wall could be 245 to 255 mm. The rebate for the door may be formed in the lining, where the thickness is adequate, or by means of a planted (nailed) stop. Linings can be broadly classified into three types

Plain. These are usually prepared from boards 38 mm thick which are either single rebated (figure 9.4.14) or double rebated (figures 9.4.11 and 9.4.13). A cheaper but less satisfactory alternative is to use a 25 mm thick lining with a 12 mm stop planted on (figure 9.4.5); this, however, economises in timber and can be adjusted to take up irregularities in the door.

Skeleton. This type of lining (figure 9.4.10) is particularly well suited for wider jambs and consists of a skeleton timber framework made up of two jambs or uprights, often 75 x 32 mm, with cross-rails tenoned to them at intervals. A board 10 or 12 mm thick is planted on to the framework to form a door stop and to give the appearance of a double

rebated lining. The framework is fixed to rough grounds which are plugged to the brickwork.

Framed. This is the best form of lining for an opening in a thick wall (figure 9.4.8). It consists of framed panels to the jambs and soffit similar to a panelled door, and the mouldings should match those on the adjoining door.

Linings can be fixed in a variety of ways and one of the more common is to use rough grounds, often 50 x 19 mm, plugged at about 900 mm intervals in the height of the opening (figures 9.4.8 and 9.4.11). The architrave, which is an ornamental timber member masking the gap between the plaster and the lining, is attached to a continuous vertical ground adjoining the lining. For walls of 215 mm thickness or more, the grounds are normally framed together (Figure 9.4.8). A cheaper form of fixing is to nail the lining to wood pads or pallets inserted in brick joints and projecting from the brickwork to give a 'plumbed' face for the lining (figure 9.4.13). Another alternative is to nail the lining to three fixing bricks on each side of the opening. Figure 9.4.14 shows the use of a cover fillet to replace the architrave and rough ground; the architrave is mitred at the top angles of the opening and is continued down to floor level, where the skirting stops against it.

METAL DOORS

Metal doors are used for two principal purposes in domestic construction, namely patio and garage doors. Aluminium sliding patio doors are becoming increasingly popular, being of very attractive appearance with a minimum obstruction of vision and with excellent weatherproofing qualities and little maintenance. One popular variety[10] contains large sliding glass panels of 6 mm polished plate or 19 mm factory sealed double glazed units, adjustable rollers, continuous stainless-steel track and an integral weathering of neoprene and siliconised wool pile. Flush glazing is obtained through neoprene gaskets fitted into the frame which is extruded from aluminium alloy with a natural anodised finish. The aluminium unit is often set in a rebated hardwood frame. Acceptable standards for single and double glazed aluminium framed sliding doors for general purposes are prescribed in BS 5286.[4]

'Up-and-over' garage doors are often manufactured from steel or aluminium and consist of pressed panels, frequently ribbed to provide extra strength, mounted on a suitably braced framework, and supplied complete with suspension and locking systems. Suspension systems may incorporate torsion springs or counterbalancing weights and are very efficient in operation. Most steel doors are hot-dip galvanised but should still be painted, using a suitable metal primer, to prevent corrosion and improve appearance. Alternatively, doors can be

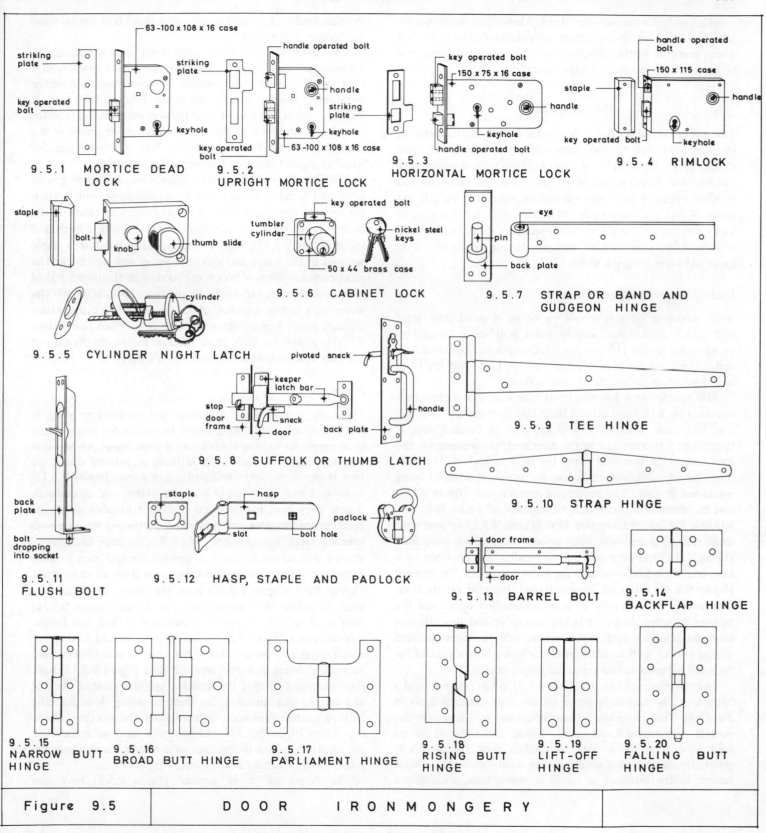

9.5.1 MORTICE DEAD LOCK

9.5.2 UPRIGHT MORTICE LOCK

9.5.3 HORIZONTAL MORTICE LOCK

9.5.4 RIMLOCK

9.5.5 CYLINDER NIGHT LATCH

9.5.6 CABINET LOCK

9.5.7 STRAP OR BAND AND GUDGEON HINGE

9.5.8 SUFFOLK OR THUMB LATCH

9.5.9 TEE HINGE

9.5.10 STRAP HINGE

9.5.11 FLUSH BOLT

9.5.12 HASP, STAPLE AND PADLOCK

9.5.13 BARREL BOLT

9.5.14 BACKFLAP HINGE

9.5.15 NARROW BUTT HINGE

9.5.16 BROAD BUTT HINGE

9.5.17 PARLIAMENT HINGE

9.5.18 RISING BUTT HINGE

9.5.19 LIFT-OFF HINGE

9.5.20 FALLING BUTT HINGE

Figure 9.5   DOOR   IRONMONGERY

supplied in 'colorcoated' steel sheet. Aluminium doors may be left unpainted as on oxidation a protective skin is formed which prevents further deterioration, but a painted finish may be preferred on grounds of appearance.

IRONMONGERY

The term *ironmongery* includes locks, latches, bolts, furniture (handles and coverplates), suspension gear and closing and check gear. BS 1331[11] covers hardware or ironmongery suitable for housing and gives the range of materials and finishes, essential minimum dimensions, minimum weights and gauge of fixing screws, while BS 455[12] prescribes the sizes of locks and latches for doors. The requirements as to materials, workmanship, construction, dimensions and weight of a wide range of hinges are given in BS 1227.[13]

Locks, Latches, Bolts and Furniture

When selecting locks it should be borne in mind that larger locks can house stronger working parts, bolt 'shoots' should be of adequate length (16 mm for a locking or key-operated bolt and 13 mm for latch or handle-operated bolt) and the latch bolt mechanism should be of 'easy action'.

The *mortice lock* has two bolts, one a dead or locking bolt operated by a key and other a latch bolt operated by a handle. The lock case, usually of wrought iron or pressed steel, is fitted into a mortice cut in the door and, in consequence, the lock cannot be removed when the door is locked. The door must be of sufficient thickness to receive it without being weakened unduly. The *horizontal mortice lock* (figure 9.5.3) can be inserted in the middle or lock rail of a panelled door, whereas the *upright mortice lock* (figure 9.5.2) is used with single panel, glazed and flush doors. The *mortice dead lock* (figure 9.5.1) has only one bolt, a locking bolt operated by a key, and is especially suitable for hotel bedrooms. The *rimlock* (figure 9.5.4), often of japanned steel, is fitted on the door face and is used mainly with matchboarded doors and the thinner panelled doors. It is less attractive and less efficient than the mortice lock. *Rim latches* with a finger-operated sliding bolt or snib in addition to the spring bolt are useful for WCs and bathrooms and are called *pulpit latches*.

The *cylinder night latch* (figure 9.5.5), often referred to as a 'Yale lock', is frequently fitted on the main external doors to dwellings. The spring-loaded bolt is opened by a key from the outside or a round knob on the inside, and the bolt can be held in or out by a slide. *A cabinet lock* (figure 9.5.6) generally consists of a small mortice dead lock with cylinder action. In the interests of safety a second lock, preferably a

mortice dead lock, should be provided about 600 mm up from the bottom of the door.

Door knobs are available in nonferrous, alloy and plastics in a variety of shapes, and are generally supplied in pairs with a steel spindle. *Lever door handles* in nonferrous, plastics or aluminium, usually have a minimum length of 90 mm. A *lockset* consists of a pair of lever handles attached to plates for screwing to the face of the door. *Finger plates*, made in the same range of materials and normally 300 x 63 mm in size, are fitted to prevent the door finish from being dirtied.

The Suffolk or thumb latch (figure 9.5.8) is used on garden gates and matchboarded doors to outbuildings, with a latch bar on one side operated by a pivoted sneck or catch on the other. Store doors are sometimes fastened with a *hasp and staple* (figure 9.5.12), with a hasp bolted to the door, a staple screwed to the frame and locked together with a padlock. The most common form of bolt is the *barrel bolt* (figure 9.5.13) of mild steel or brass varying in length from 75 to 300 mm. The *tower bolt* is less enclosed than the barrel bolt and passes through rings. A more attractive bolt is the *flush bolt* (figure 9.5.11), where the back plate finishes flush with the face of the door and the bolt is recessed into the door.

Hinges

There are various types and sizes (63 to 150 mm long) of hinges available for hanging doors. In general one pair of butts is adequate for hollow flush internal door leaves, while wider and heavier leaves, such as solid flush or external doors, are best hung on one-and-a-half pairs, *Butt hinges* (figure 9.5.15) in pressed steel, cast iron or brass are widely used on domestic doors. They have narrow flaps that are concealed within the thickness of the door and they require letting-in flush on both meeting faces. *Rising butts* (figure 9.5.18) have knuckles so shaped that when the door is opened the leaf rises to clear carpets, and this action also causes the door to close itself. *Falling butts* (figure 9.5.20) have the reverse effect and are used in public WC apartments. *Pin hinges* (figure 9.5.16) have a loose pin to permit separation of leaf and frame. *Lift-off hinges* (figure 9.5.19) have a pin attached to one flap which engages in the knuckle of the other so that the two may be readily disengaged. *Parliament hinges* (figure 9.5.17) have flaps extended so that the knuckle projects clear of the leaf and frame, thus enabling the door to swing through 180°, clearing obstructions such as architraves and skirtings. *Backflap hinges* (figure 9.5.14) have square or rectangular flaps and are usually screwed to the face of leaf and frame where the leaf is too thin to take a butt hinge.

Tee hinges or 'cross garnets' (figure 9.5.9) have one

elongated flap, 150 to 450 mm long, and are used where the weight must be carried over a large area, such as with ledged and braced leaves. *Strap, band and gudgeon,* or *hook and eye hinges* (figure 9.5.7) are used with heavy gates and garage doors, and are usually about 50 x 6 mm in section and are supplied in lengths varying from 250 to 1050 mm. Another form of *strap hinge* (figure 9.5.10) has two elongated flaps for use between folding leaves.

REFERENCES

1. BS 1186: Quality of timber and workmanship in joinery; Part 1: 1971 Quality of timber, Part 2: 1971 Quality of workmanship
2. BS 459: Doors, Part 1: 1954 Panelled and glazed wood doors; Part 2: 1962 Flush doors; Part 3: 1951 Fire-check flush doors and wood and metal frames (half-hour and one-hour types); Part 4: 1965 Matchboarded doors
3. BS 2911: 1974 Letter plates
4. BS 5286: 1978 Aluminium framed sliding glass doors
5. BS 476: Fire tests on building materials and structures, Part 7: 1971 Surface spread of flame test for materials; Part 8: 1972 Test methods and criteria for the fire resistance of elements of building construction
6. BS 1203: 1963 Synthetic resin adhesives for plywood
7. *BRE Digest 182*: Natural finishes for exterior timber. HMSO (1975)
8. BS 1567: 1953 Wood door frames and linings
9. BS 1245: 1975 Metal door frames (steel)
10. Hillaldam Coburn. *Solair aluminium sliding doors* (1979)
11. BS 1331: 1954 Builders' hardware for housing
12. BS 455: 1957 Locks and latches for doors in buildings
13. BS 1227: Hinges: Part 1A: 1967 Hinges for general building purposes
14. *BRE Digest 220*: Fire doors. HMSO (1978)
15. BS 4787: Internal and external wood doorsets, door leaves and frames, Part 1: 1972 Dimensional requirements

10 STAIRS AND FITTINGS

This chapter completes the study of joinery work by examining the design and construction of staircases and simple joinery fitments.

STAIRCASE DESIGN

Stair Types

Stairs consist of a series of steps with accompanying handrails; a *staircase or stairway* is the complete system of treads, risers, strings, landings, balustrades and other component parts, in one or more successive flights of stairs. The space occupied by a staircase is termed a *stairwell*, with the vertical distance between the floors served by a staircase described as the *lift*.

The simplest form of stair is the *straight flight* stair (figure 10.1.1) consisting of a straight flight or run of parallel steps. A *quarter-turn* stair is one containing a flight with a landing and a right-angle turn to left or right; the landing is termed a quarter-space landing (figure 10.1.3). A *dogleg* or *half-turn* stair (figure 10.1.2) has one flight rising to an intermediate half-space landing, with the second flight travelling in the opposite direction to the first flight. An *open well* or *open newel* stair (figure 10.1.3) contains a central well, with newels at each change of direction and two or more flights of steps around the outside of the well. A *geometrical stair* (figure 10.1.4) takes the form of a spiral, with the face of steps radiating from the centre of a circle which forms the plan of the outer string, and incorporates an open well. A *spiral* stair is a form of geometrical stair without a well.

Requirements for Domestic Staircases

The Building Regulations[1] prescribe certain minimum requirements for both stairways serving single dwellings and those stairways which are intended for common use in connection with two or more dwellings. The Regulations define a *parallel* tread as one having a uniform width throughout that part of its length within the width of the stairway (figure 10.1.1), whereas a *tapered* tread (figure 10.1.4) is one which has a greater width at one side than at the other and a going which changes at a constant rate throughout its length. *Pitch line* (figure 10.1.6) is a notional line which connects the nosings of all treads in a flight with the nosing of the landing at the top of the flight down to the ramp or landing at the bottom of the flight. The *going* of a tread (figure 10.1.6) is measured on plan between the nosing of the tread and the nosing of the tread, ramp or landing next above it. For the purpose of the Regulations, the width of a stairway is its unobstructed width, clear of handrails and other obstruction (figure 10.1.5).

A stairway must be designed to provide a safe, serviceable and commodious means of access from one floor to another of a building, and the Building Regulations[1] have been framed to secure this objective. The Regulations prescribe that between consecutive floors, every step or landing shall have an equal rise and every parallel tread shall have an equal going. There are also minimum headroom requirements over the whole width of the stairway: not less than 2 m clear headroom measured vertically from the pitch line (figure 10.1.6). The nosing of any tread which has no riser below it, shall overlap (on plan) the back edge of the tread below it by not less than 15 mm (figure 10.1.6).

The dimensions of the going and rise have to be in a satisfactory relationship to one another and ensure that the stairway is neither too flat or too steep. The Building Regulations[1] require that on a stairway the aggregate of the going of a parallel tread plus twice its rise shall be not less than 550 mm and not more than 700 mm (G + 2R = 550 to 700). This permits a number of combinations as shown in the hatched area in figure 10.1.7. A common arrangement is a rise of 175 mm and 225 mm going, which gives a sum of 225 + 350 = 575 and is well within the range prescribed in the Building Regulations. Furthermore, in stairways serving single dwellings the rise of a step shall not exceed 220 mm and the going shall be at least 220 mm.

For stairways for common use for two or more dwellings the rise shall not exceed 190 mm and the going shall not be less than 240 mm. The pitch of a stairway serving a single

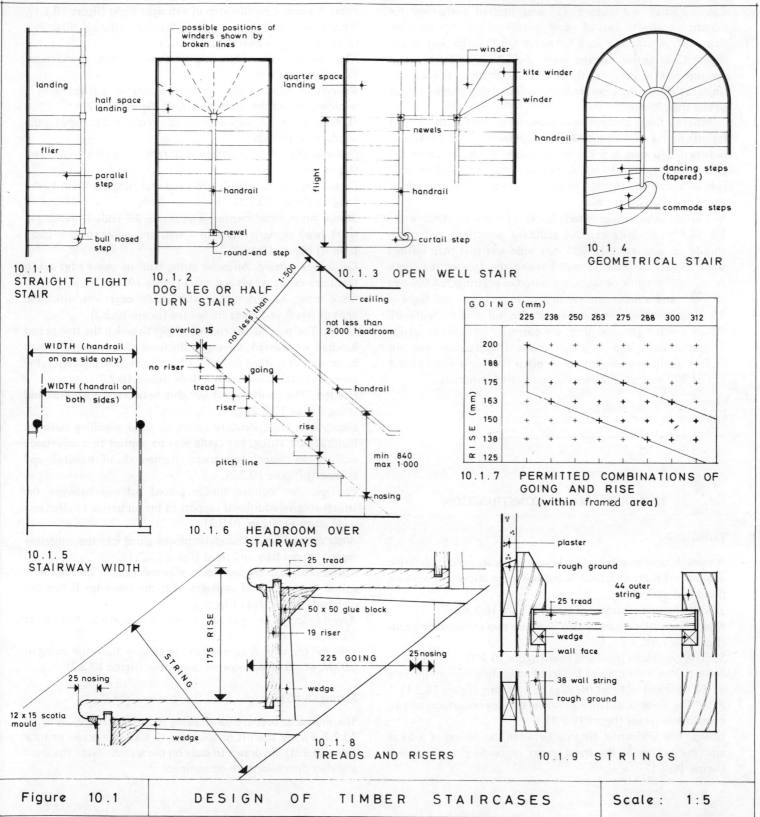

| Figure 10.1 | DESIGN OF TIMBER STAIRCASES | Scale: 1:5 |

R = 200
g = 300; 2R + g = 700 OK!

dwelling shall not exceed 42° and that of a stairway for common use with two or more dwellings shall not be more than 38°. A stairway shall be not more than 16 rises in any flight. Consecutive tapered steps shall have uniform goings measured at the centre of the length, and those on stairways in single dwellings shall not have the going of any part of a tread of less than 75 mm.

Where a flight of steps has a total rise of more than 600 mm it must have a continuous handrail fixed securely at a height of between 840 mm and 1.0 m, measured vertically above the pitch line (figure 10.1.6). A handrail shall be fixed on each side of a stairway if it is 1 m wide or more, and on one side of the stairway in other cases.

Certain other design criteria for wood stairs are contained in BS 585.[2] For instance, this standard prescribes that treads should be not less than 235 mm wide and that their nosings shall project at least 15 mm beyond the face of the risers. A minimum width of stairs, measured over strings, of 860 mm is given, and a maximum number of risers in any one flight of 15 (one less than for stairways in the Building Regulations). Where winders are used, three are normally arranged to occupy a quarter space and six a half space. The standard also prescribes a minimum height above pitch line for a sloping handrail of 840 mm and 900 mm for horizontal handrails.

TIMBER STAIRCASE CONSTRUCTION

Terminology

A considerable number of technical terms are used to describe component parts of staircases and the more important ones are listed and defined.

Tread. The upper surface of a step (figure 10.1.8).
Riser. The vertical part of step between two consecutive treads (figure 10.1.8).
Step. A combined tread and riser (figure 10.2.1).
Nosing. The front edge of the tread projecting beyond the face of the riser and includes the edge of a landing (figure 10.2.1).
Rise. The vertical distance between the upper surfaces of two consecutive treads (figure 10.2.1).
Going. The horizontal distance between the nosing of a tread and the nosing of the tread, ramp or landing next above it (figure 10.2.1).

Flier. A normal parallel step in a straight flight (figure 10.1.1).
Winder. A tapering step where the stair changes direction, radiating from a newel (figure 10.1.3).
Kite winder. The middle step of three winders at a quarter turn (figure 10.1.3).
Round-end step. A step at the bottom of a flight with a semicircular end the width of a tread (figure 10.1.2).
Half-round step. A step with a semicircular end occupying the width of two treads.
Bullnose step. A step with a quadrant or quarter-round end (figure 10.1.1).
Curtail step. A step with a scroll end matching the finish to the handrail (figure 10.1.3).
String. An inclined member supporting the ends of treads and risers (*wall* string is fixed to a wall and *outer* string is away from it) (figure 10.1.9).
Cut or open string. An outer string with its upper edge cut to the shape of the treads and risers (figure 10.2.2).
Close string. An outer string with parallel edges into which the ends of treads and risers are housed (figure 10.2.2).
Newel. The post at the end of a flight to which the strings and handrail are framed; the cap is the head of the newel and may be applied or worked on solid, while the drop or pendant is the lower end projecting below a floor (figure 10.2.2).
Baluster. The small bars or uprights between the handrail and string (figure 10.2.2).
Balustrade. This normally refers to solid panelling between handrail and string, but could also be applied to a balustrade wall or an 'open' balustrade (framework of handrail and balusters) (figure 10.2.2).
Carriage. An inclined timber placed halfway between the strings to give additional support to the underside of steps in a wide stairway (figure 10.2.1).
Angle blocks. Small triangular blocks glued into the underside angle between riser and tread (figure 10.2.1).
Rough brackets. Shaped pieces of wood, usually 25 mm thick, nailed to the side of carriages with the top edge fitting the underside of the tread (figure 10.2.1).
Apron lining. A lining or facing to a trimmer at a landing (figure 10.2.2).
Spandrel framing. A panelled infilling below the outer string of a flight which often incorporates a door (figure 10.2.2).

Setting Out Stairs

The normal procedure for drawing stairs is illustrated in figure 10.2.2. Firstly the lift or total rise (2.400) and travel or total going (2.925) are drawn to scale on the section. Next the tread and riser dimensions are determined.

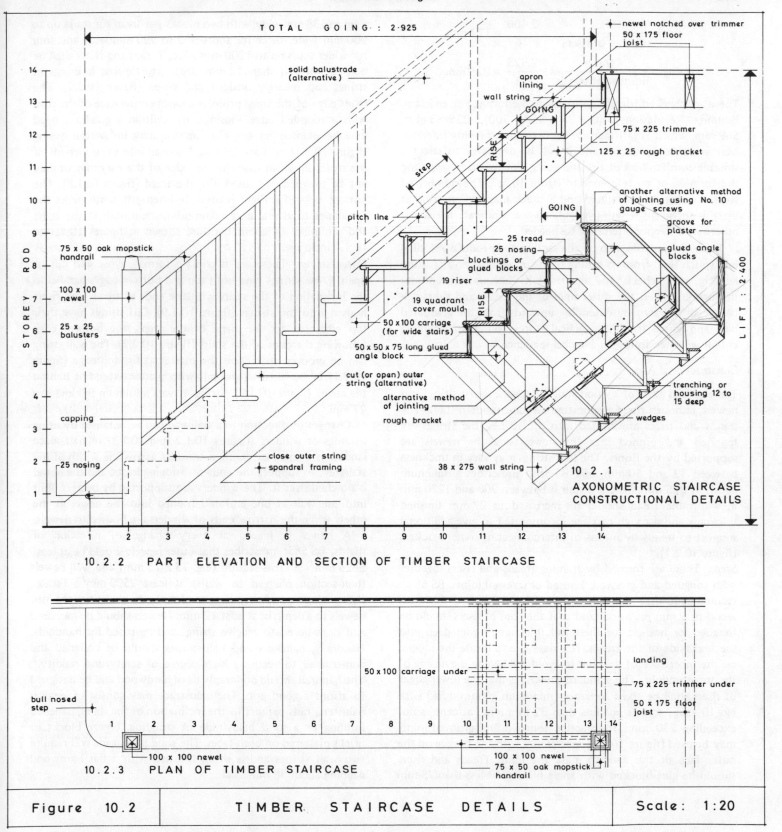

TOTAL GOING : 2·925

STOREY ROD

14
13
12
11
10
9
8
7
6
5
4
3
2
1

solid balustrade (alternative)

75 x 50 oak mopstick handrail

100 x 100 newel

25 x 25 balusters

capping

25 nosing

close outer string
spandrel framing

pitch line

step

RISE

GOING

25 tread
25 nosing
blockings or glued blocks
19 riser

19 quadrant cover mould

50 x 100 carriage (for wide stairs)

50 x 50 x 75 long glued angle block

cut (or open) outer string (alternative)

alternative method of jointing

rough bracket

38 x 275 wall string

newel notched over trimmer
50 x 175 floor joist

apron lining
wall string
GOING

75 x 225 trimmer

125 x 25 rough bracket

another alternative method of jointing using No. 10 gauge screws

groove for plaster

glued angle blocks

RISE

trenching or housing 12 to 15 deep

wedges

LIFT : 2·400

10.2.1 AXONOMETRIC STAIRCASE CONSTRUCTIONAL DETAILS

10.2.2 PART ELEVATION AND SECTION OF TIMBER STAIRCASE

bull nosed step

1 2 3 4 5 6 7 8 9 10 11 12 13 14

50 x 100 carriage under

100 x 100 newel

10.2.3 PLAN OF TIMBER STAIRCASE

landing

75 x 225 trimmer under

50 x 175 floor joist

100 x 100 newel
75 x 50 oak mopstick handrail

| Figure 10.2 | TIMBER STAIRCASE DETAILS | Scale: 1:20 |

$$\text{rise} = \frac{\text{lift}}{\text{number of risers}} = \frac{2.400}{14} = 172 \text{ mm}$$

$$\text{going} = \frac{\text{travel}}{\text{number of treads}} = \frac{2.925}{13} = 225 \text{ mm}$$

These are checked for suitability using the formula given in the Building Regulations[1] (G + 2R = 550 − 700), 225 + 344 = 569 mm, which is satisfactory. On occasions the only information available is the lift and it will be necessary to select a suitable combination of rise and going dimensions by applying the formula or by reference to figure 10.1.7. The number of steps can then be calculated and a check made to ensure that there is sufficient length available to accommodate the stairs, otherwise a steeper pitch may be needed.

Draw the bottom step on the section and produce a sloping line (pitch line) from its nosing to the uppermost nosing at the landing, and check to see there is adequate headroom throughout. Then the outer lines of the treads and risers can be speedily drawn by extending upwards from the graduated base and across from the vertical storey rod. The remaining details will then follow in a logical sequence.

Constructional Aspects

The essential parts of a wood stair are treads, risers, strings and newels, although risers are omitted on some modern stairs. The treads and risers are housed into the strings, the strings are tenoned and pinned into the newels and the newels are supported by the floors. The two strings may vary in thickness between 32 and 50 mm, and BS 585[2] prescribes a minimum depth of 225 mm. Where a stair is between 900 and 1220 mm in width the tread should be increased to 27 mm finished thickness and a rough carriage incorporated to give additional support to treads by means of interconnecting rough brackets (figure 10.2.1).

Steps. These are formed by framing treads and risers together with tongued and grooved, housed or screwed joints. BS 585[2] recommends that risers should be 14 mm finished thickness wood or 9 mm plywood, and that the tops of risers should be tongued, or housed for their full thickness, 6 mm deep into the underside of the treads. It is desirable to locate the tongue on the inner face of the riser to avoid weakening the nosing of the tread unduly. The standard prescribes that the lower edges of risers shall be fixed to treads (minimum 20 mm thick) with No 10 gauge screws not less than 32 mm long, at centres not exceeding 230 mm. In practice tongued and grooved joints may be used (figure 10.2.1), preferably with the tongue on the outer face of the riser to mask the joint. Treads and risers should be glue-blocked with angle blocks not less than 75 mm

long and 38 mm wide, with two blocks per tread for stairs up to 900 mm wide, three for stairs 900 to 990 mm wide and four for wider stairs up to 1200 mm wide. Treads and risers shall be housed not less than 12 mm deep into tapered housings in strings and securely wedged and glued (figure 10.2.1). The front edge of the tread projects over the outer face of the riser with a rounded edge (nosing). In addition a quadrant bead may be planted on the riser immediately below the nosing (figure 10.2.1) or a scotia mould housed into the underside of the tread, in which case the top edge of the riser may or may not be tongued or housed into the tread (figure 10.1.8). The bottom step of a flight is often finished differently from the remainder to give a more commodious approach to the stairs and a number of alternatives are shown in figures 10.1.1 to 10.1.4 inclusive.

Outer strings. These are two main forms − close and cut or open. Close strings have both top and bottom edges parallel to the inclination of the stairs with the ends of treads and risers housed into the strings (figure 10.1.9). Cut strings have their top edge cut to the shape of the steps, the bottom edge following the rake of the stairs (figure 10.2.2). The wall string can be grooved to receive the plaster at the top edge (figure 10.2.1) or be fixed to grounds with plaster extending behind the string (figure 10.1.9). Strings have a minimum thickness of 27 mm.[2]

Changes of direction in a stairway can be obtained by using landings or winders (figures 10.1.2 and 10.1.3). A half-space landing is formed by fixing a trimmer across the width of the stairway, to support the ends of bridging joists, and the newel is notched over it. The winders are supported by bearers built into the wall at one end and framed into the newel at the other, while the narrow ends of winders are housed to newels.

A *newel* is placed at every change of direction of flights. BS 585[2] prescribes that outer newels should be at least 5625 mm² in cross-section (e.g. 75 x 75 mm) and wall newels (half-section plugged to walls) at least 2500 mm². Treads, risers, winders and shaped ends of steps should be housed into newels to a depth of at least 12 mm. Newels should be mortised and draw-bored to receive strings and mortised for handrails. Handrails, balusters and balustrades should be designed and constructed to secure a high degree of safety and rigidity.[2] The handrail should preferably be of hardwood and be designed to afford a good grip. The balustrade may consist of vertical balusters, rails parallel to the inclination of the stairs, panelled infilling or a solid balustrade of studding, breeze blocks or bricks plastered on both faces. The solid balustrade will require two wall strings and is often finished with a flat hardwood capping about 38 mm thick.

STAIRWAYS IN OTHER MATERIALS

Apart from the traditional wood newel staircase, stairs can be constructed in various other materials or combinations of them. Some of the more common forms are examined.

Open-Riser Stairways

These consist of treads supported by strings or carriages without risers and can be constructed in a number of ways. One method, as illustrated in figure 10.3.1, incorporates hardwood treads supported on hardwood brackets bolted to a pair of hardwood strings, and with steel balustrading and hardwood handrails. Another variation is to house the ends of the treads to the strings. Alternatively, the steel brackets may be housed and screwed to the treads and welded to steel channel strings. Metal or precast concrete treads may be used in external situations. This form of stairway gives an impression of spaciousness and facilitates cleaning and the distribution of light and heat.

Stone Stairways

These generally consist of stone steps supported at their ends by brick walls. *Skeleton* stairs incorporate stone slabs usually 50 mm thick for 750 mm wide stairs and 75 mm thick for 900 mm wide stairs, with open risers. *Built-up* stairs give a stronger job with stone slab risers between the treads. *Solid* rectangular steps about 300 x 150 mm in size with a 25 mm overlap (figure 10.3.3) give a very strong job, and the inclusion of a rebated joint provides a fire check. *Spandrel* steps with splay rebated joints, similar to the precast concrete steps illustrated in figure 10.3.6, are less strong than solid steps but give a better appearance, regular soffit and increased headroom below.

Reinforced Concrete Stairways

These can be formed of beams acting as strings with the steps spanning between them, but the more common arrangement consists of a reinforced concrete slab with projections forming the steps (figure 10.3.2). Curbs may be formed *in situ* with the stairs to receive balustrades. The steps may be finished with granolithic (cement and granite chippings) or terrazzo (cement and marble chippings) placed in position after the concrete has set (figure 10.3.4). Precast inserts should be set in terrazzo faced treads, parallel with the nosings, for reasons of safety. The inserts can be formed of grooves filled with carborundum and cement or cubes of carbite, alundum or similar materials in various colours, arranged in chequerboard or other patterns.[3]

Precast Concrete Stairways

These are generally constructed of spandrel steps faced with terrazzo, granolithic or cast stone with splay rebated joints (figure 10.3.6). They have nonslip finishes incorporating carborundum or alundum, or suitable precast inserts. Another arrangement is to bed precast concrete treads and risers either separately or as combined units onto an *in situ* reinforced concrete stairway (figure 10.3.5).

Pressed Steel Stairways

These often consist of steel strings, 250 to 300 mm deep, with steel treads and risers riveted to them. Landings may be formed of flat or dovetailed sheets supported on steel angles and channels. Treads may be covered with various finishes.

WOOD TRIM

Wood trim is defined in BS 584[4] as 'products of uniform profile manufactured by linear machining only'. This standard specifies the quality, designs and dimensions of mass-produced architraves, skirtings, picture rails, cover fillets, quadrant, half-round and scotia moulds. It might be helpful at this stage to describe the more common items of wood trim even though they will follow plastering on the job as part of joinery second fixings.

Architrave. A moulding or fillet around an opening applied to the face to cover the joint between joinery and the adjoining work (figures 10.3.7 to 10.3.9).

Skirting. A finishing member fixed to a wall where it adjoins the floor to cover the joint and protect the wall finish (figure 6.2.4).

Picture rail. A plain or moulded rail fixed to walls and from which pictures and other decorative features can be hung (figure 10.3.13).

Cover fillet. A fillet to cover a joint in joinery or between joinery and adjoining work (figure 10.3.14).

The standard finishes for architraves, skirtings and picture rails are chamfered and rounded (figures 10.3.7 and 10.3.13), rounded (figures 10.3.8) and bevel-rounded (figure 10.3.9). All three mouldings are simple in form and reduce dust collection to a minimum. Quadrant and half-round moulds (figure 10.3.11) are frequently used to mask the joint between plaster and joinery. Typical sizes are

architraves and picture rails — 13 x 45 and 20 x 70 mm
skirtings — range from 13 x 70 to 20 x 170 mm in eight stages
scotia (figure 10.3.10) — 13 x 13, 20 x 20 and 27 x 27 mm

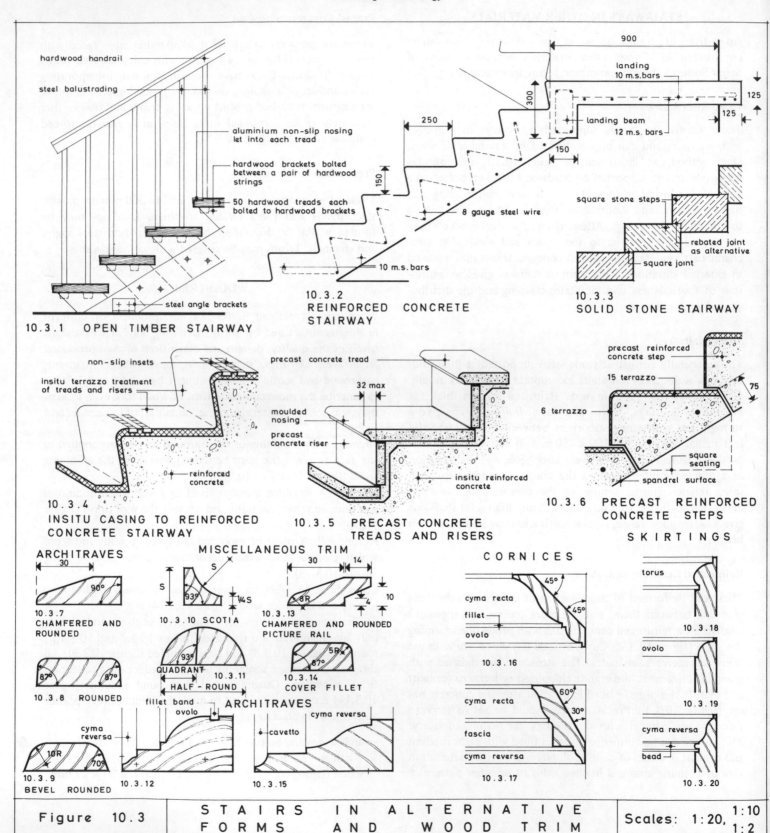

10.3.1   OPEN TIMBER STAIRWAY

hardwood handrail

steel balustrading

aluminium non-slip nosing let into each tread

hardwood brackets bolted between a pair of hardwood strings

50 hardwood treads each bolted to hardwood brackets

steel angle brackets

10.3.2   REINFORCED CONCRETE STAIRWAY

900
landing
10 m.s.bars
300
250
125
125
landing beam
12 m.s. bars
150
150
8 gauge steel wire
10 m.s. bars

10.3.3   SOLID STONE STAIRWAY

square stone steps
rebated joint as alternative
square joint

10.3.4   INSITU CASING TO REINFORCED CONCRETE STAIRWAY

non-slip insets
insitu terrazzo treatment of treads and risers
reinforced concrete

10.3.5   PRECAST CONCRETE TREADS AND RISERS

precast concrete tread
32 max
moulded nosing
precast concrete riser
insitu reinforced concrete

10.3.6   PRECAST REINFORCED CONCRETE STEPS

precast reinforced concrete step
15 terrazzo
75
6 terrazzo
square seating
spandrel surface

ARCHITRAVES

10.3.7   CHAMFERED AND ROUNDED
30
90°

10.3.8   ROUNDED
87°   87°

10.3.9   BEVEL ROUNDED
10R
70°

MISCELLANEOUS TRIM

10.3.10   SCOTIA
S
93°
S
¼S

10.3.11   QUADRANT   HALF-ROUND
93°

10.3.12
fillet band   ovolo
cyma reversa

10.3.13   CHAMFERED AND ROUNDED PICTURE RAIL
30   14
8R
4   10

10.3.14   COVER FILLET
5R
87°

10.3.15
cavetto
cyma reversa

ARCHITRAVES

CORNICES

10.3.16
45°
cyma recta
fillet
ovolo
45°

10.3.17
cyma recta
fascia
cyma reversa
60°
30°

SKIRTINGS

10.3.18
torus

10.3.19
ovolo

10.3.20
cyma reversa
bead

| Figure   10.3 | STAIRS   IN   ALTERNATIVE FORMS   AND   WOOD   TRIM | Scales: 1:20, | 1:10 1:2 |

quadrant (figure 10.3.11) — 11 x 11, 13 x 13 and 20 x 20 mm
half-round (figure 10.3.11) — 13 x 33 and 20 x 45 mm
cover fillet (figure 10.3.14) — 11 x 33 and 11 x 45 mm

More elaborate mouldings can be applied to specially designed members. Figure 10.3.12 shows an architrave incorporating a cyma reversa, ovolo and fillet band, while another (figure 10.3.15) embraces cavetto and cyma reversa moulds. Cornices at the junction of walls and ceilings generally contain a number of mouldings (figures 10.3.16 and 10.3.17), while figures 10.3.18 to 10.3.20 show some of the more ornate treatments for skirtings.

SIMPLE JOINERY FITTINGS

The design and construction of some of the more common domestic joinery fittings are now considered, with particular reference to kitchen fitments and built-in wardrobes.

Kitchen Fitments

Some measure of standardisation was essential if kitchen fitments were to be produced cheaply and efficiently. The pioneering work stems from the introduction of BS 1195[5] in 1948, which recommended a standardised range of different cupboard units, which could be assembled to produce the varying requirements for storage spaces and yet, at the same time, suit a particular kitchen plan. It was necessary to co-ordinate dimensions and BS 1195 provides a range of dimensions. The worktop normally projects 50 mm over the floor cupboard below, often giving a basic floor cupboard width of 550 mm. For hanging or wall cupboards above the worktop the width reduces to 300 mm (figure 10.4.2). Floor-cupboard units have a toe recess or recessed plinth often 50 x 80 mm high, which makes for greater comfort when standing against the cupboard and prevents damage to paintwork. The worktop level is standardised at 900 mm to produce a continuous line at the same height as the tops of cookers, sink units and other equipment. DOE design bulletin 6[6] recommends 850 mm as a more acceptable height, but design bulletin 24[7] substituted 900 mm.

Another important height dimension specified is the 'dead storage' level at 1950 to 2250 mm above the floor, being the nominal level above which a person of average height could not reach without the use of steps (highest shelf for general use — 1800 mm). These dimensions have been adopted as the maximum height of wall-cupboard space to contain articles in frequent everyday use, any cupboards above being confined to long-term storage or 'dead storage'. It is a matter of individual choice as to whether the cupboards should extend to ceiling level, but if not the tops of cupboards collect dust. The usual dimension between the worktop and the bottom of wall cupboards is 450 mm. The back and possibly sides of cupboards often consist of wall plaster (figures 10.4.3 and 10.4.4) and doors are hung from framing varying in thickness from 25 to 38 mm, closing against shelf edges without the need for rebates or stops. Vertical divisions between cupboards are often formed of plywood or hardboard. Shelves are supported on bearers plugged to side walls and screwed or housed to vertical members and divisions. The vertical distance between shelves varies from 150 mm for narrow shelves up to 450 mm for wide ones.

Doors. These are frequently framed up with plywood or hardboard on both faces with a bullnosed rebated edge which fits over the framing (figure 10.4.2). There have been objections on the grounds that this provides a ledge for dust, but it does hide the joint and it assists production by eliminating the need for individual hanging of the doors, permitting easy fixing of hinges. The concealed type of hinges can be obtained in a variety of finishes, with rust-proof steel being popular for normal work and chromiumplate on brass for good class work. D handles in anodised aluminium or chromiumplate on brass are both neat and functional, coupled with double ball catches in nickelplated steel. Doors may also be made from laminboard or blockboard with lipped edges. Plastic-faced chipboard gives a serviceable and attractive finish.

Worktops. They can be formed of solid timber, framed plywood, blockboard or laminboard, stainless steel, vitreous enamelled steel, aluminium and aluminium alloys, laminated plastics or glazed asbestos cement. Laminated plastics is the most popular finish. Solid timber worktops should be not less than 20 mm thick and if in more than one piece should be joined with machine joints or loose-tongued joints. Plywood worktops shall be not less than 6 mm thick and blockboard or laminboard not less than 15 mm thick.[5]

Drawers. These vary between 100 and 250 mm in depth and need double pulls if the length exceeds 500 mm. Drawers are often constructed of 25 mm drawer fronts, 12 or 19 mm drawer sides grooved to receive 6 mm plywood or 9 mm wood drawer bottoms, 6 mm plywood or 12 or 19 mm wood drawer backs, with the drawers sliding on 32 x 19 mm hardwood runners or slides possibly covered with plastic strips and guided by 12 x 15 mm hardwood guides (figures 10.4.1 and 10.4.2). Drawer fronts may be rebated to overlap the frame and probably the best form of joint between the wood front and sides of a drawer is the lap dovetail (figure 10.4.2).

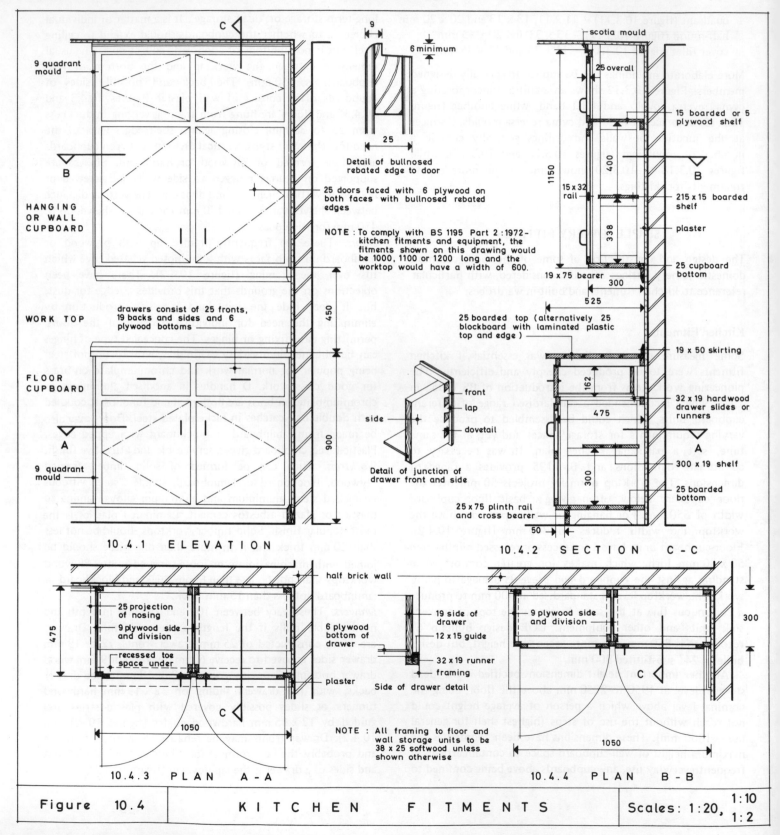

9 quadrant mould

HANGING
OR WALL
CUPBOARD

WORK TOP

FLOOR
CUPBOARD

9 quadrant mould

10.4.1   E L E V A T I O N

9

6 minimum

25

Detail of bullnosed
rebated edge to door

25 doors faced with 6 plywood on
both faces with bullnosed rebated
edges

NOTE : To comply with BS 1195 Part 2 :1972-
kitchen fitments and equipment, the
fitments shown on this drawing would
be 1000, 1100 or 1200 long and the
worktop would have a width of 600.

drawers consist of 25 fronts,
19 backs and sides and 6
plywood bottoms

front
lap
side
dovetail

Detail of junction of
drawer front and side

450

900

scotia mould

25 overall

15 boarded or 5
plywood shelf

1150

400

15 x 32
rail

215 x 15 boarded
shelf

plaster

338

25 cupboard
bottom

19 x 75 bearer

300

525

25 boarded top (alternatively 25
blockboard with laminated plastic
top and edge)

19 x 50 skirting

162

32 x 19 hardwood
drawer slides or
runners

475

300 x 19 shelf

300

25 boarded
bottom

25 x 75 plinth rail
and cross bearer

50

10.4.2   S E C T I O N   C - C

half brick wall

475

25 projection
of nosing

9 plywood side
and division

recessed toe
space under

plaster

1050

10.4.3   P L A N   A - A

19 side of
drawer

6 plywood
bottom of
drawer

12 x 15 guide

32 x 19 runner

framing

Side of drawer detail

NOTE : All framing to floor and
wall storage units to be
38 x 25 softwood unless
shown otherwise

9 plywood side
and division

300

C

1050

10.4.4   P L A N   B - B

Figure   10.4   K I T C H E N   F I T M E N T S   Scales : 1 : 20,

1 : 10

1 : 2

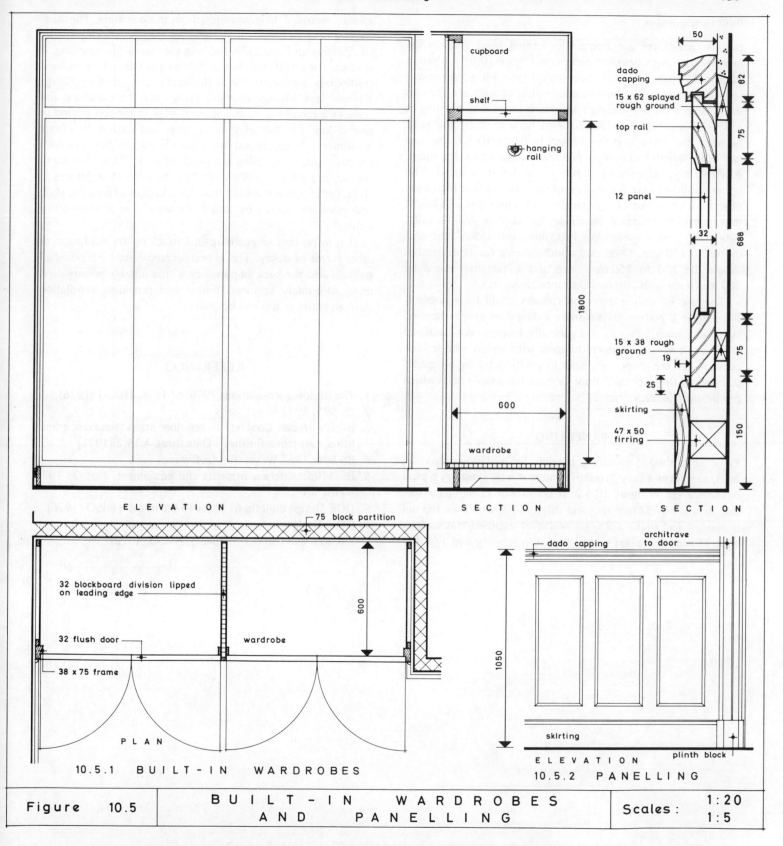

ELEVATION

cupboard

shelf

hanging rail

1800

600

wardrobe

SECTION

SECTION

50

dado capping

15 x 62 splayed rough ground

top rail

12 panel

32

688

82

75

75

15 x 38 rough ground

19

25

skirting

47 x 50 firring

75

150

75 block partition

32 blockboard division lipped on leading edge

600

32 flush door

wardrobe

38 x 75 frame

PLAN

10.5.1 BUILT-IN WARDROBES

dado capping

architrave to door

1050

skirting

plinth block

ELEVATION

10.5.2 PANELLING

| Figure 10.5 | BUILT-IN WARDROBES AND PANELLING | Scales: 1:20 1:5 |

Built-in Wardrobes

Built-in wardrobes are frequently formed in recesses in the internal partitions between bedrooms (figure 10.5.1), often in staggered formation for economical provision. A strong vertical division can be obtained with blockboard 32 mm thick, lipped on the leading edge. Flush doors may be hung to a 75 x 38 mm frame with quadrant bead masking the joint between the frame and the plaster. In figure 10.5.1, the base has been raised 150 mm to provide storage space for shoes. Alternatively, a recessed plinth could be introduced. The upper cupboards are for storage of cases and similar little-used objects. A chromiumplated hanging rail is provided at a height that provides adequate clearance for clothes (men's suits: 1000 mm, men's overcoats: 1400 mm, and ladies' coats or dresses: 1750 mm. Shoe racks and drawers for underclothes should be 100 to 150 mm deep and a hat shelf could be 300 mm wide with 200 mm clearance above it.

The use of sliding doors on runners could be valuable if floor space is restricted. Wardrobe sliding-door gear is supplied by several manufacturers and generally incorporates overhead alloy track which supports hangers with nylon wheels. The bottoms of the doors are held in position by nylon guides screwed to the cupboard floor or have retractable bolts which run in suitable track.

PANELLING

Polished hardwood panelling, possibly up to dado or chair rail level, can form a very attractive internal wall finish. A typical detail is given in figure 10.5.2. It consists of 12 mm panels let into grooves in 32 mm rails and stiles. A 105 mm wide top rail is needed to obtain a 75 mm width of exposed flat surface, with 15 mm concealed within the dado capping and a further 15 mm moulded to give emphasis to the panelling. The dado capping or chair rail is a fully moulded member which provides an effective and attractive finish to the top of the panelling. A skirting, 19 x 150 mm, has a 25 mm moulding to provide a continuous horizontal line at the foot of the panelling. Rough grounds and firrings provide fixings for the panelling and associated members, with a splayed edge to the top ground to give a key for the plaster. A large and extremely ornate architrave is used around the door opening to blend in with the panelling, terminating in a plinth block to frame the end of the skirting. The panels do not fit tightly into the grooves to allow for movement arising from shrinkage or expansion. Rails and stiles are secured by trenails or wood pins at intersecting joints.

It is important to guard against attack by dry rot fungus or other forms of decay. This is best accomplished by treating all grounds and the back of panelling with a suitable preservative, using adequately seasoned timber and providing ventilation through grilles at top and bottom.

REFERENCES

1. The Building Regulations 1976. SI 1676. HMSO (1976)
2. BS 585: 1972 Wood stairs
3. British Precast Concrete Federation. Stairs (terrazzo, granolithic, cast stone finishes): Data sheet A.15.2 (1971)
4. BS 584: 1967 Wood trim (softwood)
5. BS 1195: Kitchen fitments and equipment, Part 2: 1972 Metric units
6. DOE Design bulletin 6: Space in the home. HMSO (1976)
7. DOE Design bulletin 24, Part 2: Spaces in the home — kitchens and laundering spaces. HMSO (1972)

11 FINISHINGS

This chapter is concerned primarily with finishings to walls and ceilings. Floor finishings were considered in chapter 6. In selecting wall and ceiling finishings, probably the two most important considerations are appearance and maintenance costs. In particular situations other factors may also be important, such as resistance to condensation, acoustic properties and provision of a smooth, even surface.

INTERNAL WALL FINISHINGS

Plastering

The great range of plasters and of backgrounds to receive them that are currently available has tended to complicate the choice of plaster.[1] Before making a choice it is advisable to list the functions of plaster. It is required to conceal irregularities in the background and to provide a finish that is smooth, crackfree, hygienic, resistant to damage and easily decorated. It may also be required to improve fire resistance, to provide additional thermal and sound insulation, to modify sound absorption or mitigate the effects of condensation. These latter aspects are generally of secondary importance, and the selection is influenced mainly by the surface finish desired. The stages in the selection of a plastering specification are now listed.

Finish. Whether it should be hard or soft, smooth or textured, the choice of decorative finish and the time available for drying out of plaster before decorations are applied. Table 11.1 lists various plaster finishes and considerations affecting their choice.

Number of coats. Until quite recently, three-coat plasterwork was applied to most backgrounds, with a first levelling undercoat followed by a second undercoat to provide suction for the finishing coat. Two-coat work is generally satisfactory for use with brickwork and blockwork, provided extreme variations of suction are avoided. Some backgrounds, such as plasterboard and smooth concrete, have level surfaces and uniform suction permitting single-coat plastering.

Plaster undercoats. These need to be selected to match the suction of the background, its surface irregularities and other important properties. Undercoats based on gypsum plasters unlike those based on cement do not shrink appreciably on drying. A background of very high suction, such as aerated concrete or insulating bricks, may absorb water so rapidly from the plaster mix that its adhesion to the background is weakened and it may severely limit the time available for levelling the coat. Hence it is advisable to use mixes which retain their water in contact with backgrounds of high suction. The undercoat must have a good key or bond with the background. With backgrounds offering little or no key, such as smooth concrete, bonding agents may be used prior to plastering. Backgrounds liable to drying shrinkage should be allowed to dry out completely before plastering. Thermal movements of backgrounds are rarely sufficiently serious to cause problems, except with concrete roofs and heated concrete floors.

Type of finish. There is a wide range of plasters available and the main types with their more important characteristics are scheduled in table 11.1.

Lime plaster is weak, easily indented, slow hardening and shrinks on drying, and for these reasons is now little used. It does however absorb condensation.[2] A gauged lime plaster (cement—lime—sand) has improved qualities.

Gypsum plasters should comply with BS 1191[3] and have a number of advantages over lime plaster, requiring fewer and thinner coats, which set hard in a few hours without shrinking, providing a hard, strong surface which does not cause alkali attack on paintwork. There are four classes of gypsum plaster,[2] whose proportions and uses are listed in DOE Advisory Leaflet 2.[4]

(1) *class A* (plaster of Paris) sets so rapidly that it is unsuitable for most work apart from small repairs;

(2) *class B* (retarded hemihydrate) plasters include board finishes and give a hard surface which is sufficiently resistant to impact for normal purposes, set quickly and expand during setting;

(3) *class C* (anhydrous) plasters give a surface harder than class B plasters and are slower setting;

Table 11.1   Characteristics of plaster finishes

Plaster finish	Surface hardness and resistance to impact damage	Other surface characteristics	Restrictions on early decoration	Shrinkage or expansion	Remarks
Lime plasters lime	Weak and very easily indented	Open-textured (depending on sand), absorbs condensation	Initially only suitable for permeable finishes that are unaffected by alkali	Shrink on drying, but shrinkage is reduced by addition of fine sand	Slow hardening; only apply on dry undercoats
gauged lime	Resistance to damage increases with proportion of gypsum plaster	Similar to above, but smoother finishes obtainable		Shrinkage is restrained by the gypsum content provided that over-trowelling is avoided	Only apply on dry undercoats
Gypsum plasters class D (Keenes)	Very hard and resistant to damage	Very level and smooth; particularly suitable for low-angle lighting conditions	None, except on undercoats containing cement or lime, or unless lime is added to the finishing coat	Expand during setting. Subsequent movements usually small, but too rapid drying can lead to delayed expansion	Set slowly and so allow ample time for finishing to a smooth surface; should not be allowed to dry too quickly
class C (anhydrous)	Hard and resistant to damage	Slightly less smooth than class D			
class B (hemihydrate)	Sufficiently hard and resistant for most normal purposes, but weakened by additions of lime	Sufficiently smooth and level for most purposes		Expand during setting, though extremely slightly with board finish plasters; subsequent movements are small	Set quickly; should be allowed to dry as soon as possible
lightweight	Surface hardness similar to class B plasters. Ease of indentation varies with the type of lightweight undercoat, but resilience tends to prevent serious damage	Sufficiently smooth and level for most purposes	None, but the higher water content of lightweight undercoats makes these somewhat slower to dry than sanded gypsum plaster undercoats	Expand during setting; subsequent movements usually small and easily restrained by background	Maximum fire resistance; lightweight plaster surfaces warm up more quickly than others and so help to prevent temporary condensation
Cement–lime–sand 1:0-¼:3	Very strong and hard	Wood float finish	Initially only suitable for permeable finishes that are unaffected by alkali	Shrink on drying, but surface cracking can be minimised by avoiding overworking	Suitable for damp conditions
1:1:6	Strong and hard	Wood float finish			
1:2:9	Moderate	Wood float finish			
Single-coat finishes board finish gypsum plasters (class B)	Surface hardness similar to class B above, but resistance depends on background	On suitable backgrounds, similar to class B above	None; finish dries very quickly	Extremely small expansion on setting; subsequent movements small	
Thin-wall finishes based on gypsum	Softer than board finishes	Smooth and level on sufficiently level backgrounds	Dries very quickly: no restrictions when dry	Extremely small expansion on setting; subsequent movements small	
based on organic binders	Moderately hard, resistance depends on background	Matt surface, closely following the level of the background	Dries very quickly; no restrictions when dry	The very thin coats are restrained by the background	
gypsum projection plasters	Properties intermediate between class B and class C gypsum plasters				

Source: *Building Research Station Digest 213*[1]

(4) *class D* (Keene's) provides a very hard, smooth surface, and is slow-setting allowing ample time to bring to a fine finish that is particularly suitable for decoration with high gloss paints (the priming coat must be alkali resistant and applied when the work is dry).

Lightweight gypsum plasters are premixed, consisting of a lightweight aggregate, such as expanded perlite or exfoliated vermiculite and a class B plaster. Surface hardnesses are similar to class B plasters, but they are slower drying and so help to prevent temporary condensation.

Thin wall and special finishing plasters are based on gypsum or organic binders, and are generally applied by spraying and finishing with a broad spatula.

Although gypsum plasters give excellent results, care is needed in handling, mixing and applying. In particular, plasters belonging to different classes must not be mixed; plasters must be kept dry before use; only clean water must be used for mixing; mixes must be of the correct proportions, as too much sand in an undercoat will seriously reduce its strength; sand should comply with BS 1198[5]; cement should never be mixed in a mix with gypsum plasters; and the manufacturer's instructions should be followed regarding the suitability of plasters for different backgrounds.

Gypsum plaster mixes vary with the type of background, for instance thistle browning plaster (retarded hemihydrate gypsum plaster) is applied in proportions of one part of plaster to three parts of sand by volume to normal clay brick surfaces, but in the proportions of one to two to lightweight concrete blocks with lower suction. On the other hand Carlite browning plaster which is a premixed retarded hemihydrate gypsum plaster incorporating lightweight mineral aggregate is used neat, thus ensuring a uniform mix.[6] The work is carried out in two coats in both cases; the floating coat normally being 11 mm thick and this is lightly scratched to form a key for the finishing coat which is normally about 2 mm thick, giving a total thickness of 13 mm. The plaster should be finished truly vertical and be free from cracks, blisters and other imperfections. It is customary to round both internal and external angles of plaster to a radius of about 13 mm. Further guidance on plastering is given in BS 5492,[10] and the more common plastering defects and their prevention are described in DOE Advisory Leaflet 2.[4]

Plastering on Building Boards

Plaster can be applied satisfactorily to a number of types of board, namely: insulating board, wallboard, gypsum lath, gypsum baseboard, gypsum plank and expanded polystyrene.[7] Joints between all boards except gypsum lath should be rein-forced with a strip of jute scrim with a minimum width of 90 mm to prevent cracking of the plaster. All internal and external angles should be reinforced with scrim around the angle between the boards and the plaster. Alternatively, galvanised metal corner beads may be used to give maximum protection at external angles. Joints between walls and ceiling can be similarly scrimmed or, alternatively, a straight cut can be made through the plaster along the line of the junction. Small movements between walls and ceiling can be concealed by cornices, such as plasterboard cove.

A gypsum plaster should be used for single-coat work, preferably neat class B type b2 (retarded hemihydrate), often called board plaster, and for the undercoat in two-coat plastering, preferably sanded class B type a1 on building boards. The final coat could consist of neat gypsum plaster or a gauged lime plaster. Single-coat work is satisfactory when the boards have been fixed to give a true and level surface. Two-coat work normally finishes about 8 mm thick and single-coat work about 5 mm thick.

Dry Linings

The use of plasterboard and similar sheet materials to form a dry lining to dwellings is quite popular, as it simplifies the work of finishing trades and can lead to earlier completions. A permanent decoration can be selected in the knowledge that shrinkage cracks will not occur and painting or wallpapering can commence shortly after the lining is fixed.[8] Several methods of fixing plasterboard are now described.

Timber battening. This is a straightforward but rather expensive method, and is particularly well suited for counteracting damp penetration through the walls of a building. The battens are usually 40 mm wide and 15 mm thick, spaced at not greater than 450 mm centres, and fixed with chrome-steel nails. The plasterboard is fixed to the battens with taper-headed nails, taking care that nail heads do not puncture the plasterboard. When battens are fixed to walls which are damp or likely to become so, they should be pressure-impregnated with a timber preservative, to avoid the risk of staining the plasterboard and decorations. When the dampness is likely to be permanent, it is advisable to treat the board with a fungicidal wash before fixing, such as a one per cent solution of sodium pentachlorphenate. Plasterboard linings also provide a convenient method of improving the thermal insulation of external walls and to reduce the risk of condensation, preferably using aluminium foil-backed plasterboard.

Plasterboard battens. These consist of strips of plasterboard, stuck to the wall with a special adhesive, and can be substituted for the timber battens. Plasterboard grounds are first fixed horizontally at ceiling and floor level, and vertical grounds set between them at about 450 mm centres. The plasterboard is both stuck and nailed to the grounds.

Plaster dab and fibre pack. This is another fixing method. Pieces of bitumen-impregnated fibreboard, about 75 x 50 mm, are stuck to the walls with plaster at the top, bottom and centre of each joint, with intermediate lines of packs to support the edges of boards. After about two or three hours, plaster is applied to the wall by trowel as dabs of about 100 mm diameter spaced at 450 mm centres in both directions. As boards are tamped into position, the dabs spread to give an effective contact surface of about 200 mm diameter. The boards are then nailed to the fibre packs with corrosion-resistant hardened nails which hold the boards into position until the plaster sets.

Plaster ribbon. In this method ribbons of plaster secure the boards to the wall. The ribbons, usually about 50 mm wide and varying in thickness from 12 to 38 mm, are applied to the back of boards through a spreading box and are laid parallel to the longest edge.

Plasterboards. These are produced with square, rounded, recessed or tapered edges, and the joint finishes need to be considered in relation to the decoration. With square-edged boards it is difficult to fill the fine butt joint so as to produce a continuous plain surface to receive paint or distemper. Wallpaper with a slight texture or pattern will however mask the discrepancies. Other forms of edge can accommodate reinforcement to provide invisible joints and so give a satisfactory base for paint or distemper.

Plasterboard is essentially a building board consisting of a core of set gypsum plaster enclosed between and bonded to two sheets of heavy paper and it should comply with BS 1230.[12] There are five basic types of plasterboard as follows

(1) *gypsum wallboard*, which is used primarily for linings to framed construction, with one self-finished surface, or designed to receive decoration direct;

(2) *gypsum lath*, which is a narrow-width plasterboard with rounded longitudinal edges, less expensive than wallboard and providing an ideal base for gypsum plaster;

(3) *gypsum baseboard*, a square-edged plasterboard supplied in wider sheets than a lath as a base for gypsum plaster; unlike gypsum lath, all joints are scrimmed;

(4) *gypsum plank*, which is a relatively narrow-width board of greater thickness than other types of plasterboard; it is used as linings to framed construction and as casings to beams and columns; two types of board are available — one with two grey surfaces designed as a plaster base, and the other with one ivory surface intended for direct decoration; the former type

has square edges and the latter has tapered edges for flush jointing. There is also a moisture resistant vapour check wallboard.[9]

(5) *insulating gypsum plasterboard*, which has a bright metal veneer of low emissivity on one side.

Typical board sizes are now listed.

Type of board	Thickness (mm)	Widths (mm)	Lengths (mm)
Gypsum wallboard	9.5 and 12.7	600, 900 and 1200	1800, 2350, 2400, 2700 and 3000
Gypsum lath	9.5 and 12.7	406	1200
Gypsum baseboard	9.5	914	1200
Gypsum plank	19	600	2350, 2400, 2700 and 3000

Fixing. Plasterboard is fixed to timber studs with 2 mm galvanised nails with small, flat heads, 30 mm long for 9.5 mm boards and 40 mm for 12.7 mm boards. No nailing should take place within 13 mm of edges of boards. The recommended spacing of studs is 450 mm centres for 9.5 mm board, 600 mm for 12.7 mm and 800 mm for 19 mm board. Boards should be fixed with a 3 mm gap between cut ends and they should be nailed to every support at 150 mm centres. It is desirable to avoid joints in line with door or window jambs.

Jointing. The method of jointing depends on the edge design of the boards. For instance, tapered edge boards normally have smooth seamless jointing composed of joint filler often reinforced with joint tape and finished with a slurry coat. A cover strip joint is often used with square-edged boards, covering the joint with embossed or corrugated paper strips, or paper-faced cotton tape, fixed with suitable adhesive. When a more pronounced panel effect is required, plain or patterned cover strips of wood, metal or plastic may be fixed over the joints. With bevelled-edge boards, the base of the V-joint may be partially filled with joint filler to bridge the joint between adjacent boards. This treatment improves the appearance of the V-joint and facilitates even decoration.

Angles. Most internal angles will have at least one cut edge and should desirably be reinforced with joint tape bedded in joint finish. External angles may be reinforced with corner tape bedded in joint filler and covered with joint finish. For maximum reinforcement a metal angle bead may be used, bedded in joint filler and covered with joint finish.[6] A gypsum cove may be used to cover the junction of walls and ceilings and improve appearance. It consists of a gypsum core encased in ivory-coloured millboard and is normally fixed by sticking with adhesive or nailing or screwing to timber grounds or plugs.

Coves and cornices. The *gypsum cove* is the present day

version of the much more ornate and expensive plaster cornice of former days. Small *plaster cornices* are usually run in gauged plaster on a zinc mould. Where the projection exceeds 150 to 200 mm, the plaster must be firred out on brackets to suit the profile of the moulding. The brackets are lathed and rendered and the cornice finished with the help of a zinc mould mounted on a wooden horse running between guides fixed to wall and ceiling. Another alternative is to use *fibrous plaster* made up of plaster, glue, oil, wood wire and canvas. Fibrous plaster can be used to provide plain and moulded work and can cover large surfaces showing savings in time and weight compared with ordinary plaster. It is normally formed in gelatine moulds and fixed to grounds by specialist firms.

Plastics-surfaced Sheet Materials

These provide attractive finishes to wall surfaces on grounds of coverage, appearance, resistance to wear, durability and low maintenance costs. Probably the best type are melamine laminated thermosetting decorative sheets generally known as *decorative laminates*. They consist of layers of a fibrous material, such as paper, which has been impregnated with thermosetting synthetic resins and consolidated under heat and pressure. Upon the base lies the *print paper*, which is a layer of cellulose paper bearing the desired colour or printed pattern and impregnated with melamine formaldehyde (MF) resin. This may be covered by an overlay of MF-treated alpha-cellulose paper which becomes transparent under heat-pressure treatment. Decorative laminates are often 1.5 mm thick and are already veneered to, or prepared for veneering to, a rigid substrate such as asbestos, chipboard or plywood, to produce a composite sheet material suitable for interior cladding or free-standing constructions.[10] The quality of decorative laminated plastics sheet is prescribed in BS 3794[11] which provides for two nominal thicknesses (1.5 and 3.0 mm).

Decorative laminate surface finishes range from fine gloss to marked texture or raised patterns. They can normally withstand surface temperatures of up to 180°C for short periods without blistering or appreciable surface damage. They have been used successfully as wall linings to kitchens, bathrooms, halls, staircases, clubrooms, hospital operating theatres, public baths, lift cars, toilet cubicles and fitted furniture. Some melamine plastic laminates contain woven linens and fine quality hessians to provide the decorative feature. Plastic-faced plasterboard is also available.

Glazed Ceramic Wall Tiles

These are used extensively in kitchens and bathrooms to produce smooth, impervious and durable surfaces. There are two kinds or glaze: earthenware glazes and coloured enamels. The earthenware glazes are white and cream, while coloured enamels can be obtained in a variety of colours, either plain or mottled, and can have either a glossy or a matt surface. Wall tiles should comply with BS 1281[13] which specifies requirements for trueness of shape, standard sizes, and tolerances in dimensions and finish. Predominant sizes for tiles with cushion edges and spacer lugs are 152 x 152 x 5, 5.5, 6.5 or 8mm and 108 x 108 x 4 or 6.5 mm; and 152 x 152 x 9.5 mm for those without cushion edges and spacer lugs. The spacer lugs maintain a standard joint of 3 mm and form crushing points which will fail under compression stress without damage to the tile. There are three main types of fittings for use at angles, cappings and the like — round edge tiles which are the cheapest, attached angle tiles and angle beads. To withstand wet conditions and chemicals, it is advisable to use tiles with low porosity and a thick glaze. Wall tiles are best fixed with cement-based or organic-based adhesives meeting the requirements of BS 5385.[14]

INTERNAL CEILING FINISHINGS

The most common domestic ceiling finish is *gypsum lath plasterboard*, sometimes faced with aluminium foil on its upper face to give improved thermal insulation, all in accordance with BS 1230.[12] Each board should be nailed to each support using not less than four 40 mm clout-headed nails equally spaced across the width and driven no closer than 13 mm from the edges. End joints should be staggered in alternate courses and cut ends should be located at supports. The joints between boards are normally filled with gypsum plaster and the boards finished with a single coat of suitable neat gypsum plaster 5 mm thick.

Many years ago, *plaster* in three coats was applied to wood laths to form ceilings. The laths were usually of sawn timber 25 mm wide and 5 to 10 mm thick, in lengths varying from 900 to 1800 mm. They were spaced about 10 mm apart and fixed with flat-headed galvanised nails, butt jointed at their ends against supporting timbers and breaking joint frequently. Three-coat plasterwork in this situation was described as 'lath, plaster, float and set'. A pricking-up coat, about 10 mm thick of coarse stuff containing hair, adhered to the laths and also passed between them to form keys. The second or floating coat often contained hair and, after scratching, provided a good base for the thin final or setting coat, which was finished with a wood float or steel trowel according to the finish required. The plaster coats often consisted of lime and sand (1:3). Today *expanded metal lathing* may be used to provide a key for plaster. The metal may be dipped in black bituminous paint or be galvanised to prevent rusting, has a 6.3 mm diamond mesh and the gauge depends on the type of plaster and the spacing of the supports. The metal lathing is fixed to timber with galvanised nails or staples at 100 mm centres.

Ceilings below *solid floors*, which are mainly of reinforced concrete or precast concrete slabs and beams, are often finished with traditional plaster. One, two or three-coat work can be applied according to the shape, texture and suction of the background. Board or lath and plaster finishes can be used by fixing timber battens to the underside of the concrete, with timber fixing pads or clips embedded in the concrete when it is poured. With precast concrete slabs the fixings are usually provided in the joints.

Plastic compound finishes are sometimes applied to the decoration side of square-edge wallboard or on taper-edge board. This type of ceiling finish is applied with a brush and the texture or pattern is obtained using a comb, lacer or stippler.

Ceiling linings are available in a variety of forms ranging from plywood, blockboard, laminboard, chipboard and fibre-board to softwood and hardwood boarding. Joints often form a feature by V-jointing or rebating, or, alternatively, by covering with metal, plastic or timber strip. Some manufacturers produce panels and strips ready for use, including some faced with thin plastic or metal sheet.

In buildings with taller storey heights, ceilings may be *suspended* using timber or metal framing. In this way services can be accommodated above the ceiling and it may also contain and distribute lighting, heating and ventilation. A variety of ceiling *tiles* are available made from various materials and with a number of textures and finishes. They are often designed to absorb sound and increase thermal insulation. Common types of thermal insulating tile are made of soft mineral fibre, and thin box metal or open-textured board backed with glass or mineral wool.

Before leaving ceiling finishings mention can usefully be made of *pattern staining* on ceilings, whereby dark and light patterns appear on plaster surfaces. Where the whole surface is not at a uniform temperature due to varying rates of heat transmission, dust is deposited on the various parts of the surface at different rates, depending on the temperature. Consequently the cooler areas become coated with darker patches of dust, such as beneath ceiling joists. One of the best remedies is to provide about a 25 mm thickness of fibreglass, mineral wool or other suitable insulating material above the ceiling between the joists.

EXTERNAL RENDERINGS

An external solid wall of bricks or blocks may permit moisture penetration through haircracks between the mortar joints and the bricks or blocks and some form of external treatment then becomes necessary. Various weatherproofing applications were described in chapter 4, and in this chapter the characteristics of external renderings are examined in more detail. In selecting a suitable rendering attention should be paid to weatherproofing, durability and appearance. Renderings normally consist of two coats, the undercoat being about 13 mm thick and the final coat varying from 5 mm upwards. Brick joints should be raked out to form a key for the undercoat and the surface of the undercoat scratched to form a key for the final coat. The main types of rendering, as listed in BRE Digest 196,[15] are as follows.

Pebbledash or drydash. A rough finish of exposed small pebbles or crushed stone, thrown on to a freshly applied coat of mortar.

Roughcast or wetdash. A rough finish produced by throwing on a wet mix containing a proportion of coarse aggregate.

Scraped or textured finish. Treatment of the surface of the final coat, at the appropriate stage of setting, with different tools can produce a variety of finishes, depending on the artistry and skill of the craft operative.

Plain coat. A smooth or level surface finished with a wood, felt, cork or other suitably faced float.

Machine applied finish. The final coat is spattered or thrown on by a machine. The texture is determined by the type of machine and the mix, which is often a proprietary material such as Tyrolean.

Smooth finishes tend to suffer surface crazing, and may be patchy in appearance. Mixes rich in cement, containing fine sands or finished with a steel trowel increase the risk of crazing. Pebbledash and roughcast are least liable to change in appearance over long periods and their good watershedding properties make them well suited for use in exposed coastal areas. Dirt from atmospheric pollution shows up more on white and pale coloured finishes than on natural grey cement finishes and the periodic painting of external rendered surfaces is a costly maintenance item. Dirt also appears worse on heavily textured finishes, although rain streaks produced by projecting features will be more prominent on smooth surfaces. The three most suitable *mixes* for external renderings are

(A) 1 part Portland cement:½ part lime:4 to 4½ parts sand by volume;

(B) 1 part Portland cement:1 part lime:5 to 6 parts sand by volume;

Table 11.2 Mixes suitable for rendering (Source: BRE Digest 196[15])

Mix type	Cement : lime : sand	Cement : ready-mixed lime : sand		Cement : sand (using plasticiser)	Masonry cement : sand
		Ready-mixed lime : sand	Cement : ready-mixed material		
I	1 : ¼ : 3	1 : 12	1 : 3	—	—
II	1 : ½ : 4 to 4½	1 : 8 to 9	1 : 4 to 4½	1 : 3 to 4	1 : 2½ to 3½
III	1 : 1 : 5 to 6	1 : 6	1 : 5 to 6	1 : 5 to 6	1 : 4 to 5
IV	1 : 2 : 8 to 9	1 : 4½	1 : 8 to 9	1 : 7 to 8	1 : 5½ to 6½

NOTE. In special circumstances, for example where soluble salts in the background are likely to cause problems, mixes based on sulphate-resisting Portland cement or high alumina cement may be employed. High alumina cement should not be mixed with lime, ground limestone, ground chalk; silica flour or other suitable inert filler should be employed instead.

Table 11.3 Recommended mixes for external renderings in relation to background materials, exposure conditions and finish required (Source: BRE Digest 196[15])

NOTE. The type of mix shown underlined is to be preferred.

Background material	Type of finish	First and subsequent undercoats			Final coat		
		Exposure			Exposure		
		Severe	Moderate	Sheltered	Severe	Moderate	Sheltered
(1) Dense, strong, smooth	Wood float	II or III	II or III	II or III	III	III or IV	III or IV
	Scraped or textured	II or III	II or III	II or III	III	III or IV	III or IV
	Roughcast	I or II	I or II	I or II	II	II	II
	Dry dash	I or II	I or II		II	II	II
(2) Moderately strong, porous	Wood float	II or III	III or IV	III or IV	III	III or IV	III or IV
	Scraped or textured	III	III or IV	III or IV	III	III or IV	III or IV
	Roughcast	II	II	II	as undercoats		
	Dry dash	II	II	II			
(3) Moderately weak, porous *	Wood float	III	III or IV	III or IV			
	Scraped or textured	III	III or IV	III or IV			
	Dry dash	III	III	III	as undercoats		
(4) No fines concrete †	Wood float	II or III	II, III or IV	II, III or IV	II or III	III or IV	III or IV
	Scraped or textured	II or III	II, III or IV	II, III or IV	III	III or IV	III or IV
	Roughcast	I or II	I or II	I or II	II	II	II
	Dry dash	I or II	I or II	I or II	II	II	II
(5) Woodwool slabs *‡	Wood float	III or IV	III or IV	III or IV	IV	IV	IV
	Scraped or textured	III or IV	III or IV	III or IV	IV	IV	IV
(6) Metal lathing	Wood float	I, II or III	I, II or III	I, II or III	II or III	II or III	II or III
	Scraped or textured	I, II or III	I, II or III	I, II or III	III	III	III
	Roughcast	I or II	I or II	I or II	II	II	II
	Dry dash	I or II	I or II	I or II	II	II	II

* Finishes such as roughcast and dry dash require strong mixes and hence are not advisable on weak backgrounds.
† If proprietary lightweight aggregates are used, it may be desirable to use the mix weaker than the recommended type.
‡ Three-coat work is recommended, the first undercoat being thrown on like a spatterdash coat.

(C) 1 part Portland cement:2 parts lime:8 to 9 parts sand by volume.

Masonry—cement—sand mixes may be used as alternatives to mixes B and C. The higher sand contents should only be used if sand is well graded and a higher cement content is preferable when applied under winter conditions. The mix for a following coat should not be richer in cement than the preceding coat.

Table 11.2 gives the different mixes and table 11.3 shows how the choice of mix is influenced by the nature of the background, the type of finish and the degree of exposure. The mixes that are underlined are generally to be preferred. Dense, strong and smooth materials include dense clay bricks or blocks or dense concrete. Moderately strong and porous materials embrace most bricks and blocks and some medium-density concretes. Moderately weak and porous materials include lightweight concretes, aerated concretes and some low-strength bricks.

PAINTING

Types of Paint

Paint consists essentially of a pigment, a binding medium and a thinner to make the mixture suitable for application by brush, roller, spray or dipping. After application, the paint undergoes changes which convert it from a fluid to a tough film which binds the pigment. The nature of these changes varies with different types of paint. Some such as size-bound distemper and chlorinated rubber paint lose the thinner by evaporation. With most paints containing drying oils, part of the change on drying results from reaction of the oil with oxygen from the air. In emulsion paints and oil-bound distempers, the binding material is emulsified or dispersed as fine globules in an aqueous liquid. After application the water evaporates and the globules coalesce to form a tough, water-resistant film. CP 231[16] gives general guidance on the painting of different materials and the selection of paint types.

Painting Walls

Walls are painted to provide colour or surface texture, to waterproof them, to reflect or absorb light, and to facilitate cleaning and hygiene.[17] Problems may arise through dampness; drying out of new construction often takes an unacceptably long time and there is often a reluctance to accept temporary decoration. A coat of emulsion paint is an economical temporary decoration which, unlike distemper, need not be removed later. On internal work, heat and ventilation can accelerate the drying process. Difficulties may also arise through the presence of salts or alkali, although neat gypsum plasters are only slightly alkaline and do not usually affect paints. Surfaces to be painted must be sound and certainly not powdery or crumbly. The drying of paints is retarded by low temperatures, high humidity and lack of ventilation.[18]

Matt finishes are sometimes preferred because they avoid reflection of light sources and minimise surface irregularities. They are however less resistant to wear, more difficult to clean, and tend to absorb more condensation than glossy surfaces. Gloss finishes provide maximum washability and exterior durability and hard-gloss paints based on alkyd resins are popular. Chlorinated rubber paints provide a moderately glossy finish which is alkali-resistant. Emulsion paint (polyvinyl acetate plasticised) is very suitable for application to walls and ceilings but normally avoiding impervious surfaces, kitchens and bathrooms, and is better able to withstand scrubbing and weathering than oil-bound distemper, and is less costly than oil paint.[24] Special paints are available for specific purposes, such as anti-condensation paints containing cork or other absorbent filler, fire-retardant paints to reduce fire risk, heat-resistant paints on surfaces that become hot, and quick-drying paints for use where time is short.[16]

Asbestos cement sheeting and masonry can be painted with emulsion, chlorinated rubber and cement paints,[23] but asbestos cement sheets require back painting.[17]

Wallpaper also forms an attractive finish to walls and ceilings. A variety of finishes including hand or machine printed, embossed and plastic faced are available. Lining papers are generally applied to walls prior to hanging heavyweight or handprinted wallpapers. The two principal forms of paste for paperhanging are starch/flour and water-soluble cellulose pastes.

Painting Woodwork

On new wood, a four-coat system has been accepted as good practice, with a potential life of at least five years; early failure at susceptible areas (sills and bottom rails) is common.[19] CP 231[16] permits a three-coat paint system for economy. BRE Digest 106[19] recommends a water-repellent preservative treatment in conjunction with three coats — primer, undercoat and finish, or a combined primer/undercoat and two finishing coats. The preparatory treatment includes knotting, priming and stopping. The sealing of knots to prevent resin bleeding through the paint is done with shellac knotting to BS 1336[27] or leafing aluminium primer. A good priming coat is essential for a durable paint system and should preferably be applied in a paint-shop or factory to ensure thorough treatment of all surfaces.

The traditional pink primer for softwood contains a high proportion of white lead and some red lead, together with some titanium dioxide to improve opacity and a resin to shorten the drying time. It has good filling, water resistance and durability properties but is expensive and toxic. Aluminium wood primers are durable and are generally preferred for resinous softwoods and most hardwoods. Acrylic emulsion primers are used as shop primers by some joinery manufacturers.

Stopping and filling should follow priming and should ideally be done with hard stopping such as white lead and gold size or a mixture of hard stopping and linseed oil putty. Water-mixed cellulosic fillers or powder fillers based on gypsum plaster are only suitable for internal use. Linseed oil putty is not satisfactory as it usually shrinks and requires at least 48 hours to harden sufficiently to take paint.

Before applying undercoat, the primer should be checked and if thin, a further coat of primer applied. Undercoats should be of the same brand as the finishing coat. Better protection is often provided to exterior woodwork by using two finishing coats, but on interior work where rubbing down is specified, a smoother finish and better appearance will be obtained with two undercoats. Not all gloss finishes are suitable for use as consecutive coats and the manufacturer's instructions must be followed.

Conventional alkyd gloss and flat paints with appropriate undercoats form the main finish on wood, as they combine decorative and protective properties with the necessary elasticity to accommodate wood movement. White lead paint is still favoured by some, although it usually has some alkyd resin and titanium oxide pigment added to improve its gloss, flow and opacity. It has the double disadvantages of darkening in polluted atmospheres and of toxicity. Polyurethane paints, of the single-pack type, are useful internally for hard wear and resistance to washing with soap, detergents or bleach. Clear varnishes are effective externally for about three years, when based on alkyd or tung oil/phenolic.[19]

BRE Digest 106[19] emphasises that present painting costs rarely allow for the proper sanding down of wood and undercoats or use of fillers. Hence joinery specifications should require freedom from imperfections. Extensive thinning of paint still occurs on site, to ease brush application and speed up the work, resulting in loss of durability and standard of finish. The moisture content of joinery at time of painting should not exceed 18 per cent for exterior work and 12 per cent for interior work, and exterior painting must not be undertaken on damp surfaces.

On repainting it is ill-advisable to strip paint which is adhering well, chalking only slightly and which is free from other defects. It should be washed with a detergent solution or proprietary wash and rubbed with wet abrasive paper. Small areas of loose or defective paint must be scraped down to primer, if sound, or bare wood, and brought forward with primer and undercoat as necessary. On sound existing paint, one undercoat and one finishing coat is normally sufficient. Decayed timber should be cut out and replaced, with both new and old timber being treated with preservative. The blow lamp is the quickest and most effective way of removing paint from wood, but where impracticable, for instance near glass, paint removers of the organic solvent type may be used, taking care to remove all traces of paint remover.[19]

The main causes of failure in paintwork are

(1) crazing: undercoat has not hardened sufficiently before finishing coat applied;

(2) chalking: destruction of oil paint by chemical or physical changes;

(3) blistering and peeling: painting on base with high moisture content, inadequate preparation or subject to excessive heat;

(4) wrinkling: too thick a paint film;

(5) discolouration: absence of knotting or painting over bitumen;

(6) blooming: delayed or defective drying;

(7) saponification: alkali attack on paint, converting oil into soap.

Other causes of paint failure and suggested remedies are contained in BRE Digest 198.[18] Wood stains, properly used, show cost advantages over paint on cladding and outdoor timbers.[26]

Painting Metalwork

Metals in building work are painted mainly for protection against corrosion and only secondarily for decoration. Sites subject to sea spray or in industrial areas with heavily polluted atmospheres require special attention. All metal surfaces to be painted must be free of mill scale, rust, oil, grease, moisture and dirt. Suitable cleaning processes and associated priming treatments are

(1) grit or shot-blasting: etch and/or inhibitive primers;

(2) flame cleaning: inhibitive tolerant primers (red lead/metallic lead);

(3) phosphating: zinc chromate;

(4) hand or power tools: inhibitive tolerant primers.

The paint system normally consists of primer, undercoats and finishing coat. Guidance as to the suitable choice of a protective system for steel is given in BRE Digest 70.[20] On rough surfaces, primers are best applied by brush. The best primers contain pigments which chemically inhibit the corrosion of iron and steel. Examples of such pigments include red lead, zinc chromate, zinc phosphate and calcium plumbate. Priming paints are intended for particular metals; hence red lead primer should not be used on aluminium. Where a single primer is to be used on both steel and aluminium, a zinc chromate primer is preferable. For zinc and steel together, calcium plumbate or zinc chromate is best. Red lead, zinc-rich and metallic lead primers are especially well suited for steelwork to be left exposed on site. Advice on the protection of iron and steel structures against corrosion is given in BS 5493.[25]

The undercoats and finishes provide additional film thickness, water resistance and possibly decoration. Bituminous paints, often used for cast iron rainwater gutters, give good inexpensive protection against water, salt and some chemicals, although they are not particularly durable in sunlight. Thick, tough and resistant plastics are also being applied to steel usually in the factory (PVC) but sometimes on site by flame spraying (nylon).[20]

REFERENCES

1. *BRE Digest 213*: Choosing specifications for plastering. HMSO (1978)
2. DOE Advisory leaflet 9: Plaster mixes for inside work. HMSO (1975)
3. BS 1191: Gypsum building plasters; Part 1: 1973 Excluding premixed lightweight plasters, Part 2: 1973 Premixed lightweight plasters
4. DOE Advisory leaflet 2: Gypsum plasters. HMSO (1975)
5. BS 1198: 1976 Sands for internal plastering with gypsum plasters
6. British Gypsum. White book (1978)
7. DOE Advisory leaflet 21: Plastering on building boards. HMSO (1976)
8. DOE Advisory leaflet 64: Plasterboard dry linings. HMSO (1975)
9. British Gypsum. Gyproc moisture resistant board (1979)
10. BS 5492: 1977 Code of practice for internal plastering
11. BS 3794: 1973 Decorative laminated plastics sheet
12. BS 1230: 1970 Gypsum plasterboard
13. BS 1281: 1974 Glazed ceramic tiles and tile fittings for internal walls
14. BS 5385 Code of practice for wall tiling; Part 1: 1976 Internal ceramic wall tiling and mosaics in normal conditions, Part 2: 1978 External ceramic wall tiling and mosaics
15. *BRE Digest 196*: External rendered finishes. HMSO (1976)
16. CP 231: 1966 Painting of buildings
17. *BRE Digest 197*: Painting walls; Part 1 Choice of paint. HMSO (1977)
18. *BRE Digest 198*: Painting walls; Part 2 Failures and remedies. HMSO (1977)
19. *BRE Digest 106*: Painting woodwork. HMSO (1979)
20. *BRE Digest 70*: Painting iron and steel. HMSO (1973)
21. DOE Advisory leaflet 25: Painting woodwork. HMSO (1973)
22. DOE Advisory leaflet 11: Painting metalwork. HMSO (1974)
23. DOE Advisory leaflet 28: Painting asbestos cement. HMSO (1976)
24. DOE Advisory leaflet 37: Emulsion paint. HMSO (1976)
25. BS 5493: 1977 Code of practice for protective coating of iron and steel structures against corrosion
26. *BRE Information Paper 34/79*: Exterior wood stains. HMSO (1979)
27. BS 1336: 1971 Knotting

12 WATER SERVICES AND SANITARY PLUMBING

This chapter examines the methods of supplying water to buildings, cold and hot water supply arrangements, sanitary appliances, waste systems and hot water heating. These aspects all relate to 'sanitation and services' but are included here as it was felt that a study of 'building technology' would not be complete without them.

COLD WATER SUPPLY

A local authority has a duty to ensure that every dwelling house in its area is provided with, or has reasonably available, a sufficient supply of wholesome water for domestic purposes. Statutory water undertakers normally make byelaws for preventing waste, excessive consumption, misuse or contamination of water supplied by them. Sources of water include rainwater, surface water, lakes and rivers, and underground water in the form of springs, wells and boreholes. Some purification of the water is often necessary to remove organic or inorganic impurities and possibly to reduce the hardness of the water. Public water supplies are sterilised before being passed to the consumer by treatment with chlorine or chlorine and ammonia.

The connection to the water main is made by the water authority who normally drill and tap the main and lay a 'communication' pipe, often 13 mm in size, up to the boundary of the site, finishing with a stopvalve or stopcock in a suitable box or chamber fitted with a hinged cast iron cover. This part of the water supply installation remains the property of the water authority, even although the property owner may contribute towards it. The supply pipe is laid from the stopvalve into the building and should be at least 750 mm deep for adequate protection from frost. A second stopvalve should be fitted at the first accessible position after the service enters the building, and is usually accompanied by a draincock to permit the cold water system to be drained down.[35]

Cold Water Supply Systems

There are two principal cold water supply systems although variations of either are possible. They are often described as *direct* and *indirect*, but have also been referred to as *storage*

and *non-storage* and as *downfeed* and *upfeed*. Typical arrangements for both systems are illustrated in figures 12.1.1 and 12.1.2. In the direct system all cold water drawoff points are fed directly from the rising service pipe with the cistern, where provided, serving the sole purpose of a feed cistern (supplying cold water to the hot water system). In the indirect system the rising service pipe normally serves only one drawoff point, often over the kitchen sink, from which fresh water can be obtained. All other cold water drawoff points are supplied from a storage cistern. The cistern may serve a dual purpose — feeding a hot water system and supplying cold water for other purposes.

With the direct system, potable water is available at all outlets, and it reduces the storage capacity and often the size of cold water distribution pipes and the length of pipe runs. To be satisfactory, there must be a constant supply of water of sufficient pressure during periods of peak demand. Furthermore, where the mains pressure is high it may result in excessive noise and damage to seatings of taps and valves.

All pipes should be located in positions protected from frost or be adequately insulated;[34] suitable insulating materials are detailed in BS 5422.[37]

Cisterns

As far as possible cisterns should be located below the top floor ceiling and service pipes should be positioned on internal walls, as protection against frost. A storage cistern in a dwelling should have a minimum capacity of 90 litres[1] to cover twenty-four hours' interruption of supply. Water level in the cistern is maintained by means of a suitable ballvalve, whereby the pressure exerted by a float and increased by a lever arm is used to force a washer against the inlet orifice. Two types of ballvalve are in general use:

(1) BS 1212[2] piston or Portsmouth type, with a horizontally sliding body of brass, bronze or gunmetal and nylon seat holding a rubber washer (figure 12.1.3);

(2) Garston valve or diaphragm type with brass body and copper or plastics spherical float (reduced noise and wear)[2] (figure 12.1.4).

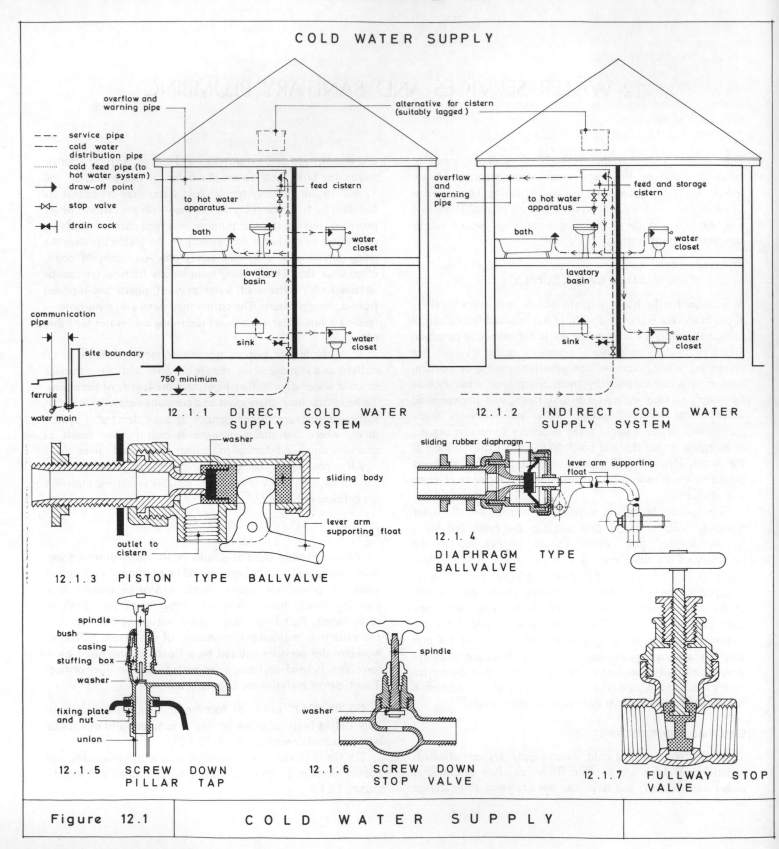

COLD WATER SUPPLY

service pipe
cold water distribution pipe
cold feed pipe (to hot water system)
draw-off point
stop valve
drain cock

overflow and warning pipe

alternative for cistern (suitably lagged)

feed cistern

to hot water apparatus

bath

water closet

lavatory basin

sink

water closet

communication pipe

site boundary

ferrule

water main

750 minimum

12.1.1   DIRECT COLD WATER SUPPLY SYSTEM

overflow and warning pipe

feed and storage cistern

to hot water apparatus

bath

water closet

lavatory basin

sink

water closet

12.1.2   INDIRECT COLD WATER SUPPLY SYSTEM

washer

sliding body

lever arm supporting float

outlet to cistern

12.1.3   PISTON TYPE BALLVALVE

sliding rubber diaphragm

lever arm supporting float

12.1.4   DIAPHRAGM TYPE BALLVALVE

spindle
bush
casing
stuffing box
washer

fixing plate and nut

union

12.1.5   SCREW DOWN PILLAR TAP

spindle

washer

12.1.6   SCREW DOWN STOP VALVE

12.1.7   FULLWAY STOP VALVE

| Figure   12.1 | COLD   WATER   SUPPLY |

Ballvalves are made in three categories to withstand varying conditions of pressure: high pressure (1400 kN/m^2), medium pressure (700 kN/m^2) and low pressure (280 kN/m^2). Each cistern is also provided with an overflow and warning pipe with a diameter of not less than 19 mm and with a larger diameter than the supply pipe to the cistern, discharging in a visible position on an external wall of the building.

Cisterns can be obtained in a variety of materials. Galvanised mild steel cisterns to BS 417[3] are quite common. Their life can be extended by painting internally with one coat of non-toxic bituminous composition,[4] particularly where water is aggressive and there is likelihood of electrolytic action due to the use of dissimilar metals in the same system. In these situations there are advantages in using cisterns made of non-metallic materials, such as asbestos cement or glass-fibre reinforced plastics, although they are more expensive. Where a cistern is placed in the roof space, it should be located centrally and be cased in timber or rigid sheet material, leaving a 50 mm gap for a layer of insulating material.[35] Performed insulating slabs, made to fit cisterns of different sizes, are also available. The cistern should be provided with a well-fitting cover, adequately insulated.[4]

Service and Distribution Pipes

Stopvalves should be provided on each branch pipe to enable repairs to drawoff points to be carried out without closing down the whole system. Pipe sizes are affected by the water pressure and the total maximum demand which is likely to occur at any one time.

A variety of materials is available for water supply pipes, of which probably the most popular is *light gauge copper* tube, largely on account of its durability, flexibility, smooth bore, neat appearance and ease of jointing. The pipe dimensions in BS 2871[5] are now given in external diameters in place of the former nominal bores.

external diameter in mm	15	22	28
nearest equivalent internal bore in inches	½	¾	1

Common sizes are 15 mm for wash basins and WC flushing cisterns and 22 mm for sinks and baths. Thicker pipes in accordance with BS 2871[5] (table Y) are recommended for underground services. Copper tubes and fittings are jointed either with capillary or compression joints. *Capillary* joints are made with solder contained in a groove formed in the wall of the socket. After cleaning and fluxing, the solder is melted by a blowlamp and fills the space between the tube and fitting. *Compression* joints are more expensive and generally consist of a compression ring compressed into the socket of a tube with a coupling nut. The bore of the compression ring corresponds to the outside diameter of the tube.

Polythene and UPVC pipes. These are useful in corrosive soils, are flexible, and their smooth bore speeds water flow and prevents the formation of scale. Compression joints probably form the best method of jointing. The pipes are rather soft, not completely resistant to ground gases, need ample support and cannot be used for earthing electrical installations. They should conform to BS 1972.[6] Unplasticised (UPVC) pipes to BS 3505[30] are stronger and more rigid than polythene, but neither is recommended for hot water systems.[31] Small diameter pipes can be hot bent but it is a skilled operation. Pipe joints can be either flexible or solvent welded.[12] The requirements for glass fibre reinforced plastics (GRP) pipes and fittings are contained in BS 5480.[39]

Lead pipes. These are now little used on account of their high cost, weight and unsuitability with soft or acid water. Soldered joints are normally used with lead pipes.

Mild steel pipes. They are relatively strong and inexpensive, and should comply with BS 1387;[7] they are made in three categories — light, medium and heavy. The most common form of joint is screwed and socketed. Stainless steel tubes can be used as an alternative to copper and should comply with BS 4127.[8] They are jointed with capillary and compression fittings similar to those used for light gauge copper tubes. Copper, polythene and steel pipes are usually supported with metal or plastics clips.

Noise in pipes (water hammer) may arise from any of a number of causes, including inadequately fixed pipework, sharp bends, defective valve seatings and large volumes of water entering the storage cistern. A Portsmouth ballvalve can be very noisy in operation unless provided with a dip pipe.

Taps and Stopvalves

A pillar tap suitable for use with sinks, baths and wash basins is illustrated in figure 12.1.5, although the form of handle varies considerably and some are very sophisticated. Stopvalves on branches to WC flushing cisterns and the like are often of the screwdown variety (figure 12.1.6), complying with BS 1010,[9] but those on the main distribution pipes may, if permitted, be of the fullway type (figure 12.1.7), complying with BS 1952,[10] as screwdown valves offer some resistance to flow.

HOT WATER SUPPLY

Before choosing a hot water system, various criteria must be considered including required performance, characteristics of local water and need, if any, for space heating. The heating source for domestic hot water supply is often an independent boiler burning solid fuel, gas or oil. Alternatives are gas or electric hot water heaters (storage or instantaneous). In smaller dwellings use is still made of back boilers behind open fires. The essential features of a hot water supply system should be to produce sufficient hot water to meet all demands, be economical in running costs, be suitable for the type of building and relatively easy to install and maintain. There are two main systems of hot water supply — direct and indirect.

Direct System

In a direct system (figure 12.2.1), hot water circulating between the boiler and storage cylinder or tank is drawn off as required for domestic use and replaced by fresh, cold water fed directly into the same circuit.[11] A totally enclosed cylinder or tank, often placed in an airing cupboard, is located as close as possible to the boiler, and connected by primary flow and return pipes. A short primary circuit ensures rapid discharge of the hot water and reduces the heat loss from pipework. The cold water feed is taken from a cold water storage cistern to the hot water cylinder or tank located below it. The water to drawoff points is delivered from the top cylinder or tank where the hottest water is stored, often through an expansion pipe. This minimises interference with the primary flow and return during periods of drawoff. The main drawback is that distribution pipes are non-circulatory and so their water content has to be run to waste before hot water is discharged at the taps. Hence water authorities normally restrict the length of non-circulatory delivery pipes to 12 m for pipes not exceeding 19 mm, 7.5 m for 25 mm pipes and 3 m for pipes exceeding 25 mm internal diameter.

Indirect System

In an indirect system (figure 12.2.2), hot water also circulates between the boiler and storage cylinder or tank, but the storage vessel (an indirect cylinder or calorifier) is so designed that the hot water in the primary circuit from the boiler is used only to raise the temperature of the stored water; it does not mix with it nor is it drawn off for domestic use. Hot water for domestic use is drawn from the secondary side of the system, which may be a complete circuit, and is replaced by cold water fed into the secondary circuit. Apart from the

replacement of water lost by expansion, the same water circulates continuously in the primary circuit.[11]

In some parts of the country the cold water supply contains calcium and magnesium bicarbonates and when heated above 68°C, the bicarbonates are converted into insoluble carbonates, which form a hard lime deposit called 'fur'. With recirculation of hot water, as in the direct system, there will be an initial production of scale but it will not be repeated. Water will expand by about 4½ per cent (1:23) when heated from 10°C to 93°C, hence a small expansion tank is necessary on an indirect system to absorb the expansion water from the primary circuit and to make up the voids when it contracts on cooling.

The water in the boiler is heated first and becomes less dense, permitting the colder, heavier water from the storage cylinder or tank to fall and force the warmer, lighter water up the primary flow into the top of the cylinder. Similarly, due to the difference in weight of the water in the feed and secondary flow, circulation of water in the secondary circuit will occur once the contents of the cylinder become heated. Loss of heat from the circulatory pipework to the surrounding air will create a temperature difference between the flow and return pipes, ensuring continuous movement.

Although both systems can be extended to serve hot water radiators for space heating, there are disadvantages with the direct system, as radiators tend to run cold when domestic water is drawn from the storage cylinder or tank and the rate of circulation of heated water is slow. Hot water for space heating in an indirect system is taken from the primary circuit and is less likely to cool when water is drawn from the secondary side of the hot water cylinder.[11]

Cistern and Pipe Sizes (internal diameters)

Common sizes for domestic hot water supply installations are as follows

cold water storage cistern	250 litres
hot water storage cylinder or tank	125 litres
flow and return pipes (primary circuit)	25 mm or 32 mm on a direct system
cold feed pipe	25 mm
delivery pipes	19 mm minimum with 19 mm branch to bath and 13 mm branches to sink and wash basin
expansion and vent pipe	19 mm

(Note: pipe sizes in different materials are expressed in different ways)

Boiler Mountings

The most common mountings for hot water supply boilers are now listed.

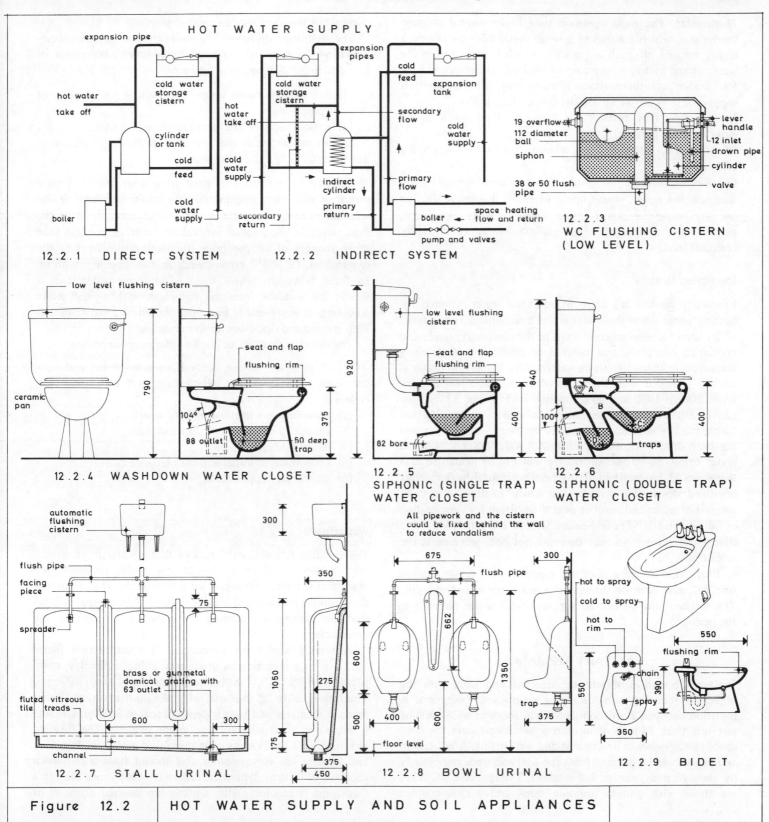

HOT WATER SUPPLY

12.2.1 DIRECT SYSTEM

expansion pipe
hot water take off
cold water storage cistern
cylinder or tank
cold feed
boiler
cold water supply
secondary return

12.2.2 INDIRECT SYSTEM

expansion pipes
cold water storage cistern
hot water take off
indirect cylinder
cold water supply
primary flow
primary return
cold feed
expansion tank
secondary flow
cold water supply
space heating flow and return
boiler
pump and valves

12.2.3 WC FLUSHING CISTERN (LOW LEVEL)

19 overflow
112 diameter ball
siphon
38 or 50 flush pipe
lever handle
12 inlet
drown pipe
cylinder
valve

12.2.4 WASHDOWN WATER CLOSET

low level flushing cistern
ceramic pan
seat and flap
flushing rim
790
375
104°
88 outlet
50 deep trap

12.2.5 SIPHONIC (SINGLE TRAP) WATER CLOSET

low level flushing cistern
seat and flap
flushing rim
920
840
400
82 bore

12.2.6 SIPHONIC (DOUBLE TRAP) WATER CLOSET

A
B
C
D
100°
400
traps

12.2.7 STALL URINAL

automatic flushing cistern
flush pipe
facing piece
spreader
fluted vitreous tile treads
channel
75
brass or gunmetal domical grating with 63 outlet
600
300
1050
175

300

12.2.8 BOWL URINAL

All pipework and the cistern could be fixed behind the wall to reduce vandalism

350
675
flush pipe
662
600
275
500
400
600
375
1350
300
trap
floor level
375
450

12.2.9 BIDET

hot to spray
cold to spray
hot to rim
chain
spray
550
550
390
350
flushing rim

Figure 12.2   HOT WATER SUPPLY AND SOIL APPLIANCES

Thermostat. The most common type is the vapour pressure thermostat, which consists of a small metal bulb containing a highly volatile oil, such as paraffin, which is immersed in the water of the boiler. In the case of oilfired, gasfired or gravity feed boilers, the thermostat will either shut off the gas or oil supply, or stop the air fan on the gravity feed boiler.

Drain valve. This enables the boiler and pipework to be emptied when required.

Thermometer. Fitted to show whether the boiler is operating effectively.

Check-draught valve. A simple pivoted valve inserted in the flue near the boiler, which opens when the draught in the flue becomes excessive due to high winds or gusty conditions, permitting the draught to be maintained at a reasonably constant level.

Immersion Heaters

Immersion heaters are often fitted in hot water cylinders for heating water when the boiler is not functioning. They consist of a resistance element connected to the electricity supply and contained in a protective material or sheath, the whole being immersed in the water to be heated. The immersion heater is usually provided with thermostatic control and lengths vary from 300 to 1050 mm with loadings of 1, 2 and 3 kW. They can be fitted vertically or horizontally in cylinders. Vertical fixing has the advantage of speedy recovery after some water has been drawn off, making more hot water quickly available. Water below an immersion heater will not be heated so that there would be merit in fixing one horizontal heater about one-third down the cylinder for quick heating of a small amount of water and another near the bottom for slow heating of all the water. These heaters should be controlled by a change-over switch so that they are not both working at the same time.

Every new dwelling shall be supplied with a minimum of one WC, one bath or shower, one wash basin and one sink. The kitchen sink should be supplied with water direct from the mains.[32]

SANITARY APPLIANCES

Sanitary appliances should be selected and sited with the object of reducing noise in living rooms and bedrooms. A partition between a bedroom and a bathroom or WC should be not less than 75 mm of insulating blockwork and the door should be as heavy as practicable and well-fitting. A WC cistern and pan should be isolated from the wall and floor respectively by resilient pads, sleeves and washers. The quietest WC suites are those with double siphonic traps and a close-coupled cistern, but these are also the most expensive.

The Building Regulations[13] require that no sanitary accommodation (a room or space which contains watercloset or urinal fittings) shall open directly into:

(1) a habitable room unless used solely for sleeping or dressing purposes;

(2) a room used for kitchen or scullery purposes;

(3) a room in which any person is habitually employed in any manufacture, trade or business.

Sanitary accommodation must have a window or similar means of ventilation opening directly into the external air and with an opening area of not less than one-twentieth of the floor area; or mechanical ventilation securing not less than three changes of air per hour discharging directly into the external air. CP 305[14] recommends that in dwellings with up to three bedrooms where there is only one watercloset it should be separate from the bathroom with a wash basin adjoining. It is preferable that there should be a watercloset at both ground and first-floor levels in houses.

The essential qualities of good sanitary appliances are:

(1) cleanliness — strong, smooth, non-absorbent and non-corroding surfaces, largely self-cleansing and permitting easy cleaning;

(2) durability — withstanding hard wear;

(3) simplicity of design and construction;

(4) accessibility;

(5) economical in initial and maintenance costs;

(6) satisfactory appearance.

Waterclosets

The Building Regulations[13] require that a watercloset receptacle or pan shall have a smooth and readily cleansed non-absorbent surface, discharging through an effective and suitably dimensioned trap to a soil pipe or drain. The flushing apparatus shall be capable of securing the effective cleansing of the receptacle.

A typical washdown watercloset is illustrated in figure 12.2.4 with a ceramic pan of vitreous china or fireclay, complying with BS 1213,[16] with a normal height of about 400 mm. The trap is 'S' or 'P' in shape, integral with the pan, of 88 mm internal diameter and with a depth of seal of 50 mm. The seat is normally of plastics for maximum hygiene. The flushing cistern can be of a variety of materials, of which ceramic ware and plastics are very popular and should have a minimum capacity of 9 litres. Where the cistern is raised above the pan a flush pipe is required, with a minimum internal diameter of

38 mm for a low-level suite, discharging into the flushing rim around the top of the pan.

Siphonic waterclosets are more efficient and quieter than the washdown variety but there is greater risk of blockage if misused. In the single trap variety (figure 12.2.5), the waterway is designed to produce full bore flow and the siphonic action assists in cleansing the pan. The double trap is illustrated in figure 12.2.6. When flushed the patent device A reduces the air pressure in chamber B; seal C is broken allowing the escape of flush water and setting up siphonage. Siphonic waterclosets are sometimes provided with flushing cisterns with capacities of 11 and 14 litres.

The older type high level flushing cisterns were usually of the siphon pipe or the dome pattern.[17] In the dome type, the dome is lifted by a handle forcing water over the open top of the flush pipe, relieving air pressure and starting siphonic action which draws all the water out of the flushing cistern. In low level WC suites, flushing cisterns of the type illustrated in figure 12.2.3 in ceramic ware or plastics are often used. The bent siphon tube, one leg of which is connected to the flush pipe, has an enlarged cylinder at the bottom of the other leg. The lever handle lifts a valve in the cylinder, which forces the water in the cistern over the siphon and into the flush pipe.

Urinals

Under the Building Regulations[13] a urinal shall have one or more slabs, stalls, troughs, bowls or other suitable receptacles, which

(1) have a smooth and readily cleansed non-absorbent surface;

(2) have an outlet fitted with an effective grating and trap;

(3) are so constructed as to facilitate cleansing.

In addition, it shall be provided with an effective flushing apparatus.

Urinals can be formed of ceramic materials, stainless steel or perspex (acrylic resin). A typical stall urinal is illustrated in figure 12.2.7 with a width of 600 mm per stall and heights of 900 and 1050 mm. A 50 or 62 mm diameter end outlet to the channel is adequate for four stalls, a central outlet for five to seven, and a minimum of two outlets where there are more than seven stalls. The flushing capacity is normally 4½ litres per stall. Automatic flushing cisterns, often set to discharge at twenty-minute intervals, should comply with BS 1876.[18] The cistern is fed with water at a steady rate and siphonic action takes place automatically and rapidly when water reaches the designed level. The water passes through 25 or 32 mm flush pipes and spreaders to thoroughly flush each stall. The bowl type urinal (figure 12.2.8) has a smaller surface to be fouled

and flushed and the bowls are normally fixed at 600 to 675 mm centres, each with its own separate trap.

Bidets

Bidets are used for cleansing the lower excretory organs of the body in a thorough and convenient manner by sitting astride the appliance. Hot and cold water can be delivered either to the rim or to the ascending spray as shown in figure 12.2.9. Bidets have a secondary use as a footbath. They are normally formed of ceramic materials with 32 mm outlet and separate trap.[32]

Wash Basins

Most wash basins are in white or coloured ceramic ware; BS 1188[19] recognises both fireclay and vitreous china and the latter is available in two grades (heavy and ordinary). Wash basins provide facilities for personal ablutions in bathrooms, dressing rooms, bedrooms and cloakrooms. They can be supported by a pedestal or by cantilever brackets built into or screwed to a wall. The normal size of wash basin is 635 x 455 mm (figure 12.3.1) although 560 x 405 mm may be used in confined spaces. A common waste size is 32 mm and a separate trap is provided. BS 1188[19] prescribes requirements for tap holes, soap sinkings, chain stay and overflow. A typical overflow detail is shown in figure 12.3.3 and a corner wash basin in figure 12.3.2. The waste pipe will be taken from a ground floor appliance to connect with a back-inlet gully. Metal wash basins are available in cast iron or pressed steel (vitrified porcelain enamelled), or stainless steel in accordance with BS 1329.[20]

In recent years it has become quite common practice to incorporate steel or ceramic wash basins in bedrooms and ladies powder rooms into a melamine-faced phenolic laminate surround, when the fittings are called *vanitory units*. This arrangement permits a counter-type surface for make-up and similar materials to be placed around the wash basin without the risk of being splashed. The laminate can be bent into two dimensional curves under heat, permitting the provision of vanitory units with rounded front edges and swept back upstands. Some vanitory units also incorporate shallow drawers.

Baths

The majority of baths are made of porcelain enamelled cast iron in accordance with BS 1189[21] or vitreous enamelled pressed steel conforming to BS 1390,[22] although baths are also manufactured in plastics, fireclay and cast acrylic sheets.[29] Bath panels, for enclosing the bath, are available in a variety of materials, including polystyrene, super hardboard, steel, asbestos cement, preslabbed ceramic tiles and plastics. A

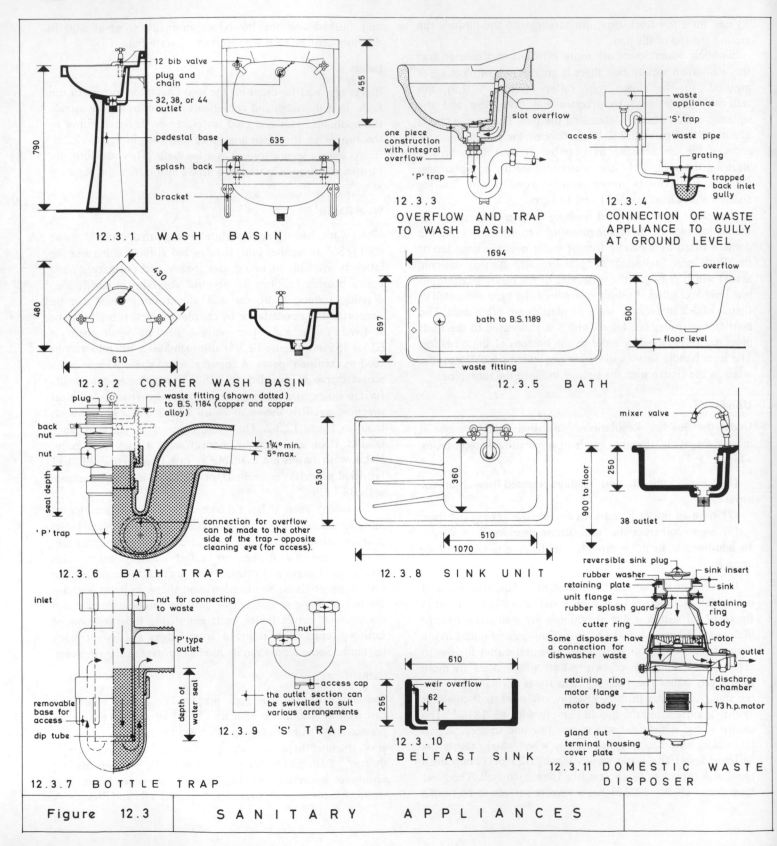

12.3.1   WASH   BASIN

12 bib valve
plug and chain
32, 38, or 44 outlet
pedestal base
splash back
bracket

790
635
455

12.3.3   OVERFLOW AND TRAP TO WASH BASIN

slot overflow
one piece construction with integral overflow
'P' trap

12.3.4   CONNECTION OF WASTE APPLIANCE TO GULLY AT GROUND LEVEL

waste appliance
'S' trap
access
waste pipe
grating
trapped back inlet gully

12.3.2   CORNER WASH BASIN

430
480
610

12.3.5   BATH

1694
697
500
bath to B.S.1189
waste fitting
overflow
floor level

12.3.6   BATH TRAP

plug
back nut
nut
seal depth
'P' trap
waste fitting (shown dotted) to B.S. 1184 (copper and copper alloy)
1¼° min.
5°max.
connection for overflow can be made to the other side of the trap – opposite cleaning eye (for access).

12.3.8   SINK UNIT

mixer valve
250
900 to floor
38 outlet

360
510
1070
530

12.3.7   BOTTLE TRAP

inlet
nut for connecting to waste
'P' type outlet
removable base for access
dip tube
depth of water seal

12.3.9   'S' TRAP

nut
access cap
the outlet section can be swivelled to suit various arrangements

12.3.10   BELFAST SINK

610
255
weir overflow
62

12.3.11 DOMESTIC WASTE DISPOSER

reversible sink plug
rubber washer
retaining plate
unit flange
rubber splash guard
cutter ring
Some disposers have a connection for dishwasher waste
retaining ring
motor flange
motor body
gland nut
terminal housing cover plate
sink insert
sink
retaining ring
body
rotor
outlet
discharge chamber
1/3 h.p.motor

Figure   12.3   S A N I T A R Y   A P P L I A N C E S

typical overall size for a rectangular bath is 1694 x 697 x 500 mm high, as illustrated in figure 12.3.5. Other bath types, specified in BS 1189,[21] include the rectangular shallow pattern and the tub (parallel) pattern. The bottom of the bath should be as flat as possible and hand grips are desirable for reasons of safety. The normal rectangular bath has a capacity of about 120 litres when filled to 225 mm above the waste fitting, a common waste size being 38 mm. A suitable bath trap is shown in figure 12.3.6; the depth of seal varies from 38 to 75 mm.

Sinks

Sinks used for culinary, laundry and other domestic purposes are normally made either of white glazed fireclay to BS 1206[23] or of vitreous porcelain enamelled cast iron or pressed steel, or stainless steel to BS 1244.[24] Fireclay sinks can take various forms and sizes, with or without back shelves and with and without integral fluted drainers. A typical reversible sink (610 x 455 x 255 mm deep), without shelves (Belfast type), is illustrated in figure 12.3.10 and a suitable S-trap for use with it in figure 12.3.9. Outlets are similar to baths, while plain sinks have weir-type overflows (figure 12.3.10) and sinks with shelves have slot overflows (figure 12.3.8). A combined sink and drainer unit is illustrated in figure 12.3.8. Metal sinks are made in round bowls, normally 430 mm diameter, as well as the rectangular pattern. More attractive forms of exposed trap include the chromium-plated or stainless steel 'bottle' trap illustrated in figure 12.3.7. Butler's, pantry or 'wash-up' sinks used for cleaning utensils that are liable to breakage may be made of teak, timber lined with lead or stainless steel.

Waste Disposers

Waste disposers are fitted to sinks and are designed to dispose of organic food waste quickly, hygienically and electro-mechanically and to flush the residue into the drain. Metal, rags and plastic objects should not be placed in a waste disposer. A typical disposer is illustrated in figure 12.3.11 and is operated by turning on cold water, switching on the disposer and feeding waste into the unit. The waste falls on to a high speed rotor and is flung against a stationary cutting ring with great force, shredding the waste into very small particles. The partially liquefied waste filters through the rotor into the waste pipe, joining the flow of water which keeps the apparatus clean and free from unpleasant smells. A thermal overload device cuts off the power in the event of jamming.

Showers

Showers enable washing to be performed quickly, in limited space and with the minimum quantity of water. They may be fitted in separate enclosures or be used in conjunction with baths. They normally consist of an overhead or shoulder height rose or spray nozzle. The shower base or tray may be in porcelain enamelled cast iron or fireclay, the most common sizes being 750 and 900 mm square by 150 and 175 mm deep overall, drained by a 38 mm trapped gully with bars not exceeding 6 mm apart. One-piece cubicles in acrylic plastics and built-up panels of ceramic tiles are available, and water-proof curtains may be used to give privacy.

Spray nozzles are of various patterns. Some contain a control tap to vary the shape or volume of the spray, and all may be mounted on a swivel joint to vary the direction. Showers may be controlled by stopcocks on the hot and cold supplies, by mechanical mixers or by thermostatic mixing valves. Thermostatic mixing valves are to be preferred as they reduce the danger from scalding.

Materials used in Sanitary Appliances

Some of the more important characteristics of the more common materials used in sanitary appliances are now listed. *Ceramic materials.* Fireclay is semi-porous yellow or buff refractory clay of great strength especially suitable for large appliances. Vitreous china is a white non-porous clay of very fine texture, which is very strong with high resistance to crazing and staining. The final coating of ceramic glaze resembles glass coating fused at high temperature and can be transparent or coloured.
Vitreous enamelled cast iron or steel. This is sometimes referred to as porcelain enamel and consists of opaque glass fused to the metal forming a permanent bond. It produces a smooth, even, very durable and high-gloss finish in a variety of colours which are permanent and non-fading.
Stainless steel. A high alloy steel containing a large proportion of chromium. Satin finish has a high resistance to marking, scratching and corrosion. Stainless steel is light in weight, but the cost is much higher than vitreous enamelled steel.
Plastics. Perspex, polypropylene, nylon and glass reinforced plastic require no protective coating and are generally self-coloured, homogeneous, free from ripples or blemishes, very tough, lightweight, warm to the touch and resilient. They can however be damaged by abrasion, lighted cigarettes and hot utensils.

SANITARY PIPEWORK ABOVE GROUND

Traps

The Building Regulations[13] require that every soil and waste appliance must be adequately trapped with a satisfactory water seal and have means of access for internal cleansing. These aspects are reiterated in BS 5572[15]: 'The entry of foul air from the drainage system into the building is prevented by the installation of suitable traps which should be self-cleansing. A trap which is not an integral part of the appliance should be attached to and immediately beneath its outlet and the bore of the trap shall be smooth and uniform throughout. All traps shall be accessible and be provided with adequate means of cleaning. There is advantage in providing traps which are capable of being readily removed or dismantled'. Typical traps are illustrated in figures 12.3.3, 12.3.6, 12.3.7 and 12.3.9.

Waste outlets and traps should have a minimum internal diameter of 30 mm for wash basins and bidets, and 38 mm for sinks, baths and shower trays. Minimum depths of seals should be 50 mm for water closets, and for other appliances — 75 mm for traps up to and including 50 mm diameter, and 50 mm for traps over 50 mm diameter.

The traps to the various appliances must remain sealed in all conditions of use, otherwise there is a risk of unpleasant smells entering the building. A seal will be broken if the pressure changes in the branch pipe are of sufficient intensity and duration to overcome the head of water in the trap itself. One way of restricting the changes in air pressure during discharge is to provide an extensive system of vent piping with the object of equalising pressures throughout the system. Research conducted by the Building Research Establishment has shown that this may also be achieved by appropriate design of the pipework using the single-stack system. Another but less-certain alternative is to use special resealing traps. The main causes of loss of seal are *self-siphonage* due to flow in the branch waste pipe to the main stack, and *induced siphonage* due to flow down the main stack, and *back pressure* due to flow of water down the main stack in conjunction with a sharp bend at the foot of the stack.[25] Other causes are *evaporation*, *wind effect* across the top of tall stacks and *leakage* from the trap.

Discharge Pipes

BS 5572[15] introduced the term *discharge pipe* in place of 'soil and waste pipes'. A discharge pipe may also convey rainwater. The primary function of discharge pipes is to convey discharges from appliances to the underground drains and they can be arranged in three different ways: two-pipe, one-pipe

and single-stack systems. An efficient system should satisfy the following requirements

(1) effective and speedy removal of wastes;

(2) prevention of foul air entering building;

(3) ready access to interior of pipes;

(4) protection against extremes of temperature;

(5) protection against corrosion and erosion of pipes;

(6) restriction of siphonage and avoidance of liability to damage, deposition or obstruction;

(7) obtaining economical and efficient arrangements, which are assisted by compact grouping of sanitary appliances.

Two-pipe System

This is the traditional system in the United Kingdom whereby soil (discharge from water closets, urinals, slop-hoppers, and similar appliances) is conveyed through soil pipes directly to the drain, and waste (discharge from sinks, wash basins, baths and similar appliances) is conveyed to the drain by a waste pipe discharging through a trapped gully or directly into a drain. This arrangement is straightforward and effective but costly, as it results in a large number of pipes. It is especially suitable, however, where the sanitary appliances are widely dispersed. The Building Regulations[13] require all soil and waste pipes serving buildings over three storeys to be internal and hopper heads are no longer permitted on waste pipes. A typical arrangement of a two-pipe system for a two-storey house is shown diagrammatically in figure 12.4.1, with separate soil and waste pipes serving first floor sanitary appliances and with both pipes extended above roof level as ventilating pipes. The Building Regulations[13] prescribe that a soil pipe shall normally have a diameter of not less than 75 mm. Ventilating pipes shall terminate so as not to become prejudicial to health or a nuisance and shall be fitted with a durable wire cage or other suitable cover. Branches may need anti-siphonage pipes if unsealing is otherwise possible. Ground floor appliances are best taken into the drains; the water closet is shown with a direct connection and the wash basin discharges under the grating of a back inlet gully.

One-pipe System

The one-pipe system is illustrated in figure 12.4.2, whereby all sanitary appliances discharge into one main stack or discharge pipe, relying upon appliance trap seals to act as the foul air barrier. The risk of loss of trap seal was overcome by using deeper seal traps and fully ventilating the system. This system requires the close grouping of appliances around the main stack to be economical and has now been largely superseded by the simpler and more economical single-stack system,

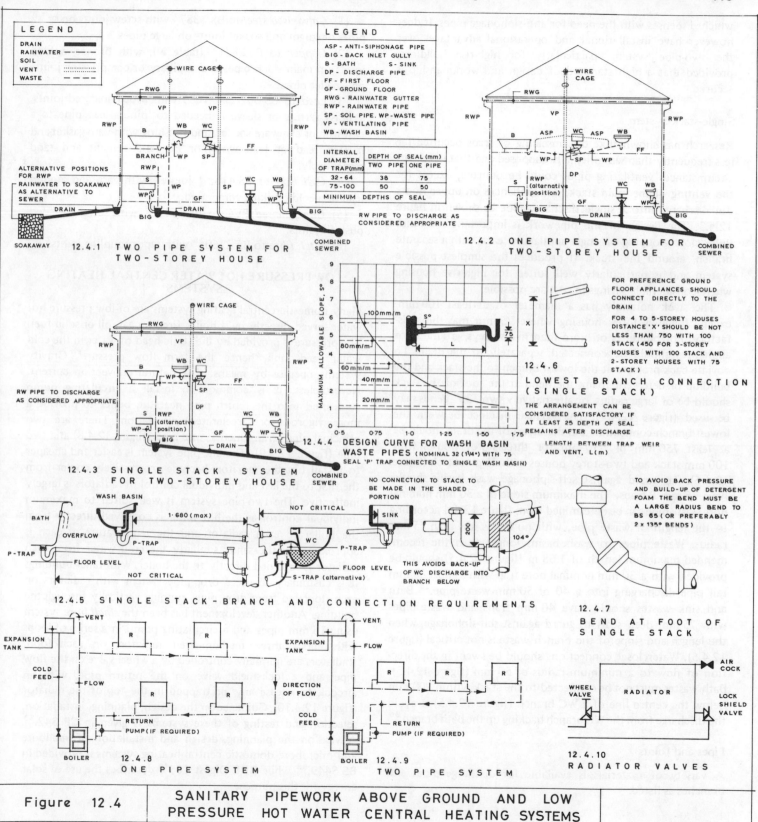

LEGEND

DRAIN	▬▬▬
RAINWATER	— · — · —
SOIL	· · · · · ·
VENT	· · · · · ·
WASTE	

LEGEND

ASP - ANTI-SIPHONAGE PIPE
BIG - BACK INLET GULLY
B - BATH   S - SINK
DP - DISCHARGE PIPE
FF - FIRST FLOOR
GF - GROUND FLOOR
RWG - RAINWATER GUTTER
RWP - RAINWATER PIPE
SP - SOIL PIPE. WP - WASTE PIPE
VP - VENTILATING PIPE
WB - WASH BASIN

INTERNAL DIAMETER OF TRAP (mm)	DEPTH OF SEAL (mm)	
	TWO PIPE	ONE PIPE
32 - 64	38	75
75 - 100	50	50
MINIMUM DEPTHS OF SEAL		

12.4.1 TWO PIPE SYSTEM FOR TWO-STOREY HOUSE

ALTERNATIVE POSITIONS FOR RWP
RAINWATER TO SOAKAWAY AS ALTERNATIVE TO SEWER

12.4.2 ONE PIPE SYSTEM FOR TWO-STOREY HOUSE

RW PIPE TO DISCHARGE AS CONSIDERED APPROPRIATE

12.4.3 SINGLE STACK SYSTEM FOR TWO-STOREY HOUSE

RW PIPE TO DISCHARGE AS CONSIDERED APPROPRIATE

12.4.4 DESIGN CURVE FOR WASH BASIN WASTE PIPES (NOMINAL 32 (1¼") WITH 75 SEAL 'P' TRAP CONNECTED TO SINGLE WASH BASIN)

LENGTH BETWEEN TRAP WEIR AND VENT, L (m)

FOR PREFERENCE GROUND FLOOR APPLIANCES SHOULD CONNECT DIRECTLY TO THE DRAIN

FOR 4 TO 5-STOREY HOUSES DISTANCE 'X' SHOULD BE NOT LESS THAN 750 WITH 100 STACK (450 FOR 3-STOREY HOUSES WITH 100 STACK AND 2-STOREY HOUSES WITH 75 STACK)

12.4.6 LOWEST BRANCH CONNECTION (SINGLE STACK)

THE ARRANGEMENT CAN BE CONSIDERED SATISFACTORY IF AT LEAST 25 DEPTH OF SEAL REMAINS AFTER DISCHARGE

12.4.5 SINGLE STACK - BRANCH AND CONNECTION REQUIREMENTS

NO CONNECTION TO STACK TO BE MADE IN THE SHADED PORTION

THIS AVOIDS BACK-UP OF WC DISCHARGE INTO BRANCH BELOW

TO AVOID BACK PRESSURE AND BUILD-UP OF DETERGENT FOAM THE BEND MUST BE A LARGE RADIUS BEND TO BS 65 (OR PREFERABLY 2 x 135° BENDS)

12.4.7 BEND AT FOOT OF SINGLE STACK

12.4.8 ONE PIPE SYSTEM

12.4.9 TWO PIPE SYSTEM

12.4.10 RADIATOR VALVES

Figure	12.4	**SANITARY PIPEWORK ABOVE GROUND AND LOW PRESSURE HOT WATER CENTRAL HEATING SYSTEMS**

which dispenses with the need for anti-siphonage pipes. It does however have installational and operational advantages over the two-pipe system, particularly for high-rise buildings, provided that a high standard of design and workmanship is secured.

Single-stack System

Research has shown that the unsealing of traps occurred far less frequently than was originally supposed and that in certain circumstances ventilating pipes could be omitted, except for the venting of the main stack. All appliances on upper floors can discharge into a single discharge pipe as shown in figure 12.4.3. The design of the pipework is important and relies upon close grouping of single appliances, each with a separate branch, around the stack. It produces the simplest possible system and is particularly well suited for high-rise housing where considerable savings in cost are possible.

The stack normally has a diameter of at least 100 mm, except for two-storey housing where 75 mm may be satisfactory. For buildings of more than five storeys, ground floor appliances should be connected separately to the drain. To obviate back pressure at the lowest branch to the stack and the build-up of detergent foam, the bend at the foot of the stack should be of large radius, or alternatively two 135° bends may be used (figure 12.4.7). The vertical distance between the lowest branch connection and the invert of the drain should be at least 750 mm or 450 mm for three-storey houses with 100 mm stack and two-storey houses with 75 mm stack (figure 12.4.6). To guard against self-siphonage wash basins should have 75 mm P-traps. The maximum slope of a 30 mm nominal bore waste pipe can be determined from figure 12.4.4 according to the length of waste pipe, with no bends less than 75 mm radius. Waste pipes to wash basins longer than the recommended maximum length of 1.68 m (figure 12.4.5) should be provided with a 30 mm nominal bore trap with a short 30 mm tail pipe discharging into a 40 or 50 mm waste pipe.[25] Bath and sink wastes should have 40 mm traps and waste pipes with 75 mm deep seals to guard against self-siphonage, when the length and slope of the branch waste is not critical (figure 12.4.5). Watercloset connections should be swept in the direction of flow to a minimum radius of 50 mm (figure 12.4.5). Bath wastes must not be connected to the stack within 200 mm below the centre line of a WC branch (figure 12.4.5) to avoid the discharge from the WC branch backing up the bath branch.[36]

Pipes and Joints

A variety of materials is available for discharge pipes and branches as listed.

(1) *Galvanised steel* to BS 1387[7] with screw joints on sn ‖ pipes or spigot and socket joints on large pipes.

(2) *Copper* to BS 2871[5] (table X), with bronze welded joints on main stack pipes and capillary or compression joints on smaller pipes.

(3) *Lead* to BS 602 and 1085 with wiped soldered joints; a brass ferrule or sleeve is needed to joint a lead pipe to a cast iron or clayware socket, and finishing with yarn gasket and molten lead or lead wool, or Portland cement and sand respectively.

(4) *Cast iron* to BS 416,[26] jointed with yarn and molten lead or a cold caulking compound.

(5) *Unplasticised* PVC to BS 4514[27] with solvent welded push-fit joints.

(6) *Pitch fibre* to BS 2760[28] with taper coupling joints.

LOW PRESSURE HOT WATER CENTRAL HEATING SYSTEMS

Many domestic central heating systems are of low pressure hot water served directly by a boiler fired by gas, oil or solid fuel. The pressure is provided by the static head of water in the cold water feed tank, hence the term 'low pressure'. Gravity systems operate by means of a natural convection current. Nevertheless, it is common to install a circulating pump connected into the return pipe near the boiler to provide a more vigorous or 'accelerated' circulation. There are two principal pipe arrangements: one pipe (figure 12.4.8) and two pipe (figure 12.4.9). The one-pipe system is easier and cheaper to install but the radiators become progressively cooler from the first to the last and individual control of radiators is largely ineffective. The two-pipe system is well suited to maximum individual control as the hot water is conveyed directly from the boiler to each radiator and the temperature of each is approximately the same; similarly the cold water from each radiator is returned directly to the boiler. With taller buildings it is customary to use a drop-pipe system with radiators on each floor fed from drop pipes running vertically through the building. Another development has been the *small bore* system using 15 mm pipes and a circulating pump in a series of loops with about three medium-sized radiators on each loop. Radiators are normally controlled by a wheel valve on the flow pipe and a lockshield valve on the return pipe, with an aircock to release any air trapped in the top of the radiator (figure 12.4.10). Guidance on the design, planning, installation, selection and testing of these systems is given in CP 342.[33] Advice on the planning, design and installation of smallbore and microbore domestic central heating systems is provided in BS 5449,[38] while BRE Digest 205[40] describes the use of solar energy for this purpose.

REFERENCES

1. CP 310: 1965 Water supply
2. BS 1212: Ballvalves, Part 1: 1977 Piston type, Part 2: 1978 Diaphragm type (brass body)
3. BS 417: Galvanised mild steel cisterns and covers, tanks and cylinders, Part 2: 1973 Metric units
4. *BRE Digest 15*: Pipes and fittings for domestic water supply. HMSO (1977)
5. BS 2871: Copper and copper alloys – tubes, Part 1: 1971 Copper tubes for water, gas and sanitation
6. BS 1972: 1967 Polythene pipe (type 32) for cold water services
7. BS 1387: 1967 Steel tubes and tubulars suitable for screwing to BS 21 pipe threads
8. BS 4127: Light gauge stainless steel tubes, Part 2: 1972 Metric Units
9. BS 1010: Draw-off taps and stopvalves for water services (screwdown pattern), Part 2: 1973 Draw-off taps and above ground stopvalves
10. BS 1952: 1964 Copper alloy gate valves for general purposes
11. *BRE Digest 146*: Modernising plumbing systems. HMSO (1972)
12. DOE Advisory leaflet 76: Plastic pipes for cold water supply. HMSO (1977)
13. The Building Regulations (1976). SI 1676. HMSO (1976)
14. CP 305: Sanitary appliances, Part 1: 1974 Selection, installation and special requirements
15. BS 5572: 1978 Code of practice for sanitary pipework
16. BS 1213: 1974 Ceramic washdown WC pans
17. BS 1125: 1973 WC flushing cisterns
18. BS 1876: 1972 Automatic flushing cisterns for urinals
19. BS 1188: 1974 Ceramic wash basins and pedestals
20. BS 1329: 1974 Metal hand rinse basins
21. BS 1189: 1972 Cast iron baths for domestic purposes
22. BS 1390: 1972 Sheet steel baths for domestic purposes
23. BS 1206: 1974 Fireclay sinks, dimensions and workmanship
24. BS 1244: Metal sinks for domestic purposes, Part 2: 1972 Metric units
25. *BRE Digest 80*: Soil and waste-pipe systems for housing. HMSO (1975)
26. BS 416: 1973 Cast iron spigot and socket soil, waste and ventilating pipes (sand cast and spun) and fittings
27. BS 4514: 1969 Unplasticised PVC soil and ventilating pipe, fittings and accessories
28. BS 2760: 1973 Pitch-impregnated fibre pipes and fittings for below and above ground drainage
29. BS 4305: 1972 Baths for domestic purposes made from cast acrylic sheet
30. BS 3505: 1968 Unplasticised PVC pipe for cold water services
31. CP 312: Plastics pipework (thermoplastics material), Part 1: 1973 General principles and choice of material; Part 2: 1973 Unplasticised PVC pipework for the conveyance of liquids under pressure
32. BS 5505: 1977 Bidets
33. CP 342: Centralised hot water supply, Part 1: 1970 Individual dwellings
34. CP 99: 1972 Frost precautions for water services
35. DOE Advisory leaflet 41: Frost precautions. HMSO (1975)
36. DOE Advisory leaflet 73: Single stack plumbing. HMSO (1968)
37. BS 5422: 1977 The use of thermal insulating materials
38. BS 5449: Code of practice for central heating for domestic premises, Part 1: 1977 Forced circulation hot water systems
39. BS 5480: Glass fibre reinforced plastics (GRP) pipes and fittings, Part 1: 1977 Dimensions, materials and classification
40. *BRE Digest 205*: Domestic water heating by solar energy. HMSO (1977)

13 DRAINAGE

This chapter is concerned with building drainage from initial design and statutory requirements to the constructional techniques and materials used and methods of disposal of the effluent. As in previous chapters reference will be made to appropriate Codes of Practice and British Standards.

DESIGN OF DRAINS

Drainage Systems

Drainage systems must be designed to provide an efficient and economical method of carrying away waterborne waste, in such a way as to avoid the risk of pipe blockage and the escape of effluent into the ground. A drainage system normally consists of a network of pipes laid from a building to fall to a local authority sewer, although in some cases it may be necessary to install a plant to treat the effluent or a pump to raise it to a sewer at a higher level.

The layout of the drainage system is also influenced by the sewer arrangements, which can be one of three types.

(1) *Combined system* whereby foul water from sanitary appliances and surface water from roofs and paved areas discharge through a single drain to the same combined sewer. This simplifies and cheapens the house drainage system, ensures that the drains are well flushed in time of storm and that the house drain cannot be connected to the wrong sewer. On the other hand silting may occur in large pipes and it may entail storm overflows on sewers and high costs of pumping and sewage disposal.

(2) *Separate system* in which foul wastes pass through one set of drains to a foul sewer, whereas surface water is conveyed to a separate surface water sewer or soakaways. This arrangement reduces pumping and sewage treatment costs to the main drainage authority but results in additional expense with the house drainage system, eliminates the flushing action of the surface water in foul drains and permits the possibility of an incorrect connection.

(3) *Partially separate system* in which the foul sewer takes some of the surface water (possibly that from the rear roof slopes and paved areas) and another sewer takes surface water only. This arrangement is a compromise and lessens both the advantages and disadvantages of the previous systems.

The layout of a drainage system should be as simple and direct as possible, and drains should be laid in straight lines between points where changes of direction or gradient occur. The drains should be of sufficient strength and be constructed of sufficiently durable materials with watertight joints.[1] The Building Regulations[1] also require that a drain under a building should be adequately supported throughout its length without restricting thermal movement, and that it should be reasonably accessible for maintenance and repair. The Code on Building Drainage[2] recommends that pipes under buildings should have flexible joints and that lintels or relieving arches should be provided in walls over pipes to prevent wall loads bearing on them.

Pipe Sizes and Gradients

Pipes must be laid to falls which ensure a sufficiently rapid flow to prevent the settlement of solid matter, which might lead to a blockage. The pipe size and fall must also be chosen to ensure that the pipe will not run full which might, by suction, unseal traps either within the building or in gullies. BS Code of Practice 301[2] recommends that at peak flow, pipes should be designed to run at 90 per cent capacity, which is equivalent to a depth of flow not exceeding three-quarters of the pipe diameter. For most housing contracts, pipes of 100 or 150 mm diameter will be used for which the Code[2] recommends a gradient of not less than 1 in 80 for a 100 mm pipe when serving not less than five housing units or dwellings and 1 in 150 for a 150 mm pipe when serving not less than ten units. A 100 mm pipe serving the first housing unit should have a fall of not less than 1 in 70.

In difficult situations the Code[2] permits gradients of 1 in 130 for a 100 mm pipe and 1 in 200 for a 150 mm pipe, but emphasises the need for a high standard of workmanship and supervision with these flat gradients. In practice estimates of peak flow per person are prepared, often $\frac{1}{80}$ litre/second, and pipe sizes are then checked using flow charts such as those supplied by the pipe manufacturers. The aim is to secure a self-cleansing velocity and with small flows, continuous flows containing solid matter at less than one litre per second, or long drains, a gradient of not less than 1 in 40 will be required for a 100 mm pipe.

The Code[2] recommends that a maximum of 20 housing units can be connected to a 100 mm pipe and 150 housing units to a 150 mm pipe. The Building Regulations[1] prescribe that the internal diameter of a soil or foul drain should be not less than 100 mm, whilst a surface water drain should be not less than 75 mm. For normal house drains serving two or three houses, the flow in dry weather is very small and usually intermittent, and to be self-cleansing the pipe diameter should be kept to the smallest practicable (100 mm). The volume of surface water is usually calculated on the basis of a rainfall intensity of 50 mm per hour.

Structural Design of Drains

An important aspect in the design of drainage work is to avoid damage resulting from settlement, particularly in made ground and in areas subject to mining subsidence. The Code[2] prescribes two alternative approaches to the structural design of drains.

The first method (computed load method) relates the load-carrying capacity of the drain to the maximum vertical loads likely to be imposed upon it by fill and surface surcharges and aims to meet these loads by flexible construction. It is necessary to select a rigid pipe with an appropriate safe crushing test strength and a suitable class of bedding which, together with an adequate factor of safety, will provide the required load-bearing capacity. It is advisable to use flexible joints with the pipes.

The second method relies upon the provision of additional support to the pipes in the form of concrete haunching or surround. The pipes, often of vitrified clayware or unreinforced concrete with either rigid or flexible joints, are not designed to carry specific loads as in the first method. The form of concrete protection varies with the amount of cover to the pipe and should conform to the following recommendations as contained in the Code[2]

(1) Pipes with 6 m or more of cover are to be surrounded with at least 150 mm of concrete (figure 13.1.9).

(2) Pipes with 4.30 m to 6 m of cover are to be bedded on and haunched with at least 150 mm of concrete to at least the horizontal diameter of the pipe. The splaying of concrete above that level is to be tangential to the pipe (figure 13.1.9).

(3) Subject to (1), all pipes of 300 mm diameter and over are to be bedded on and haunched with at least 150 mm of concrete in the method described in (2).

(4) Subject to (5), all pipes under 300 mm diameter and with less than 4.30 m cover may be laid without concrete, if the joints are of the socket or collar type.

(5) All pipes with less than 1.20 m of cover under roads (except concrete roads) or 900 mm when not under roads, are to be surrounded with at least 150 mm of concrete.

Traditional practice has tended to follow the second method, but there is quite a high risk of failure due to the forces resulting from earth settlement. Hence modern practice favours the first method using flexible joints or flexible pipes and joints and laying the pipes in compacted granular fill in a trench of limited width, which will absorb some movement and distribute pressure uniformly over the surface of the pipe (figure 13.1.1). In areas subject to major subsidence, expansion joints should be incorporated.

Access to Drains

Access to drains is required for inspection and rodding to clear blockage. Rodding is normally done with flexible canes, or metal or plastic rods or tubes to which a suitable ram is attached. The traditional means of access is the inspection chamber (figure 13.3.5) which becomes a manhole, according to the Code,[2] when the depth exceeds 900 mm. These chambers are described in more detail later in the chapter. Inspection chambers or manholes are mainly provided at changes of direction and gradient. Other desirable locations are listed in the Code[2] and described later in the chapter.

Manholes and inspection chambers are, however, expensive to build and rodding eyes may therefore be substituted at the head of shallow branch drains. A rodding eye consists of a vertical or inclined length of pipe with a suitable cover or cap at the top end at ground level, while at its base it makes a curved junction with the drain. It is likely that the extra cost of clearing blockages, resulting from the greater time needed because of the more restricted form of access of the rodding eye, is considerably less than the greater initial cost of providing a manhole or inspection chamber. A typical rodding eye as produced by Marley is illustrated in figure 13.1.2.

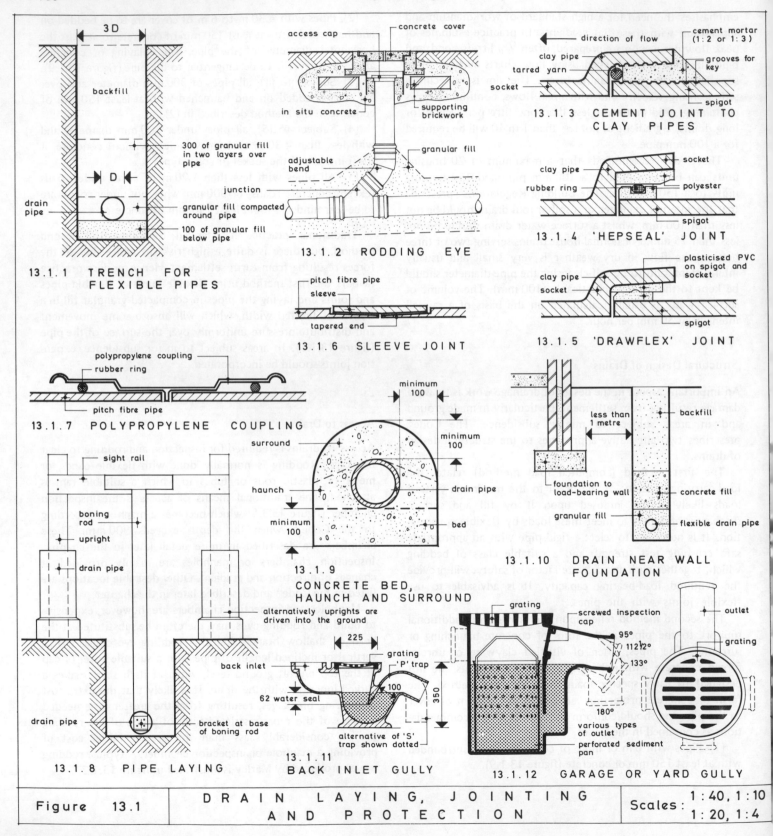

13.1.1   TRENCH FOR FLEXIBLE PIPES

13.1.2   RODDING EYE

13.1.3   CEMENT JOINT TO CLAY PIPES

13.1.4   'HEPSEAL' JOINT

13.1.5   'DRAWFLEX' JOINT

13.1.6   SLEEVE JOINT

13.1.7   POLYPROPYLENE COUPLING

13.1.8   PIPE LAYING

13.1.9   CONCRETE BED, HAUNCH AND SURROUND

13.1.10   DRAIN NEAR WALL FOUNDATION

13.1.11   BACK INLET GULLY

13.1.12   GARAGE OR YARD GULLY

| Figure   13.1 | DRAIN LAYING, JOINTING AND PROTECTION | Scales: 1:40, 1:10 1:20, 1:4 |

CHOICE OF PIPES

All the pipes subsequently described are suitable for use below ground, but the beam strength of a pipeline may become a limiting factor under difficult loading or ground conditions. In these situations, rigid pipes with flexible joints should be used, and short lengths of pipe in ground subject to severe settlement. Where pipes are laid above ground, special attention should be paid to structural support and protection against mechanical damage, frost and corrosion.[2] The various pipe materials are now described and compared.

Rigid Pipes

Vitrified clay pipes. Manufactured to BS 65 and 540[5] with nominal bores of 75 to 900 mm and lengths of 300 mm to 1.50 m, in two classifications of 'British Standard' and 'British Standard Surface Water'. The strength of larger sizes is obtainable from the manufacturer. The omission of salt glaze is permitted under these standards. Clay pipes are resistant to attack by a wide range of substances, both acid and alkaline. These pipes are still very popular although the traditional joint made of two rings of tarred yarn and with the remainder of the gap between spigot and socket filled with cement mortar to a 1:2 or 1:3 mix (figure 13.1.3) is becoming increasingly displaced by mechanical or flexible joints, of which two types are illustrated in figures 13.1.4 and 13.1.5. Cement mortar shrinks on setting thus making the traditional form of cement joint vulnerable and, being rigid, it is liable to damage by settlement. Furthermore the short pipe lengths produce a large number of joints. Flexible joints incorporating natural rubber or plastics are recommended in DOE Advisory Leaflet 24.[3]

Concrete pipes. These are suitable for use with normal effluents but may be attacked by acids or sulphates in the effluent, or in the surrounding soil or groundwater. Concrete pipes are used mainly for larger pipes of 225 mm diameter and upwards, and with these sizes external wrappings of glass-fibre laminate are available which reinforce the pipes and protect them from external attack. Concrete pipes to BS 556[6] are supplied either reinforced or unreinforced in lengths of 900 mm to 5 m with socket and spigot joints and are often fitted with push-fit flexible rubber ring seals. Concrete pipes to BS 4101[9] have ogee joints which are jointed in cement mortar and their use is normally restricted to the carriage of surface water. Prestressed concrete pipes are also available complying with BS 5178.[7]

Asbestos cement pipes. These are made to BS 3656,[10] often dipped in bitumen and provided with flexible joints. They are occasionally used for drainage purposes and have the same shortcomings as concrete pipes.

Cast iron pipes. These can be supplied with spigot and socket joints to BS 437[11] for caulking with lead or a proprietary material, or as pressure pipes with flexible joints to BS 1211,[12] which are much more satisfactory for use in difficult or waterlogged conditions or ground subject to large movement. The coating on these pipes gives good protection against corrosion and a reasonable life with average ground conditions and normal effluents, although care is needed during handling; they can be laid at any depth on account of their great strength. BS 4622[13] gives metric sizes for pressure grey iron pipes and fittings. Cast iron pipes are made in varying lengths, but the most commonly used length is around 3.60 m. Ductile iron pipes are covered by BS 4772.[8]

Flexible pipes

Pitch-impregnated fibre pipes. Made to BS 2760[14] they are becoming increasingly popular due to their suitability for use with normal domestic and most trade wastes. They are manufactured in nominal bores of 50 to 225 mm and standard lengths are 1.7, 2.5 and 3 m. They are more economical than clay pipes where long lengths are involved and in bad ground conditions. There are two principal forms of joint: (1) The pipes have tapered ends and an internally tapered sleeve is driven onto the pipes to connect them (figure 13.1.6). This gives flexibility but there is no provision for longitudinal expansion or contraction. (2) A better alternative is to use a polypropylene moulded coupling with snap ring rubber seals (figure 13.1.7) which provides longitudinal flexibility.[4]

Unplasticised PVC pipes. Manufactured to BS 4660,[15] to 110 and 160 mm nominal sizes (external), are golden brown in colour and are suitable for domestic installations and surface water drainage. UPVC pipes are available in 1, 3 and 6 m lengths. They should not be used for effluents at high temperatures and they become brittle at low temperatures and then require handling with care. Push-in joints are available for UPVC pipes either as loose couplings or integral sockets and these provide flexibility.[4] Although relatively expensive, they are available in long lengths, are lightweight and readily handled and assembled.

Reinforced plastics pipes. These are made of thermosetting resin and have advantages of light weight and resistance to corrosion and effluents with high temperature.

PIPE LAYING

Pipes should always be laid with the sockets pointing uphill, commencing from the point of discharge (sewer connection or treatment works). Painted sight rails, as illustrated in figure

13.1.8, should be fixed across the trench, usually at manholes or inspection chambers, at a height equal to the length of the boning rod to be used above the invert level of the drain. A line sighted across the tops of two adjacent sight rails will represent the gradient of the drain at a fixed height above invert level. At any one time there should desirably be at least three sight rails erected on a length of drain under construction.

Wooden pegs or steel pins are driven into the trench bottom at intervals of at least 900 mm less than the length of straight-edge in use. The use of a boning rod (figure 13.1.8) will enable each peg or pin to be driven until its head represents the pipe invert at that point. The underside of the straight-edge resting on the tops of the pegs or pins will give the levels and gradient of the pipe. The pegs or pins are withdrawn as the pipes are laid.

To obtain a true line in a horizontal plane, a side line is strung tightly between steel pins at half pipe level, with the pipe sockets just free of the side line. Pins will normally be located at each manhole or inspection chamber, but intermediate pins will also be needed on very long lengths.

In firm ground, rigid pipes may be laid upon the carefully trimmed trench bottom or formation, with socket and joint holes formed where necessary to give a minimum clearance of 50 mm between the socket and the formation. Where poor ground or groundwater is involved, the trench should be excavated below the pipe invert to give a minimum thickness of 100 mm of granular bed as in figure 13.1.1.

Where rigid pipes are to be laid on a concrete bed, pipes should be supported clear of the trench bottom by placing blocks or cradles under each pipe, after which concrete is placed solidly under the barrel of each pipe to give continuous support. The concrete bed shall be not less than 100 mm thick for pipes up to 300 mm diameter, and the bed or haunch should extend at least 100 mm on either side of the pipe (figure 13.1.9). Where a pipe trench is within one metre of a loadbearing wall foundation, the trench shall be filled with concrete to the level of the underside of the foundation (figure 13.1.10).[1]

In general, flexible pipes should be laid on a granular bed not less than 100 mm thick. With certain types of flexible pipe, such as pitch-impregnated fibre and UPVC pipes, it is necessary to place selected bedding material up to a level of 100 mm above the top of the pipe, between the pipe and the undisturbed soil of the trench sides and to ensure that the side fill is well compacted.

DRAINAGE LAYOUT

When designing a drainage layout the primary objectives should be to secure an efficient and economical arrangement. Wherever practicable, sanitary appliances should be grouped together to reduce the lengths of drain and to obtain steeper gradients. A drainage layout for a bungalow and ancillary buildings is illustrated in figure 13.2.1, and consists of separate foul and surface water drains each connected to a separate sewer. The foul drain picks up the sinks, wash basins and bath through back inlet gullies of the type illustrated in figure 13.1.11, which act as a water seal and the pipes discharge below gratings to prevent fouling, whilst water closets are connected direct to the drains and are normally fitted with a 'S' trap. Gullies prevent undesirable matter entering drains but gratings and traps need cleansing periodically. A ventilating pipe must be provided at the head of the foul drainage system.

Rainwater pipes discharge through back inlet gullies or direct into the surface water drains. These drains also pick up the yard gullies in the drive and paved patio. Care has to be taken to obtain adequate difference in level between the two drains to allow surface water connections to cross the line of the foul drain, to provide adequate falls to ensure self-cleansing velocities and yet, at the same time, to avoid excessive excavation. Inspection chambers are provided at the head of each main pipe run, at changes of direction and at main intersections. Other forms of disposal of surface water are soakaways (figure 1.1.6 in chapter 1) and watercourses, where permitted by Regional Water Authorities to drain large areas and where there are impervious soils, with a non-return flapvalve at the outlet end of the pipe to prevent backflow or debris blocking it in times of flood.

Sewer Connection

Except where junctions have been provided on the sewer, the drain connection is made with a saddle of the type illustrated in figure 13.2.3. A hole is formed in the top half of the sewer and carefully trimmed to fit the saddle. The saddle, incorporating a socket piece, is jointed all round in cement mortar and finally surrounded in 150 mm of concrete after the mortar has set. The saddle connection should be carried out by the local authority's employees or under their supervision. Where the sewer and drain are of similar size or the drain is laid to a very steep gradient, it is advisable to construct a manhole at the point of connection. Another alternative when connecting to a 150 mm sewer is to take out three sewer pipes and replace with two straight pipes and a junction pipe, or to use a double spigot pipe and loose collar.

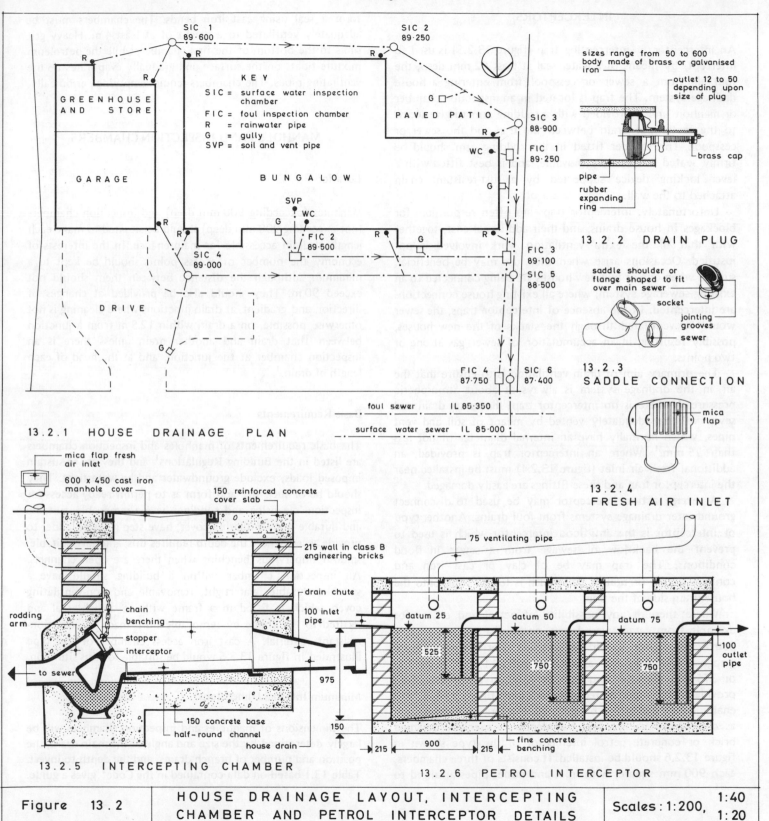

SIC 1
89·600

SIC 2
89·250

R

R

R

**GREENHOUSE
AND STORE**

PAVED PATIO

G

SIC 3
88·900

R

WC

FIC 1
89·250

KEY

SIC = surface water inspection
chamber
FIC = foul inspection chamber
R = rainwater pipe
G = gully
SVP = soil and vent pipe

GARAGE

BUNGALOW

G

SVP
WC

G

FIC 2
89·500

R

R

G

R

FIC 3
89·100

R

SIC 5
88·500

R

SIC 4
89·000

DRIVE

G

sizes range from 50 to 600
body made of brass or galvanised
iron

outlet 12 to 50
depending upon
size of plug

brass cap

pipe

rubber
expanding
ring

13.2.2 DRAIN PLUG

saddle shoulder or
flange shaped to fit
over main sewer

socket

jointing
grooves

sewer

**13.2.3
SADDLE CONNECTION**

FIC 4
87·750

SIC 6
87·400

foul sewer   IL 85·350

surface water sewer IL 85·000

13.2.1 HOUSE DRAINAGE PLAN

mica
flap

**13.2.4
FRESH AIR INLET**

mica flap fresh
air inlet

600 x 450 cast iron
manhole cover

150 reinforced concrete
cover slab

75 ventilating pipe

215 wall in class B
engineering bricks

drain chute

100 inlet
pipe

datum 25

datum 50

datum 75

rodding
arm

chain
benching

stopper

interceptor

to sewer

525

750

750

100
outlet
pipe

975

150 concrete base

half-round channel

house drain

150

215   900   215

fine concrete
benching

13.2.5 INTERCEPTING CHAMBER

13.2.6 PETROL INTERCEPTOR

| Figure | 13.2 | HOUSE DRAINAGE LAYOUT, INTERCEPTING CHAMBER AND PETROL INTERCEPTOR DETAILS | Scales: 1:200, | 1:40 1:20 |

INTERCEPTORS

An interceptor or intercepting trap (figure 13.2.5) is used to intercept, by means of a water seal at least 62 mm deep, the foul air from a sewer or cesspool from entering a house drainage system. The trap is located in an inspection chamber or manhole and is provided with a rodding arm to give access to the section of drain between the trap and the sewer or cesspool. The stopper fitted in the rodding arm should be firmly seated but easily removable, and is best fitted with a lever locking device supported by a rust-resistant chain attached to the wall.

Unfortunately, interceptor traps are often responsible for blockages in house drains, and their additional cost, together with that of the extra ventilation work involved is not justified. Occasions arise when their use may be beneficial, such as when one or two new houses are being connected to an existing sewerage system, where all existing house connections are intercepted. In the absence of interceptor traps, the sewer would be ventilated through the stacks of the new houses, possibly resulting in an accumulation of sewer gas at one or two points.

The primary aim of drain ventilation is to ensure that the air in the drainage system is always at about atmospheric pressure. If there is no interceptor trap, the house drain and sewer will be adequately vented by means of soil and vent pipes, which normally have an internal diameter of not less than 75 mm.[1] Where an interceptor trap is provided, an additional fresh air inlet (figure 13.2.4) must be installed near the interceptor trap and these fittings are easily damaged.

A reverse action interceptor may be used to disconnect groundwater drainage systems from foul drains. Another type of interceptor is the anti-flood interceptor which is used to prevent the backflow of sewage from a sewer in flood conditions. The trap may be of clay or cast iron and contains a flap or hollow ball which is forced back onto the house drain side of the trap.

Where there is any possibility of petrol and oil, as for example from garage washdowns, entering a drain, a suitable petrol interceptor should be installed. For a single private garage, a deep gully trap as detailed in BS 1130,[16] being 300 or 375 mm diameter and 600 or 750 mm deep, should be provided with a perforated sediment pan (figure 13.1.12), to enable debris and grit to be removed. For larger premises, such as commercial garages and bus depots, a specially designed brick or concrete petrol interceptor of the type shown in figure 13.2.6 should be installed. It consists of three chambers, each 900 mm square, with inlet and outlet pipes arranged to form a seal using cast iron bends. The chambers must be adequately ventilated to a height of at least 4 m. Heavy grit sinks to the bottom of the first chamber, while the petroleum mixture floats on the surface and gradually evaporates up the ventilating pipes. The chambers require emptying periodically.

MANHOLES AND INSPECTION CHAMBERS

Location

Manholes (exceeding 900 mm deep) and inspection chambers (not exceeding 900 mm deep) should be situated to make each length of drain accessible for maintenance. In the interests of economy the number of access points should be kept to a minimum, although the distance between them should not exceed 90 m. They should also be provided at changes of direction and gradient, at drain junctions where cleaning is not otherwise possible, on a drain within 12.5 m from a junction between that drain and another drain unless there is an inspection chamber at the junction, and at the head of each length of drain.[1]

Basic Requirements

The basic requirements of manholes and inspection chambers are listed in the Building Regulations[1] and they are to sustain imposed loads, exclude groundwater and be watertight. They should be of such size and form as to permit ready access for inspection, cleansing and rodding; have a removable, suitable and durable non-ventilating cover; have step irons or ladder to provide access where the depth requires this; and have suitable, smooth impervious benching when there are open channels. An inspection chamber within a building should have a suitable, durable, watertight, removable and non-ventilating cover, which is fitted in a frame with an airtight seal and secured to the frame by removable bolts made of corrosion-resistant material. A cast iron access chamber of the type illustrated in figure 13.3.6 would be suitable for this situation.

Minimum Internal Dimensions

The dimensions of manholes and inspection chambers will be largely determined by the size and angle of the main drain, the position and number of branch drains, and the depth to invert. Table 13.1 based on data contained in the Code[2] gives a guide.

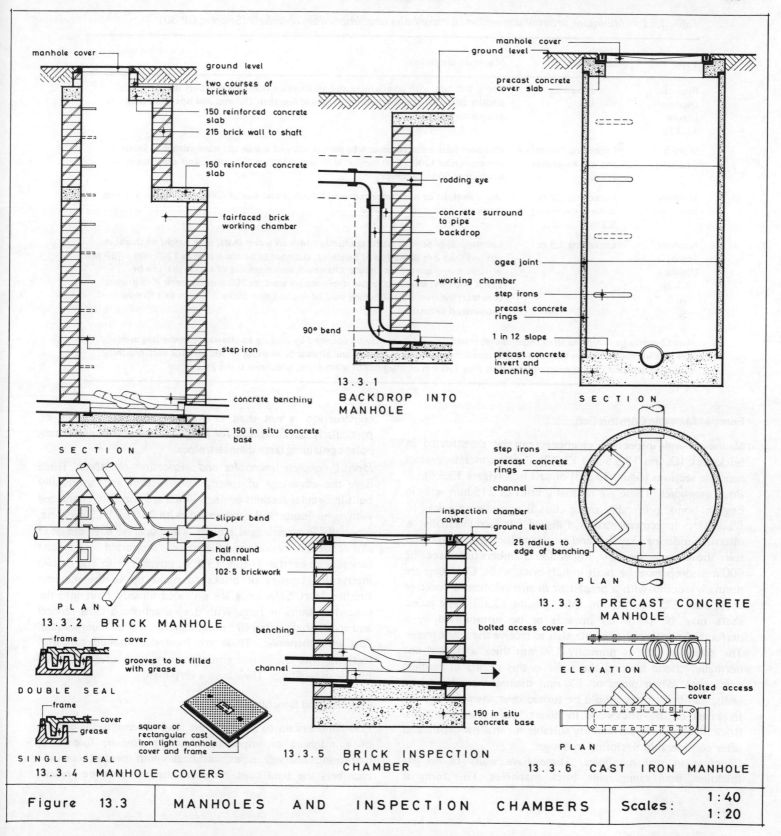

manhole cover

ground level

two courses of brickwork

150 reinforced concrete slab

215 brick wall to shaft

150 reinforced concrete slab

fairfaced brick working chamber

step iron

concrete benching

150 in situ concrete base

SECTION

slipper bend

half round channel

102·5 brickwork

PLAN

13.3.2 BRICK MANHOLE

frame

cover

grooves to be filled with grease

DOUBLE SEAL

frame

cover

grease

SINGLE SEAL

13.3.4 MANHOLE COVERS

square or rectangular cast iron light manhole cover and frame

manhole cover

ground level

rodding eye

concrete surround to pipe

backdrop

working chamber

90° bend

step iron

13.3.1 BACKDROP INTO MANHOLE

inspection chamber cover

ground level

benching

channel

150 in situ concrete base

13.3.5 BRICK INSPECTION CHAMBER

manhole cover

ground level

precast concrete cover slab

ogee joint

step irons

precast concrete rings

1 in 12 slope

precast concrete invert and benching

SECTION

step irons

precast concrete rings

channel

25 radius to edge of benching

PLAN

13.3.3 PRECAST CONCRETE MANHOLE

bolted access cover

ELEVATION

bolted access cover

PLAN

13.3.6 CAST IRON MANHOLE

| Figure 13.3 | MANHOLES AND INSPECTION CHAMBERS | Scales: | 1:40 1:20 |

Table 13.1   *Minimum internal dimensions for manholes and inspection chambers* (Source: CP 301[2])

Type	Depth to invert	Minimum dimensions
Inspection chamber (figure 13.3.5)	Not exceeding 900 mm	750 x 700 mm, with length increased where more than two branches on either side; smaller chambers may be used for depths of less than 750 mm and not more than one branch on either side
Manhole (shallow)	Exceeding 900 mm and not exceeding 2.7 m	Chamber must be large enough to permit entry of a man and have minimum internal dimensions of 1200 x 750 mm or 1100 mm diameter, if circular, and a minimum cover opening of 900 x 600 mm
Manhole (shallow)	Exceeding 2.7 m and not exceeding 3.3 m	Man can stand up in chamber covered by slab; cover size of 600 x 600 mm or 550 mm diameter, if circular, is suitable
Manhole (deep) (figure 13.3.2)	Exceeding 3.3 m	Economical to provide working chamber with an access shaft, with height of chamber not less than 2 m above top of benching; chamber to be not less than 1200 mm x 750 mm or 1200 mm diameter for circular chambers; minimum size of access shaft to be 750 x 700 mm, where malleable step irons are used, or 700 mm diameter if in precast concrete; the clear cover opening shall be not less than 600 x 600 mm or 550 mm diameter, if circular

Note: Subject to minimum sizes listed above, internal dimensions may be obtained by making an allowance in the length of 300 mm for each 100 mm branch and 375 mm for each 150 mm branch, and allowance in width of 300 mm for each benching with branches or 150 mm without branches, plus 150 mm or diameter of main drain, whichever is the greater.

Form of Manhole Construction

Manholes and inspection chambers can be constructed in brickwork (figures 13.3.5 and 13.3.2), *in situ* concrete, precast concrete sections (figure 13.3.3) or cast iron (figure 13.3.6).

Brick manholes. These are normally built of 215 mm walls in English bond, preferably using class B engineering bricks to BS 3921[17] in cement mortar, finished fair face internally, as internal rendering may fail and result in blockages. In granular soils above the water table, inspection chambers not exceeding 900 mm deep may be built in half-brick walls. Chambers are normally roofed with a precast or *in situ* reinforced concrete cover slab, 125 or 150 mm thick (figure 13.3.2). The access shaft may be corbelled inwards or be surmounted by a perforated reinforced concrete slab to receive the cover frame. The concrete base is normally 150 mm thick and need not normally extend beyond the walls as this results in needless extra cost. When pipes of 300 mm diameter are built into walls, half-brick rings should be turned over them for the full thickness of the brickwork to divert loads from the pipes. Brick manholes are particularly suitable for shallow depths and offer considerable flexibility in design.

Concrete in situ *manholes*. These have walls of no less thickness than comparable brick manholes. This form of construction is not used extensively although it could be particularly advantageous for irregularly shaped shallow manholes containing large diameter pipes.

Precast concrete manholes and inspection chambers. These have the advantage of speedy construction and are usually built in circular sections or rings which are normally connected with ogee joints and surrounded with 150 mm of concrete, especially in waterlogged ground. The concrete base can be *in situ* or precast concrete and the cover is supported on a precast concrete cover slab (figure 13.3.3), possibly with one or two intervening courses of bricks to make up to the required finished level. Step irons are provided already built into the precast sections or rings. With deep manholes a taper section and access shaft, often 675 mm diameter, will be incorporated.

Cast iron manholes. These are formed of bolted cast iron sections[16] for use in bad ground or cast iron access chambers for use in buildings. Their cost is very high.

Channels and Benchings

For diameters up to 300 mm, the open channel in the bottom of a manhole or inspection chamber may be formed of half-round channel pipes, although with precast concrete chambers the base containing the channels may be precast.

Side branches are often in the form of three-quarter section standard branch bends, bedded in cement mortar and discharging in the direction of flow in the main channel. Larger diameter channels are often formed of *in situ* concrete finished with a granolithic rendering to a mix of 1:1:2.

The benching should rise vertically from the top edge of the channel pipe to a height not less than that of the soffit of the outlet and then be sloped upwards to meet the wall of the manhole at a gradient of about 1 in 12, to control the flow and to allow a man to stand on the benching. It is usually floated to a smooth, hard surface with a coat of cement mortar (1:2), laid monolithic with the benching.

Access to Manholes

The most usual form of access is galvanised malleable cast iron step irons to BS 1247.[18] They are normally built into manhole walls at 300 mm vertical intervals and, unless of the straight-bar corner type, set staggered in two vertical runs at 300 mm centres horizontally. The top step iron should be 450 mm below the top of the manhole cover and the lowest not more than 300 mm above the benching, to be of practical use. On very deep manholes it is customary to provide mild steel ladders, often made up of 65 x 12 mm stringers supporting 22 mm diameter rungs at 300 mm centres.

The majority of manhole covers and frames are made of cast iron complying with BS 497[19] and coated with a tar or bitumen-based composition. They are supplied in four grades: grade A for use in carriageways, grade B1 for use in minor residential roads, grade B2 for use in pedestrian precincts (occasional vehicular access) and grade C in positions inaccessible to wheeled vehicles. Covers and frames are also made in various shapes – mainly circular (500 to 600 mm diameter) for grades A and B and rectangular for grade C. Light-duty cast iron covers and frames with both single and double seals are illustrated in figure 13.3.4. Sizes of cover frame openings vary between 450 x 450 and 600 x 600 mm. The covers are usually bedded in grease and the frames in cement mortar. Covers are also manufactured in steel, or with a steel frame and steel mesh bottom filled with concrete and surfaced with a material to match the surrounding finish.

Backdrop Manholes

These are provided to accommodate a significant difference in invert levels, which may arise where a domestic drain approaches a sewer. The usual method is to construct a vertical drop pipe outside the manhole, surrounded in concrete or granular fill, as in figure 13.3.1. The upper drain is carried through the manhole wall to form a rodding eye and the vertical drop terminates at its lower end with a bend discharging into an open channel at the bottom of the manhole. Alternatively, the vertical drop pipe may be constructed inside the manhole where there is adequate space. The backdrop arrangement economises in drain trench excavation and avoids the need for excessively steep drain gradients.

TESTING OF DRAINS

The Building Regulations[1] require that all drains after laying, surrounding with concrete, where appropriate, and backfilling of trenches, shall withstand a suitable test for watertightness. The same regulations also prescribe that soil, waste and ventilating pipes should be able to withstand a smoke or air test for a minimum period of three minutes at a pressure equivalent to a head of not less than 38 mm of water. The principal objective in testing a length of drain is to ensure that it is watertight, as defective drains can create serious problems.

Water Test

The most effective test is the water test whereby suitably strutted plugs of the type illustrated in figure 13.2.2 are inserted in the lower ends of lengths of drain, which are then filled with water before the trenches are backfilled. For house drains, a knuckle bend and length of vertical pipe may be temporarily jointed at the upper end to provide the required test head. A drop in the level of water in the vertical pipe may be due to one or more of the following causes:

(1) absorption by pipes or joints;
(2) sweating of pipes or joints;
(3) leakage from defective pipes, joints or plugs;
(4) trapped air.

Hence it is advisable to fill the pipes with water for one hour before testing and then to measure the loss of water over a 30-minute period, normally applying a test pressure of 1.2 m head of water at the upper end and not more than 2.4 m at the lower end. It may thus be necessary to test steeply graded drains in stages, to avoid exceeding the maximum head. The Code[2] recommends that the average quantity of water added in the 30-minute period for drains up to 300 mm nominal bore should not exceed 0.06 litre per hour per 100 m per mm of the nominal bore of drain. Where there is a trap at the upper end of a branch drain, a rubber or plastics tube should be inserted through the trap seal to draw off the confined air as the pipes are filled with water. A second and similar test is generally undertaken after the completion of bedding or concrete surround, backfilling, compaction and reinstatement.

Air Test

An air test may be used to test the watertightness of drains, in which an appreciable drop in pressure will indicate a defective drain. The length of drain is plugged in the manner previously described and air pumped into the pipes, usually by means of a hand pump, until a pressure of 100 mm head of water is shown in an attached U-tube. The air pressure should not fall to less than 75 mm head of water during a period of five minutes after a period for stabilisation. For trapped drains the corresponding values are 50 and 47.5 mm respectively. Where there is an appreciable drop in pressure, it may still be difficult to detect the location or rate of leakage. Hence failure to satisfy this test should be followed by a water test to provide more conclusive information. An alternative to the water test is the smoke test with the smoke generated by a smoke-testing machine consisting of a double-action leather bellows and a copper fire-box enclosed within a copper tank and floating dome.

Infiltration Test

This test should be applied after backfilling where any part of the soffit of the drain (underside of crown of pipe) is 1.2 m or more below the water table. All inlets are sealed and inspection of manholes or inspection chambers will show any flow.

Tests for Straightness and Obstruction

Tests for line, uniform gradient and freedom from obstruction can be carried out using a mirror at one end of the drain and a lamp at the other.

CESSPOOLS

A cesspool is an underground chamber constructed for the reception and storage of foul water. The Building Regulations[1] require that it should be impervious to both liquid from the inside and subsoil water from the outside; should not pollute any source of domestic water supply; be properly covered with a suitable manhole cover; be adequately ventilated; have a minimum capacity of 18 m^3; and be sufficiently far from occupied buildings. When siting a cesspool attention should be paid to the slope of the ground, direction of the prevailing wind, access for emptying and possibility of future connection to sewer.

In general cesspools should only be used when no other alternative is available because of the problems associated with ensuring watertightness and emptying; a typical cesspool emptier only has a capacity of 4550 litres. A small sewage treatment works, if practicable, will be cheaper in the long-term and much more satisfactory. Capacity limitations will restrict the use of cesspools to single properties with not more than eight inhabitants. The Code for cesspools[20] prescribes a minimum capacity of 45 days dry weather flow (DWF) based on 136 litres per head per day. With four persons this would require a capacity of 24 600 litres, but it would need emptying eight times per annum and might well prove prohibitive in cost.

The depth below cover should not exceed 4.3 m to facilitate pumping, and the most satisfactory shape is circular with a diameter equal to the depth below the incoming drain. A typical cesspool is illustrated in figure 13.4.4. The walls may be of 215 mm brickwork, minimum of 150 mm concrete, or large diameter precast concrete pipes surrounded with a minimum of 100 mm of concrete, built from a concrete base. The drain inlet enters through a bend and a 100 mm fresh air inlet (750 mm above ground) and 100 mm vent pipe (not less than 3 m above ground) must be provided to ensure adequate ventilation.

SMALL DOMESTIC SEWAGE TREATMENT WORKS

In country areas where no public sewers are available, sewage from single houses or groups of houses, with a total population up to 300, is often treated in small sewage treatment works, consisting essentially of a septic tank and a biological filter. In the septic tank heavier solids settle at the bottom as sludge and lighter solids rise and form a scum which acts as a surface seal and permits anaerobic decomposition by bacteria. Oxidation of the organic matter contained in the tank effluent takes place by aerobic bacteria in the filter medium. The Code for small sewage treatment works[21] prescribes a minimum distance of 15 m from habitable buildings of works serving up to ten persons, increasing to 90 m for works serving 100 persons or more. Other siting requirements are similar to those listed for cesspools.

Septic tanks. These should have a capacity in litres of (180P + 2000) where P is the design population.[21] Alternatively it can be calculated on the basis of 48 hours DWF for ten persons reducing to 12 hours DWF for 100 persons or more, with a minimum capacity of 1365 litres. The Building Regulations[1] prescribe a minimum capacity of 2.7 m^3. A simple tank is illustrated in figure 13.4.2 consisting of a rectangular chamber of from 1.2 to 1.8 m water depth with its length about three times its width. Precast concrete slabs, set slightly apart for ventilation purposes and fitted with lifting

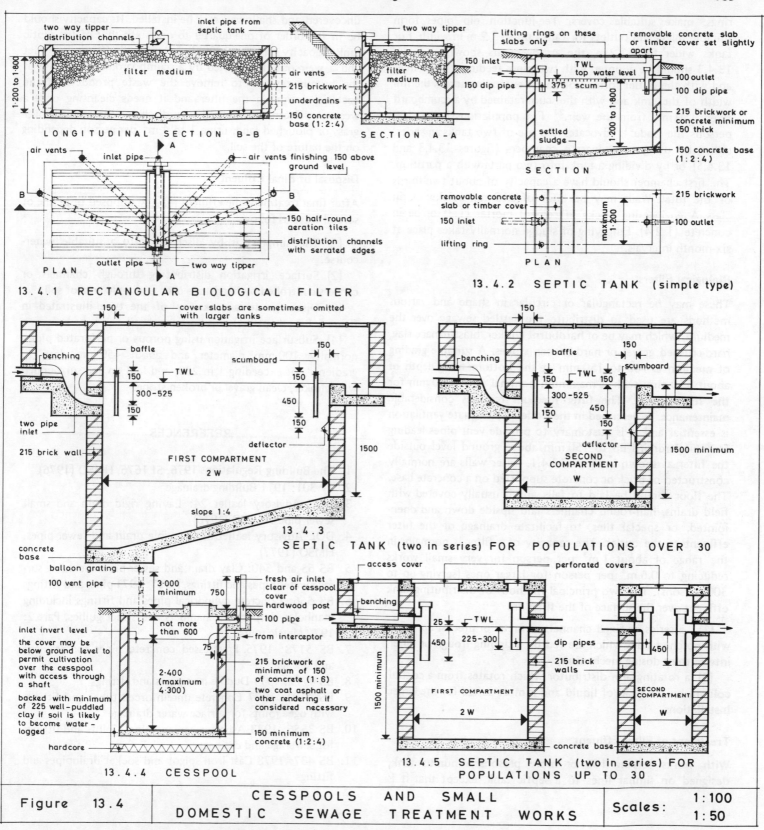

13.4.1 RECTANGULAR BIOLOGICAL FILTER

13.4.2 SEPTIC TANK (simple type)

13.4.3 SEPTIC TANK (two in series) FOR POPULATIONS OVER 30

13.4.4 CESSPOOL

13.4.5 SEPTIC TANK (two in series) FOR POPULATIONS UP TO 30

| Figure 13.4 | CESSPOOLS AND SMALL DOMESTIC SEWAGE TREATMENT WORKS | Scales: 1:100 1:50 |

rings, make suitable covers. Tee-junction dip pipes form suitable inlets and outlets for tanks up to 1.2 m in width. For tanks wider than 1.2 m, the arrangements shown in figure 13.4.3 should be adopted with two pipes feeding a baffle inlet and with the outlet formed of a weir extending across the full width of the tank and with the scum retained by a scumboard fixed 150 mm from the weir.[21] For populations below 100 persons the Code[21] advocates the use of two tanks in series, either by using two separate chambers (figures 13.4.3 and 13.4.5) or by dividing a tank into two parts with a partition. The first chamber should have a capacity of about two-thirds of the total. Walls may be built of 215 mm brickwork in class A engineering bricks in cement mortar (1:3) or be in concrete (1:2:4). Emptying of sludge normally takes place at six-month intervals.

Biological Filters

These may be rectangular or circular in shape and various methods are used to distribute the settled sewage over the medium, which may be of hardburnt clinker, blastfurnace slag, hard-crushed gravel or hard-broken stones. A suitable grading of medium is 100 to 150 mm at the bottom for a depth of about 150 mm, with a maximum nominal size of 50 mm for the remainder.[21] The filter medium requires considerable maintenance, as described in the Code.[21] Adequate ventilation is essential and it is customary to provide vent pipes leading from underdrains up to 150 mm above ground level outside the filter as shown in figure 13.4.1. Filter walls are normally constructed of brick or concrete supported on a concrete base. The floor should be laid to falls and is usually covered with field drains, half-round channels laid upside down and open-jointed, or special tiles, to facilitate drainage of the filter effluent to the outlet. The capacity of a filter is normally in the range of about 1 m^3 per person for very small works reducing to 0.6 m^3 per person for larger ones (serving up to 300 persons). The two principal methods of distributing tank effluent over the surface of the filter are

(1) a series of fixed channels of metal or precast concrete with notched sides which are fed by a tipping trough or other intermittent dosing mechanism;

(2) a rotating arm distributor which rotates from a central column by the head of liquid and is more suited for the larger installations.

Treatment of Filter Effluent

With works serving 100 or more persons a humus tank, designed on similar lines to a septic tank, except that it is uncovered and shallower, may be installed. Its capacity should be in the order of one-quarter to one-sixth that of the septic tank (capacity = (30P + 1500) litres) with a baffle or weir inlet and a weir outlet protected by a scumboard. The purpose of the humus tank is to remove the waste products of the bacterial action in the filter and it needs cleansing once a week. An alternative method is to discharge the effluent over grass or ploughed land, using 1 to 4 m^2 per person depending on the nature of the soil.

Disposal of Final Effluent

After final treatment the effluent may be disposed of in one of several ways as follows

(1) The simplest is by discharge into a suitable water-course.

(2) Surface irrigation distributing through channels or carriers controlled by handstops over a suitable area of soil.

(3) Discharge to a soakaway of the type illustrated in figure 1.1.6 (chapter 1), in a porous subsoil.

(4) Subsurface irrigation using porous or perforated pipes, normally 100 mm diameter and about 450 mm deep to gradients not exceeding 1 in 200, and laid on and surrounded with clinker, clean gravel or broken stone.

REFERENCES

1. The Building Regulations 1976. SI 1676. HMSO (1976)
2. CP 301: 1971 Building drainage
3. DOE Advisory leaflet 24: Laying rigid drain and small sewer pipes. HMSO (1976)
4. DOE Advisory leaflet 66: Flexible drain and sewer pipes. HMSO (1977)
5. BS 65 and 540: Clay drain and sewer pipes including surface water pipes and fittings, Part 1: 1971 pipes and fittings
6. BS 556: Concrete cylindrical pipes and fittings including manholes, inspection chambers and street gullies, Part 2: 1972 Metric units
7. BS 5178: 1975 Prestressed concrete pipes for drainage and sewage
8. BS 4772: 1971 Ductile iron pipes and fittings
9. BS 4101: 1967 Concrete unreinforced tubes and fittings with ogee joints for surface water drainage
10. BS 3656: 1973 Asbestos cement pipes, joints and fittings for sewerage and drainage
11. BS 437: 1978 Cast iron spigot and socket drainpipes and fittings

12. BS 1211: 1958 Centrifugally cast (spun) iron pressure pipes for water, gas and sewage
13. BS 4622: 1970 Grey iron pipes and fittings
14. BS 2760: 1973 Pitch-impregnated fibre pipes and fittings for above and below ground drainage
15. BS 4660: 1973 Unplasticised PVC underground drain pipe and fittings
16. BS 1130: 1943 Schedule of cast iron drain fittings
17. BS 3921: 1974 Clay bricks and blocks
18. BS 1247: 1975 Manhole step irons
19. BS 497: Manhole covers, road gully gratings and frames for drainage purposes, Part 1: 1976 Cast iron and cast steel
20. CP 302, 200: 1949 Cesspools
21. CP 302: 1972 Small domestic sewage treatment works

14 EXTERNAL WORKS

The final chapter is concerned with the various external works provided in connection with buildings, apart from drainage and other underground services. They consist principally of roads, paved areas, footpaths, landscape work, fencing and gates.

ROAD DESIGN

The normal housing estate road has a width of about 4.90 m, although this is sometimes increased to 5.50 m where a large volume of traffic is to be carried. For the principal distribution roads to large housing estates, widths of 6.10 or 6.70 m would be more appropriate. At the other end of the scale, the shorter service roads with development on one side only and culs-de-sac often have a width of 4.0 m, although this is inadequate to accommodate two wide vehicles alongside one another. Private drives and accesses to garages may be 2.50 to 3.00 m wide.

Longitudinal gradients must be kept within reasonable limits, such as 1 in 20 to 1 in 250. If a road is too flat it will be difficult to remove surface water, while if it is too steep it will become difficult to negotiate in snowy or frosty weather. Vertical curves should be designed to provide a suitable parabolic curve linking the two gradients, with the levels normally determined at 6 m intervals. Roads can be constructed with a camber (figure 14.1.3) or with a single crossfall.

The selection of the form of road construction will be influenced by a number of factors, including type of subgrade, liability to subsidence, initial costs, maintenance costs, appearance, resistance to wear and non-skid qualities.

ROAD CONSTRUCTION

The construction methods can be broadly subdivided into two main groups

(1) flexible (figure 14.1.1) consisting of a stone base with a surfacing of tar or bitumen-coated stones;

(2) rigid (figure 14.1.2) constructed of a concrete road slab.

Flexible Roads

Flexible roads usually contain a sub-base of ashes, clinker, shale, gravel or hoggin, ranging from 100 to 250 mm thick. The sub-base reduces stress in the subgrade (natural formation), protects the subgrade against frost and constructional traffic, and prevents mud entering the road structure. The base may be formed of various materials — limestone or slag 63 to 38 mm in size, blinded stone pitching 200 to 250 mm long, low binder macadam of limestone, slag or gravel 50 mm in size bound with tar or bitumen, base course and single course tar or bituminous macadam 50 or 38 mm in size, or tar or bitumen-bound granular base (limestone, slag or gravel) 38 or 25 mm in size. A roller weighing 8 to 10 tonnes is used to compact the base.

Bituminous surfacings provide a smooth, non-skid, abrasion-resistant and jointless surface which possesses colour and texture, and has good cleansing and surface water runoff properties. The thickness varies from 100 mm laid in two courses for main roads to 56 mm in two courses or 50 mm in a single hot-laid course for minor roads. The binders consist of tar, bitumen or a mixture of both. Warm-laid macadams are usually cheaper but have a shorter effective life than hot-laid surfacings. They are particularly suitable for low-cost jobs and minor roads, often using local stone as aggregate and with tar or bitumen as a binder. Tar is cheaper in first cost and offers more resistance to petrol and oil droppings, whereas bitumen has a longer life, can be laid through a large range of temperatures and is less subject to deformation in hot weather.

Another suitable surfacing is *cold asphalt*[1] which is a mixture of bitumen and crushed igneous rock, limestone or slag aggregate, has a life of up to ten years and is used extensively for patching and carpet coats. Other alternatives are dense bitumen and tar-base courses and dense bitumen or tar surfacing. *Hot-rolled asphalt*[2] gives a dense, impervious and

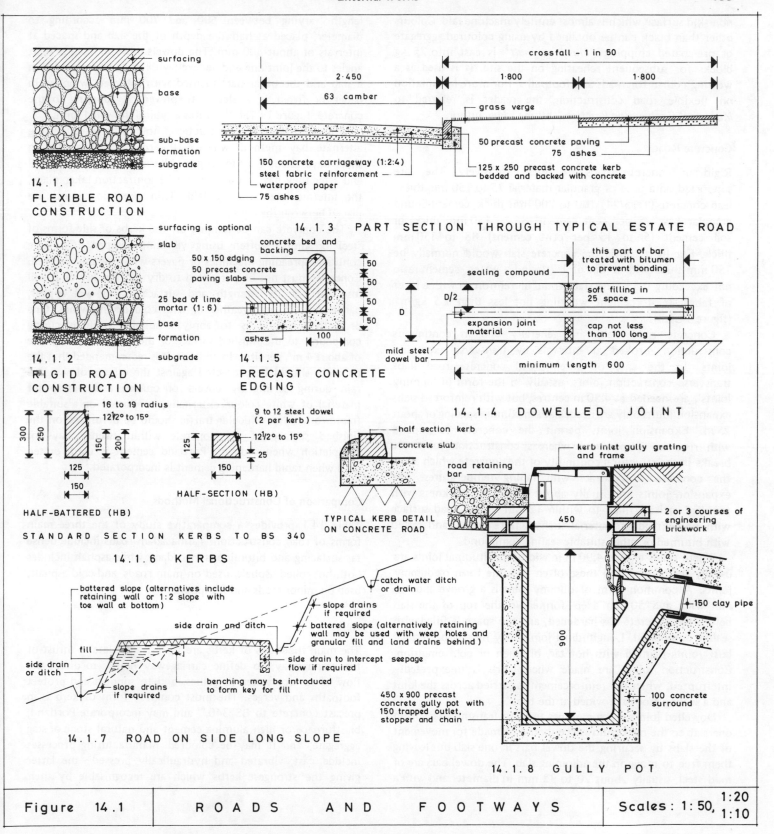

14.1.1 FLEXIBLE ROAD CONSTRUCTION

surfacing
base
sub-base
formation
subgrade

14.1.2 RIGID ROAD CONSTRUCTION

surfacing is optional
slab
base
formation
subgrade

14.1.3 PART SECTION THROUGH TYPICAL ESTATE ROAD

crossfall – 1 in 50
2·450
63 camber
1·800
1·800
grass verge
50 precast concrete paving
75 ashes
125 x 250 precast concrete kerb bedded and backed with concrete
150 concrete carriageway (1:2:4)
steel fabric reinforcement
waterproof paper
75 ashes

14.1.5 PRECAST CONCRETE EDGING

concrete bed and backing
50 x 150 edging
50 precast concrete paving slabs
25 bed of lime mortar (1:6)
ashes
50
50
50
50
100

14.1.4 DOWELLED JOINT

this part of bar treated with bitumen to prevent bonding
sealing compound
soft filling in 25 space
D/2
D
expansion joint material
cap not less than 100 long
mild steel dowel bar
minimum length 600

14.1.6 KERBS

16 to 19 radius
12½° to 15°
300
250
150
200
125
150
150
HALF-BATTERED (HB)

9 to 12 steel dowel (2 per kerb)
12½° to 15°
125
150
25
HALF-SECTION (HB)

STANDARD SECTION KERBS TO BS 340

half section kerb
concrete slab
TYPICAL KERB DETAIL ON CONCRETE ROAD

14.1.8 GULLY POT

kerb inlet gully grating and frame
road retaining bar
2 or 3 courses of engineering brickwork
450
900
150 clay pipe
100 concrete surround
450 x 900 precast concrete gully pot with 150 trapped outlet, stopper and chain

14.1.7 ROAD ON SIDE LONG SLOPE

battered slope (alternatives include retaining wall or 1:2 slope with toe wall at bottom)
catch water ditch or drain
slope drains if required
side drain and ditch
battered slope (alternatively retaining wall may be used with weep holes and granular fill and land drains behind)
fill
side drain or ditch
side drain to intercept seepage flow if required
slope drains if required
benching may be introduced to form key for fill

| Figure 14.1 | R O A D S   A N D   F O O T W A Y S | Scales: 1:50, 1:20 1:10 |

non-skid surface which is almost entirely machine laid. Colours other than black can be obtained by using coloured aggregate or pre-coated chippings. *Mastic asphalt*[3] is cast into 25 kg blocks for subsequent reheating on the site. It is used as a wearing course for roads and footpaths. For more information on flexible road construction, the reader is referred to *Municipal Engineering Practice*,[4] by the same author.

Concrete Roads

Rigid or concrete roads are used extensively. They are supported on a base of granular material 75 to 150 mm thick, lean concrete (1:15/24) 100 to 200 mm thick, cement-bound granular base (5 per cent cement) 88 to 150 mm thick, or soil cement (5 to 15 per cent cement) 88 to 150 mm thick. For light traffic the concrete slab would normally be 150 mm thick with a mix of 1:2:4 and a water–cement ratio not exceeding 0.55. The slab would be reinforced with a layer of fabric reinforcement weighing not less than 2.5 kg/m^2 (figure 14.1.3).

Concrete road slabs are generally laid in continuous construction, broken at intervals by transverse expansion joints. In the case of unreinforced concrete road slabs transverse contraction joints, usually in the form of 'dummy joints', are needed at 4.50 m centres, but with reinforced slabs expansion joints only are needed at a maximum spacing of about 25 m. Expansion joints permit the concrete to expand with rises in temperature, whereas construction joints are breaks in the structural continuity of the concrete which allow the concrete to contract with falling temperatures. The expansion joints are usually about 20 mm in thickness with rounded arrises, filled to within 12 mm of the road surface with non-extruding material such as fibre board, and sealed with bitumen or other suitable sealing compound.

Where roads exceed 4.90 m in width, longitudinal joints are usually introduced and these often take the form of dummy joints. A common form of dummy joint is a groove about 5 mm wide and 50 mm deep—formed in the top of the slab before the concrete has hardened, and the space is filled with sealing compound. Longitudinal joints may merely be vertical butt joints painted with hot tar, bitumen or cold emulsion. Construction joints are made when work is unexpectedly interrupted, when the reinforcement is carried across the joint and a sealing groove provided at the top.

Dowelled joints are sometimes used to transmit loads from one slab to the next. Allowance should be made for movement of the slabs by securing the dowel bars in one slab but leaving them free to move in the adjoining slab. The dowel bars are of mild steel, usually about 20 to 32 mm in diameter and with a

length varying between 500 and 700 mm according to diameter, placed at half the depth of the slab and spaced at intervals of about 300 mm. The dowels must be fixed at right angles to the joint line and be level; with one part embedded in a slab and the other part painted with bitumen, wrapped in paper or fitted in a sleeve, to prevent it adhering to the concrete. Figure 14.1.4 illustrates a typical dowelled joint.

Another form of construction, now little used, is the alternate bay method, whereby the road is constructed in short bays about 6 m in length. Alternate bays are first laid and after they have set and the initial contraction taken place, the intermediate bays are laid. Thin expansion joints are placed between the bays.

The concrete can be tamped off the kerbs or side forms of steel or timber, often using vibrating tampers. On large contracts spreading machines or pavers will probably be used. Concrete must not be permitted to dry out quickly or it will lose part of its strength and surface cracks will appear, particularly in hot weather or with cold drying winds. One protective method is to apply a suitable resinous curing compound to the finished surface or the concrete at the rate of about 4 m^2 per litre. In addition it is recommended that the concrete should be protected against the effects of sun and rain during setting by covers of opaque and waterproof material of white colour externally, supported on a suitable framework.[5] No vehicular traffic should be permitted on the finished surface of the concrete within twenty days of completion when ordinary Portland cement is used, or ten days when rapid hardening cement is incorporated.[13]

Comparison of Constructional Methods

Table 14.1 provides a comparative study of the three main forms of road construction. Tarmacadam roads include dense tar surfacing and bituminous macadam, while asphalt includes both hot-rolled asphalt, used on main roads, and cold asphalt, used on minor roads and footpaths.

Kerbs

The main functions of kerbs are to resist the lateral thrust of the carriageway, to define carriageway limits, to direct the flow of surface water to gullies, and to support and protect footpaths and verges. The most common form of kerb is in precast concrete to BS 340,[6] and may incorporate Portland, blastfurnace or high alumina cement and natural stone or slag aggregate, and it may be coloured. Manufacturing processes include cast, vibrated and hydraulically pressed – the latter giving the strongest kerbs which are recognisable by their

Table 14.1   Comparison of constructional methods for roads

Characteristics	Concrete roads	Tarmacadam roads	Asphalt roads
Liability to subsidence damage	Considerable	Reasonably flexible	Reasonably flexible
Initial cost	Reasonably priced in most areas; requires little skilled labour	Depends on proximity of tarmacadam plant, but not excessively high	More expensive than tarmacadam
Maintenance costs	If well constructed should be low for many years after construction	Fairly high as periodic surface dressing required every five years and carpet coat at longer intervals	Carpet coat required at about eight to ten-year intervals at much greater cost than tarmacadam
Appearance	Glare from sun	Reasonable	Good
Non-skid properties	Slightly corrugated surface gives reasonable resistance	Good	Fairly good
Resistance to wear	Good if concrete quality is carefully controlled	Fairly good to good, depending on type of filler and surface dressing stone	Good
Ease of reinstatement	Costly, difficult, requires curing time	Comparatively simple	Comparatively simple

patterned faces. Less popular kerbs include granite and sandstone.

Precast concrete kerbs are made in 900 mm lengths and in four standard sections — bullnosed (with 15 to 19 mm radius edge), splayed (75 mm wide x 75 mm deep), half-battered (100 mm deep) and half-section (half-battered). The last two sections are illustrated in figure 14.1.6 and the most commonly used sizes are 125 x 250 mm and 150 x 125 mm respectively. Radius kerbs are made to a variety of radii ranging from 1 to 12 m. Bullnosed and half-battered kerbs are used extensively for urban roads, whereas splayed kerbs are more often used on dual carriageways and rural roads.

With flexible roads, it is usual to lay full-section kerbs (100 x 250 to 150 x 300 mm) to take the lateral thrust. They are usually bedded on a 12 to 25 mm layer of cement mortar on a concrete foundation (1:3:6), 100 to 150 mm thick, and backed with similar quality concrete, 100 to 125 mm thick, to within 75 mm of the top of the kerb. Alternatively, the kerbs may be laid on a fresh semi-dry concrete foundation without an intervening mortar bed. The kerbs may be butt jointed without any mortar, or have mortared and pointed joints.

With rigid roads, it is better to use half-section kerbs on a mortar bed, 12 to 25 mm thick, laid directly on the concrete road slab. To give adequate support to the kerbs it is advisable to provide two 9 to 12 mm diameter steel dowel bars, 125 to 150 mm long, to each length of kerb and driven 75 mm into the concrete while it is still green and then backed with concrete (figure 14.1.6). This form of construction prevents water penetrating between the kerb face and the edge of the road slab.

Channels

Channels help surface water runoff, define the limits of the carriageway, facilitate road sweeping and strengthen the edges of flexible roads. The most commonly used channels are precast concrete to BS 340,[6] of 250 x 125 mm rectangular section, laid on a concrete foundation. In concrete roads channels are normally formed about 200 or 250 mm wide with a wood or steel float. Precast concrete edging (figure 14.1.5) varying in size from 50 x 150 to 50 x 250 mm, with tops finished to half-round, square, bullnosed or chamfered sections, may be used to support and define the edges of drives in place of kerbs or channels.

ROAD DRAINAGE

Groundwater Drainage

Water entering the subgrade can cause a substantial loss in strength in the soil foundation to a road. Hence the road formation should be kept a minimum of 1.20 m above the highest water level. With permeable soils it is possible to reduce the water table by installing longitudinal land drains on each side of the road, with the drain trenches filled with coarse sand or other suitable material and sealed with clay at the top of the trench. Where there is a risk of surface water from adjoining higher land penetrating into the subgrade this must be intercepted by side drains or ditches as shown in figure 14.1.7. With embankments it is customary to provide slope drains, often constructed as French drains, running down the slope to connect with a side drain or ditch at the side of the road.

Surface Water Drainage

Road surfaces are laid to specified cambers or straight crossfalls to shed surface water to channels. Minimum cross-falls vary with the form of construction, ranging from 1 in 50 for relatively smooth surfaces such as asphalt or dense bitumen—macadam to 1 in 35 for rougher surfaces such as open-textured macadam or slightly corrugated concrete. Roads in rural areas are often drained through trenches or 'grips' about 375 mm wide and 150 mm deep, leading across grassed roadside areas to ditches. Paved areas should generally have a minimum crossfall of 1 in 60.[17]

Gullies

In urban areas surface water runoff from roads is almost invariably discharged into gullies of the type illustrated in figure 14.1.8. These are often spaced at about 45 m intervals, with each gully draining about 220 m^2 of road surface. Gully pots are made in a variety of materials including precast concrete, clayware, cast iron, brickwork and *in situ* concrete. Precast concrete pots to BS 556[7] and clayware pots to BS 539[8] are both popular. They can be trapped or untrapped and supplied with or without rodding eyes, in a variety of sizes ranging, for instance, from 300 mm diameter amd 600 mm deep to 600 mm diameter and 1200 mm deep for precast concrete pots. The outlets are usually 150 mm in diameter. A typical gully pot is illustrated in figure 14.1.8, which shows brickwork above the pot to permit some flexibility in the fixing of the cover and frame.

Gully gratings, through which surface water enters the pot, are of two main forms – channel and kerb inlet, in cast iron, cast steel or a mixture of both to BS 497[9]. The kerb inlet type are only really suitable for light duty but have a number of advantages – they eliminate obstructions in the carriageway, facilitate road construction and resurfacing and there is less risk of blockage. On the other hand they obstruct a section of the path or verge and need anti-deflector plates in the channel where the gradient exceeds 1 in 50.

FOOTPATHS AND PAVED AREAS

Footpaths and paved areas can be constructed of a variety of materials, and the choice will be determined largely by such factors as initial cost, maintenance cost, appearance, wearing qualities and non-skid properties. An examination of the comparative schedule in table 14.2 may prove helpful in this connection. The width of a footpath varies from about 1.35 to 1.80 m on housing estates and may increase to as much as 6 m in shopping centres. They are normally laid to a crossfall of about 1 in 30 towards the kerb and there may be a grass verge between the path and kerb. A tree-planted grass verge improves the appearance of residential development but it increases maintenance costs.

Table 14.2   Comparison of constructional methods for footpaths

Form of construction	Advantages	Disadvantages
Precast concrete paving slabs	Good appearance; hard wearing; non-slip	Expensive in initial and maintenance costs; can soon become dangerous with slight settlement; easily damaged by vehicles mounting kerb
In situ concrete	Reasonably cheap; can be coloured; reasonably hard wearing if concrete of good quality; reasonably non-slip	Needs frequent expansion joints and time for curing; unattractive appearance and difficult to reinstate satisfactorily
Bituminous macadam or tarmacadam	Reasonably priced and hard wearing; non-slip; flexible; fairly easily maintained; reasonable appearance	Periodic surface dressing required; generally needs path edging (figure 14.1.5) at back of path
Asphalt	Good appearance; hard wearing; reasonably non-slip; flexible; fairly easily maintained	Fairly expensive; generally needs path edging (figure 14.1.5) at back of path

Precast Concrete Paving Slabs

Precast concrete paving slabs or flags are used extensively in the built-up areas of towns and to a lesser extent on residential estates. The best slabs are hydraulically pressed and all should comply with BS 368;[10] they can be made with a variety of aggregates and to various colours and surface finishes. Standard sizes are 600 x 450, 600 x 600, 600 x 750 and 600 x 900 mm, in both 50 and 63 mm thicknesses.

The slabs are usually laid on a 75 or 100 mm bed of ashes consolidated with a 2½ tonne roller or 300 to 500 kg vibratory roller, and a 12 to 25 mm bed of sand, lime mortar or gauged mortar. Where the slabs may be subject to vehicular traffic they should be laid on concrete. Slabs are often laid to a 150 mm bond and a crossfall of 1 in 30. After laying, slabs are usually grouted with lime or cement mortar to give a solid job. Although expensive, precast concrete paving slabs are attractive, durable and easily reinstated.

Alternatives to precast concrete slabs are natural stone or reconstructed stone.
Natural stone. In particular, York stone, which is hard-wearing, non-slip and attractive, but is extremely expensive and hence has a restricted use.
Reconstructed stone slabs. These contain natural stone both as fine and coarse aggregates, giving colour and non-slip qualities. They are supplied in several shapes for patterned work, including hexagonal, triangular and circular, when they are often surrounded by asphalt, tarmacadam or cobbles.

In situ Concrete

In situ concrete is very suitable for large paved areas and paths to houses. The usual mixes range from 1:2:4 to 1:3:6, using 19 mm graded aggregate and the concrete is often 75 mm thick, but may be thickened at the edges, on a clinker or hardcore base. It is advisable to provide dummy joints at about 1.80 m intervals and expansion joints, often of timber battens, about 9.0 m apart. A variety of finishes are available including tamped, rolled with a crimping roller leaving pyramidal depressions, brushed aggregate, pigmented, and surface dressed with coloured chippings lightly tamped into the concrete. It is relatively cheap but its main deficiency is the difficulty of reinstatement to gain access to underground services.

Bituminous Surfacings

A common footpath surfacing is tarmacadam *tarpaving* to BS 1242,[11] using a crushed rock or slag coarse aggregate and a binder of tar or tar—bitumen mixture. It may be laid warm or cold on a bed of clinker, hardcore or other suitable material, 50 to 75 mm in thickness. Tarpaving may be laid in a single course, 20 to 30 mm thick, or in two courses with a total thickness of 45 to 75 mm (30 to 60 mm base course and 19 to 25 mm wearing course). Tarpaving should be consolidated with a roller weighing not less than 4 tonnes to a crossfall of not less than 1 in 48. It provides a relatively economical, non-slip and fairly easily reinstated surface, but it does require periodic surface dressing.

Another alternative, which is becoming increasingly popular, is a single course of fine cold asphalt to BS 1690,[1] laid on hardcore, concrete, soil-cement or other suitable base. A tack coat of bitumen emulsion (1.8 to 2.7 m^2 per litre) should precede the asphalt, unless the asphalt is being laid on a recently laid base course. The thickness varies from 13 to 20 mm and the asphalt is consolidated with a 350 to 2500 kg roller. After initial consolidation, 10, 13 or 20 mm precoated chippings are sometimes added and rolled into the surface of the asphalt. Fine cold asphalt gives a more dense and hard-wearing surface than tarpaving.

Cobbles, Setts and Bricks

These smaller paving units are used to give interest, for demarcation purposes, to form drainage channels, to pave small irregular areas and to discourage the passage of vehicles and pedestrians.
Cobbles. These are obtained from river beds, beaches and gravel quarries, are hard wearing and have a good texture. They are often about 50 to 75 mm in size but can be 150 to 225 mm in areas subject to vehicular traffic. For lightly trafficked areas cobbles are usually laid on a fine gravel or ash bed with the intervening spaces between cobbles filled with a dry cement and sand mix brushed in and sprayed with water. For heavily trafficked areas cobbles are usually laid on a semi-dry concrete bed about 100 mm thick, with the cobbles extending above the concrete for about one-third of their depth.
Setts. These are usually of granite or whinstone and are generally either hammer dressed (left roughly rectangular) or nidged (chisel dressed to a true rectangle). They are extremely hard wearing with good textures and colours, and laid to a crossfall of about 1 in 40 on a concrete base with a 25 mm cement mortar bed, and with the joints between setts filled with 6 mm chippings and grouted with pitch or pitch—tar compound. Cement mortar can be used for jointing in pedestrian areas.
Engineering bricks and brick pavers. These are suitable for paving in a variety of bonds and patterns. They are usually laid on a concrete base with a 38 mm bed of semi-dry cement mortar (1:4) with 6 to 9 mm joints between bricks.

Gravel

Gravel is used for surfacing lightly used pedestrian areas in rural situations, landscaped areas, around trees in car parks and in shaded areas where grass will not thrive. The formation is usually treated with a weedkiller. The gravel can be 'all-in' material such as ballast or hoggin, or graded material such as river gravels and crushed rock (often 19 to 3 mm in size). It is normally laid in two layers, well watered and rolled to a consolidated depth of about 150 mm to a crossfall of about 1 in 30.

LANDSCAPE WORK

This section is mainly concerned with grassed areas and the planting of trees and shrubs.

Grassed Areas

Grassed areas and roadside verges possess great amenity value and also perform other useful functions such as absorption of sound and reduction of glare. They are relatively cheap to establish but expensive in maintenance. It is undesirable to lay verges to gradients in excess of 1 in 4 and less than 1.35 mm in width. The finished surface should project about 20 mm above adjoining surfaces such as kerbs, pavings and manhole covers. It is good practice to provide a 225 mm wide strip of gravel or ballast, treated with weedkiller, between grassed areas and walls.

Prior to soiling and seeding, the formation should be broken up and a layer of clinker or gravel provided for drainage purposes with heavy clay. The layer of soil should be at least 100 mm thick worked to give 50 mm of tilth and have a pH value just over 5.0. A suitable fertiliser is distributed at the rate of 0.05 kg/m^2 followed by a suitable mixture of grass seed at 0.06 kg/m^2, which should be lightly raked in and lightly rolled if the surface is dry. The author favours the following general purpose mix, although the choice of mix is influenced by the type of soil and location of the grassed area.

	percentage
Short-seeded perennial rye grass	60
Chewings fescue	10
Red fescue	10
Agrostis tenuis	10
Crested dogstail	10

Practical difficulties of the seasons when seeding can take place and the time taken for the grass to become established may favour more expensive turfing in some situations. Turves vary in size but are often 600 × 300 mm or 900 × 300 mm and about 38 mm thick, and are laid from boards to break bond over well-prepared soil. They should be watered after laying in dry periods.

Trees and Shrubs

The provision of trees and shrubs can do much to improve the appearance of a building project. Trees take many years to grow and so existing trees which are in good condition should be preserved wherever possible. Furthermore, existing trees and shrubs give a good indication of what new species are likely to thrive as far as climate and soil are concerned. There is rarely sufficient space on residential developments for forest-type trees, but smaller trees such as whitebeam and mountain ash, as well as many types of flowering tree, with heights of 6 to 9 m, can be used to advantage. BS 4043[16] contains recommendations for transplanting semi-mature trees.

Trees generally have great value in form as well as colour, and they should be chosen to blend the architecture and landscape into a single composition. They change with the season and give living quality to the hard surfaces and angular shapes of buildings and roads. In summer trees in leaf have great beauty in both form and colour, and in winter the silhouette and shadow of branches and twigs provide fascinating decoration to nearby walls. The irregular shapes and siting of trees can often be used deliberately as a contrast to architectural regularity. Trees are also useful as a barrier to noise and glare, and to give shade and shelter from winds. Consideration must be given to soil and climatic conditions, ultimate size of trees, and location of underground services and buildings, when selecting trees.[12] Shrubs provide a variety of colours in flowers, berries and leaves, with differing flowering seasons and heights varying from 0.60 to 2.40 m.

Small trees require holes about 1.5 m in diameter which should be filled with topsoil, preferably with some leaf mould or manure placed around the tree roots. In wet soils, brick rubble or clinker should be laid over the bottom of the hole. The normal planting season stretches from October to mid-March. The roots should be well spread out and the soil made firm around them. Young trees should be protected with stakes and guards and they need to be watered and mulched with grass cuttings or manure during a dry spring.

FENCING

Fences of a variety of materials and types are used to form boundaries between land of different occupiers. The choice may be influenced by a number of factors including appearance, durability, initial and maintenance costs, and effectiveness.

Close-boarded fences. These are both attractive and effective but have a high initial cost. They should comply with BS 1722[14] and a typical fence of this type is illustrated in figure 14.2.1. Posts may be of concrete or timber, preferably oak. Timber posts look better — they should ideally have weathered tops and mortices to receive rails, be 100 x 100 or 100 x 125 mm in section and be 600 to 750 mm in the ground. A normal spacing is 3 m. Rails are normally triangular in section with two cut from 75 x 75 mm timber — with two rails for fences up to 1.20 m high and three rails for higher fences. The boarding usually consists of vertical pales about 100 mm wide and feather edged approximately 14 to 7 mm (minimum) thick (two ex 100 x 25 mm), with a 65 x 38 mm twice-weathered capping to protect the top of the pales. The pales are nailed to the rails with 50 mm galvanised nails. A horizontal gravel board, about 32 or 25 x 200 mm, is often fixed below the pales to prevent their bottoms from being in contact with the ground with liability to decay.

Woven wood fences. These are reasonably attractive, fairly expensive and not exceptionally durable (figure 14.2.2). They should comply with BS 1722[14] and posts may be of concrete or timber. The maximum length of panels is 1.80 m and heights vary from 0.60 to 1.80 m. The panels consist of horizontal and vertical slats woven together within a frame of double vertical battens and either double horizontal battens or a single counter rail and gravel board, together with single or double vertical centre stiffeners. Battens are not usually less than 19 x 38 mm and slats not less than 75 x 5 mm. A weathered capping, about 19 x 50 mm, is usually fixed to the top of each panel. Wood posts are usually 75 x 75 mm with a weathered cap, and concrete posts are generally 100 x 85 mm with a weathered top.

Wooden post and rail fences. These are also specified in BS 1722[14] with posts often 75 x 150 mm, possibly with pointed bottoms for driving (figure 14.2.3). Rails are often rectangular, 38 x 87 mm. This type of fence is reasonably durable and attractive, of moderate cost, but does not provide a very effective division, as people, animals and objects can pass through it.

Wood palisade fences. These (figure 14.2.4) may consist of either concrete or wooden posts supporting two or three arris or triangular rails (ex 75 x 75 mm), according to the height of the fence (three rails when exceeding 1.20 m high), and vertical pales or palisades of 75 x 20 or 65 x 20 mm section, all in accordance with BS 1722.[14] This is a reasonably popular type of fence but has no particular merit.

Cleft chestnut pale fences. These (see figure 14.2.5) are also covered by BS 1722[14] with the pales normally hand-riven from sweet chestnut and a girth of not less than 100 mm, and roughly triangular or half-round in section. The pales are supported by two or three lines of 2 mm diameter galvanised mild steel wire, with each line consisting of four wires twisted together between the pales. Straining and intermediate posts can be of concrete or wood with the spacing of intermediate posts varying between 2.00 and 3.00 m according to the material used, function and height of fence. This type of fence is useful for temporary work but its poor appearance and liability to damage prevent its more widespread use.

Chain-link fences. Made to BS 1722[14], these are widely used as they are reasonably economical and form a very effective boundary, although their appearance even using plastic-coated chain link, is not very attractive (figure 14.2.6). The chain link consists of a diamond-shaped mesh with an average mesh size of 50 mm and average 3 mm diameter finished with a galvanised or plastics coating or using aluminium wire. The chain link is tied with wire to mild steel line wires of 2.5 to 4.75 mm diameter (two lines for fences up to 900 mm high and three lines for higher fences). Posts may be of reinforced concrete, steel or wood. The line wires are pulled tight with straining fittings at straining posts (figure 14.2.6) and intermediate posts are provided at not more than 3 m centres. Post sizes vary with the height of the fence, but with 1.20 m high fences typical sizes would be:

	straining posts (mm)	intermediate posts (mm)
reinforced concrete	125 x 125	125 x 125
steel (angle)	60 x 60 x 6	45 x 45 x 5
wood	100 x 100	75 x 75

Straining posts are strutted in the manner shown in figure 14.2.6.

Other types of fence. Covered by BS 1722,[14] these include woven wire; strained wire; mild steel or wrought iron continuous bar; mild-steel or wrought iron unclimbable fences with round or square verticals and flat standards and horizontals; and anti-intruder chain-link fences with extension arms to posts supporting lines of barbed wire.

GATES

Domestic front entrance gates are covered by BS 4092[15] and are of two main types — wooden and metal. Standard wooden single gates have widths of 0.81 and 1.02 m, whereas pairs of gates may have overall widths of 2.13, 2.34 and 2.64 m.

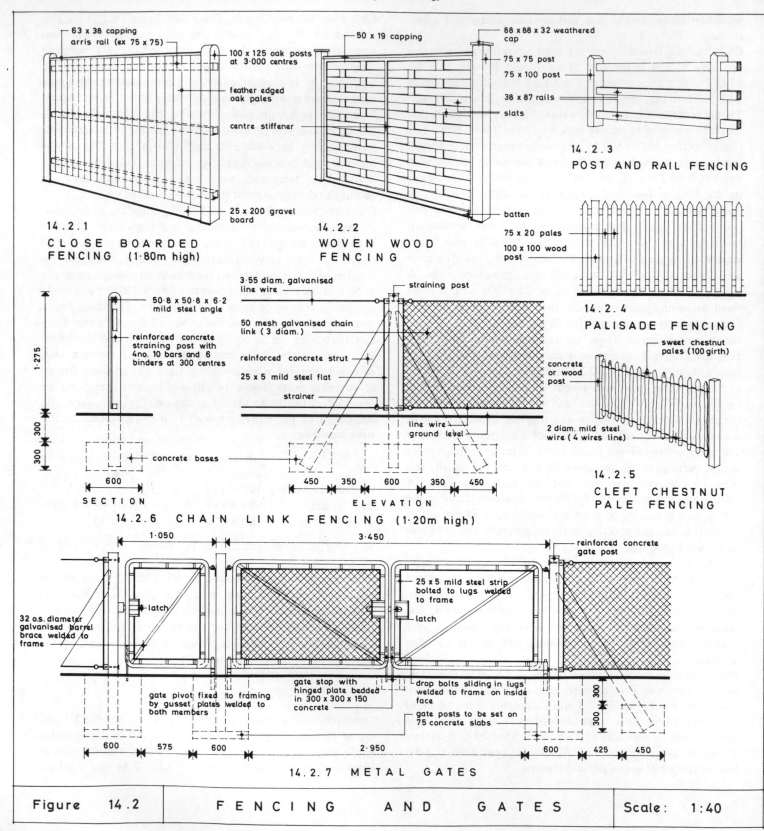

14.2.1 CLOSE BOARDED FENCING (1·80m high)
- 63 x 38 capping
- arris rail (ex 75 x 75)
- 100 x 125 oak posts at 3·000 centres
- feather edged oak pales
- centre stiffener
- 25 x 200 gravel board

14.2.2 WOVEN WOOD FENCING
- 50 x 19 capping
- slats
- batten

14.2.3 POST AND RAIL FENCING
- 88 x 88 x 32 weathered cap
- 75 x 75 post
- 75 x 100 post
- 38 x 87 rails

14.2.4 PALISADE FENCING
- 75 x 20 pales
- 100 x 100 wood post

14.2.5 CLEFT CHESTNUT PALE FENCING
- sweet chestnut pales (100 girth)
- concrete or wood post
- 2 diam. mild steel wire (4 wires line)

14.2.6 CHAIN LINK FENCING (1·20m high)
- 50·8 x 50·8 x 6·2 mild steel angle
- reinforced concrete straining post with 4no. 10 bars and 6 binders at 300 centres
- 3·55 diam. galvanised line wire
- straining post
- 50 mesh galvanised chain link (3 diam.)
- reinforced concrete strut
- 25 x 5 mild steel flat
- strainer
- line wire
- ground level
- concrete bases
- 1·275
- 300
- 300
- 600
- SECTION
- 450 350 600 350 450
- ELEVATION

14.2.7 METAL GATES
- 1·050
- 3·450
- reinforced concrete gate post
- 32 o.s. diameter galvanised barrel brace welded to frame
- latch
- 25 x 5 mild steel strip bolted to lugs welded to frame
- latch
- gate pivot fixed to framing by gusset plates welded to both members
- gate stop with hinged plate bedded in 300 x 300 x 150 concrete
- drop bolts sliding in lugs welded to frame on inside face
- gate posts to be set on 75 concrete slabs
- 300
- 300
- 600 575 600
- 2·950
- 600 425 450

| Figure 14.2 | FENCING AND GATES | Scale: 1:40 |

Minimum component sizes are specified as follows

hanging stiles for pairs of gates	95 x 45 mm
hanging stiles for single gates	70 x 45 mm
shutting stiles	70 x 45 mm
rails behind infilling	70 x 25 mm
other rails	70 x 45 mm
brace	70 x 25 mm
thickness of infilling	14 mm

Rails should be through-tenoned into stiles and each tenon should be pinned with a hardwood pin (not less than 10 mm diameter) or a non-ferrous metal star dowel (not less than 6 mm diameter).

For metal gates, the standard widths are 0.90, 1.0 and 1.1 m for single gates and 2.3, 2.4 and 2.7 m for pairs of gates. Metal gates may be constructed of mild steel or wrought iron, should have a continuous framework, and be truly square and welded at junctions. Minimum dimensions of framework and infilling are prescribed in BS 4092,[15] together with suitable hanging and latching fittings and protective treatments. Figure 14.2.7 shows suitable single and pairs of metal gates for use with chain-link fencing. Alternative forms of construction are detailed in BS 1722.[14]

REFERENCES

1. BS 1690: 1962 Cold asphalt
2. BS 594: 1973 Rolled asphalt (hot process) for roads and other paved areas
3. BS 1446: 1973 Mastic asphalt (natural rock asphalt fine aggregate) for roads and footways; BS 1477: 1973 Mastic asphalt (limestone fine aggregate) for roads and footways
4. I. H. Seeley. *Municipal Engineering Practice*. Macmillan (1967)
5. DOE. Specification for road and bridge works. HMSO (1976)
6. BS 340: 1963 Precast concrete kerbs, channels, edgings and quadrants
7. BS 556: Concrete cylindrical pipes and fittings including manholes, inspection chambers and street gullies, Part 2: 1972 Metric units
8. BS 539: 1971 Dimensions of fittings for use with clay drains and sewer pipes
9. BS 497: Manhole covers, road gully gratings and frames for drainage purposes, Part 1: 1976 Cast iron and cast steel
10. BS 368: 1971 Precast concrete flags
11. BS 1242: 1960 Tarmacadam 'tarpaving' for footpaths, playgrounds and similar works
12. BS 5837: 1979 Trees in relation to construction
13. DOE. Road note 29: A guide to the structural design of pavements for new roads. HMSO (1970)
14. BS 1722: Fences Part 1: 1972 Chain link fences, Supplement No. 1 (1974) Part 1: 1972 Gates and gateposts used in conjunction with chain link fences, Part 2: 1973 Woven wire fences, Part 3: 1973 Strained wire fences, Part 4: 1972 Cleft chestnut pale fences, Part 5: 1972 Close-boarded fences, Part 6: 1972 Wooden palisade fences, Part 7: 1972 Wooden post and rail fences, Part 8: 1978 Mild steel continuous bar fences, Part 9: 1963 Mild steel or wrought iron unclimbable fences with round or square verticals and flat standards and horizontals, Part 10: 1972 Anti-intruder chain link fences, Part 11: 1972 Woven wood fences
15. BS 4092: Domestic front entrance gates, Part 1: 1966 Metal gates, Part 2: 1966 Wooden gates
16. BS 4043: 1966 Recommendations for transplanting semi-mature trees
17. CP 308: 1974 Drainage of roofs and paved areas

APPENDIX: METRIC CONVERSION TABLE

Length

1 in = 25.44 mm (approximately 25 mm)

then $\dfrac{mm}{100} \times 4$ = inches

1 ft = 304.8 mm (approximately 300 mm)
1 yd = 0.914 m (approximately 910 mm)
1 mile = 1.609 km (approximately $1\frac{3}{5}$ km)
1 m = 3.281 ft = 1.094 yd (approximately 1.1 yd)
(10 m = 11 yd approximately)
1 km = 0.621 mile ($\frac{5}{8}$ mile approximately)

Area

1 ft^2 = 0.093 m^2
1 yd^2 = 0.836 m^2
1 acre = 0.405 ha (1 ha or hectare = 10 000 m^2)
1 mile^2 = 2.590 km^2
1 m^2 = 10.746 ft^2 = 1.916 yd^2 (approximately 1.2 yd^2)
1 ha = 2.471 acres (approximately 2½ acres)
1 km^2 = 0.386 mile^2

Volume

1 ft^3 = 0.028 m^3
1 yd^3 = 0.765 m^3
1 m^3 = 35.315 ft^3 = 1.308 yd^3 (approximately 1.3 yd^3)
1 ft^3 = 28.32 litres (1000 litres = 1 m^3)
1 gal = 4.546 litres
1 litre = 0.220 gal (approximately 4½ litres to the gallon)

Pressure

1 lbf/in^2 = 0.007 N/mm^2
1 lbf/ft^2 = 47.88 N/m^2
1 tonf/ft^2 = 107.3 kN/m^2

Mass

1 lb = 0.454 kg (kilogramme)
1 cwt = 50.80 kg (approximately 50 kg)
1 ton = 1.016 tonnes (1 tonne = 1000 kg = 0.984 ton)
1 kg = 2.205 lb (approximately $2\frac{1}{5}$ lb)

Density

1 lb/ft^3 = 16.019 kg/m^3
1 kg/m^3 = 0.062 lb/ft^3

Velocity

1 ft/s = 0.305 m/s
1 mile/h = 1.609 km/h

Energy

1 therm = 105.506 MJ (megajoules)
1 Btu = 1.055 kJ (kilojoules)

Thermal conductivity

1 Btu/ft^{2}h$^\circ$F = 5.678 W/m$^2\,^\circ$C
(where W = watt)

Temperature

x°F = $\frac{5}{9}(x - 32)^\circ$C
x°C = $\frac{9}{5}x + 32^\circ$F
0°C = 32°F (freezing)
5°C = 41°F
10°C = 50°F (rather cold)
15°C = 59°F
20°C = 68°F (quite warm)
25°C = 77°F
30°C = 86°F (very hot)

INDEX